T0192084

Graduate Texts in Physics

Graduate Texts in Physics

Graduate Texts in Physics publishes core learning/teaching material for graduate- and advanced-level undergraduate courses on topics of current and emerging fields within physics, both pure and applied. These textbooks serve students at the MS- or PhD-level and their instructors as comprehensive sources of principles, definitions, derivations, experiments and applications (as relevant) for their mastery and teaching, respectively. International in scope and relevance, the textbooks correspond to course syllabi sufficiently to serve as required reading. Their didactic style, comprehensiveness and coverage of fundamental material also make them suitable as introductions or references for scientists entering, or requiring timely knowledge of, a research field.

Series Editors

Professor Kurt H. Becker
Vice Dean for Academic Affairs
Professor of Applied Physics and
of Mechanical and Aerospace Engineering
Editor-in-Chief, European Physical Journal D
NYU Polytechnic School
of Engineering
15 Metro Tech Center, 6th floor
Brooklyn, NY 11201, USA
kurt.becker@nyu.edu

Professor Richard Needs
Cavendish Laboratory
JJ Thomson Avenue, Cambridge
CB3 0HE, UK
rn11@cam.ac.uk

Professor William T. Rhodes
Department of Computer and Electrical
Engineering and Computer Science
Imaging Science and Technology Center
Florida Atlantic University
777 Glades Road SE, Room 456
Boca Raton, FL 33431, USA
wrhodes@fau.edu

Professor Susan Scott
Department of Quantum Science
Australian National University
Canberra, ACT, 0200, Australia
susan.scott@anu.edu.au

Professor H. Eugene Stanley
Center for Polymer Studies
Department of Physics
Boston University
590 Commonwealth Avenue, Room 204B,
Boston, MA, 02215, USA
hes@bu.edu

Professor Martin Stutzmann
Technische Universität München
Am Coulombwall,
Garching, 85747, Germany
stutz@wsi.tu-muenchen.de

More information about this series at
www.springer.com/series/8431

Tânia Tomé • Mário J. de Oliveira

Stochastic Dynamics and Irreversibility

 Springer

Tânia Tomé
Institute of Physics
University of São Paulo
São Paulo, Brazil

Mário J. de Oliveira
Institute of Physics
University of São Paulo
São Paulo, Brazil

ISSN 1868-4513   ISSN 1868-4521 (electronic)
ISBN 978-3-319-36481-0   ISBN 978-3-319-11770-6 (eBook)
DOI 10.1007/978-3-319-11770-6
Springer Cham Heidelberg New York Dordrecht London

Dedicated to
Maria Roza, Wilson Tomé
Natalina, João Batista
and Pedro

Preface

Erratic or irregular movements, which we call unpredictable or random, occur spontaneously and are essential part of microscopic and macroscopic worlds. When superimposed to the predictable movements, they make up the random fluctuations sometimes called noise. Heat and temperature are macroscopic manifestations of these fluctuations at the microscopic level, and statistical physics constitutes the discipline that studies the systems affected by these fluctuations. According to this discipline, the states of thermodynamic equilibrium are described by the Boltzmann-Gibbs distribution, from which we may obtain the thermodynamic properties and, in particular, the phase transitions and critical phenomena. The states out of thermodynamic equilibrium, on the other hand, do not have such a general description. However, there are very well known laws that describe out of equilibrium or near equilibrium systems, such as the Fourier's law, the Boltzmann H theorem, the Onsager reciprocity relations and the dissipation-fluctuations theorems.

This book presents an introduction to the theory of random processes and an exposition of the stochastic dynamics as a contribution to development of non-equilibrium statistical physics and, in particular, to the irreversible phenomena. The contents of Chaps. 1–8 comprise the fundamentals of the theory of stochastic processes. In addition to the basic principles of probability theory, they include the study of stochastic motion or processes in continuous spaces, which are described by Langevin and Fokker-Planck equations, and in discrete spaces, which are described by Markov chains and master equations. Chapter 9 presents the fundamental concepts used in the characterization of the phase transitions and critical behavior in reversible and irreversible systems. Chapter 10 concerns the stochastic description of chemical reactions. Starting from Chap. 11, topics related to the non-equilibrium statistical physics are treated by means of the stochastic dynamics of models defined on lattices. Among them, we study models with inversion symmetry, which include the Glauber-Ising model; models with absorbing states, which include the contact process; and models used in population dynamics and spreading of epidemics. We analyze also the probabilistic cellular automata, the reaction-diffusion processes, the random sequential adsorption and percolation.

An important part of the book concerns the phenomenon of irreversibility, which occurs both in systems that evolve to equilibrium and in out of equilibrium stationary states. According to our point of view, the irreversible macroscopic phenomena are treated by means of a microscopic description that contains in itself the irreversibility. This means that the descriptive level which we consider is the one in which the microscopic irreversibility can be clearly displayed. We cannot refrain from being interested in irreversibility, because it may be the enigma and the key to unveil the phenomena of nature.

São Paulo, Brazil   Tânia Tomé
June 2014   Mário José de Oliveira

Contents

Chapter 1
Random Variables

1.1 Probability

Natural phenomena to which we assign a random character occur with great frequency. They are commonly characterized as unpredictable, irregular or erratic, and happen not only in the macroscopic world but also in the microscopic world. They are clearly distinguished from movements that we characterize as predictable, as with the fall of a compact solid body. An apple, for example, dropped a few meters from the ground, from the rest, will hit the ground at a point situated vertically below the point where it was released. If we observe the fall of a compact solid body, no matter how many times, we can see that the body will always fall at the same point. On the other hand, if we observe the fall of a body as light as a leaf, many times, we see that the leaf reaches different points of the ground, despite of being released from the same point, even in the complete absence of winds. The planar shape of the leaf associated with its small weight increase considerably the friction with the air, making the motion of the leaf irregular. The first trial, represented by the falling apple, is a predictable phenomenon, while the second, with the leaf, is an unpredictable erratic phenomenon, which we call random.

At first sight, one might think that a random phenomenon does not have a regularity and therefore would not be liable to a systematic study. However, after a careful observation, we see that it is possible, indeed, to find regularities in random phenomena. As an example, we examine the test in which a coin, biased or not, is thrown a number of times. The test is performed several times and for each test, that is, for each set of throws, we annotate the frequency of occurrences of heads, which is the ratio between the number of heads and the total number of throws. We can see that the frequencies obtained in different tests are very close to one another and the difference between them decreases as the number of throws in each test is increased.

Probability theory and its extensions, the theory of stochastic processes and stochastic dynamics, constitute the appropriate language for the description of

© Springer International Publishing Switzerland 2015
T. Tomé, M.J. de Oliveira, *Stochastic Dynamics and Irreversibility*, Graduate Texts in Physics,
DOI 10.1007/978-3-319-11770-6_1

random phenomena. They are based on two fundamental concepts: probability and random variable. The definition of probability is carried out by constructing the set of all outcomes of a given experiment, grouping them into mutually exclusive subsets. If, to each subset, one assigns a nonnegative real number such that their sum is equal to one, then we will be facing a probability distribution defined over the set of possible outcomes. We emphasize that this definition is very general and therefore insufficient for determining the probability to be associated with a particular experiment. The determination of the probability distribution that must be assigned to the results of a specific experiment is a major problem that must be solved by the construction of a theory or a model that describes the experiment.

The concept of probability, as well as any other physical quantity, has two basic aspects: one concerning its definition and the other concerning its interpretation. For most physical quantities, the two aspects are directly related. However, this does not happen with probability. The interpretation of probability does not follow directly from its definition. We interpret the probability of a certain outcome as the frequency of occurrence of the outcome, which constitutes the frequentist interpretation. In the example of the throw of a coin, we see that the probability of occurrence of heads is interpreted as being the ratio between the number of heads and the total number of throws. As happens to any experimental measurement, we should expect a difference between the frequency obtained experimentally and the probability assigned to the experiment. This difference however is expected to decrease as the number of throws increases and must vanish when this number grows without limits.

1.2   Discrete Random Variable

Consider a numerical variable ℓ which takes integer values and suppose that to each value of ℓ is associated a real nonnegative number p_ℓ,

$$p_\ell \geq 0, \tag{1.1}$$

such that

$$\sum_\ell p_\ell = 1, \tag{1.2}$$

which is the normalization condition. If this happens, ℓ is a discrete random variable and p_ℓ is the probability distribution of the random variable ℓ.

Example 1.1 The Bernoulli distribution refers to a random variable that takes only two values, that we assume to be 0 or 1. It is defined by

$$p_0 = a, \qquad\qquad p_1 = b, \tag{1.3}$$

where a is a parameter such that $0 < a < 1$ and $b = 1 - a$.

Example 1.2 Geometric distribution,

$$p_\ell = a^\ell b, \tag{1.4}$$

where a is a parameter such that $0 < a < 1$ and $b = 1 - a$. The random variable ℓ takes the integer values $0, 1, 2, \ldots$ The normalization follows from the identity

$$\sum_{\ell=0}^{\infty} a^\ell = \frac{1}{1-a}. \tag{1.5}$$

Example 1.3 Poisson distribution,

$$p_\ell = e^{-\alpha} \frac{\alpha^\ell}{\ell!}, \tag{1.6}$$

where α is a parameter such that $\alpha > 0$. The random variable ℓ takes the values $0, 1, 2, 3, \ldots$. The normalization is a consequence of the identity

$$\sum_{\ell=0}^{\infty} \frac{\alpha^\ell}{\ell!} = e^\alpha. \tag{1.7}$$

Example 1.4 Binomial distribution,

$$p_\ell = \binom{N}{\ell} a^\ell b^{N-\ell}, \tag{1.8}$$

where a is a parameter such that $0 < a < 1$, $b = 1 - a$, and

$$\binom{N}{\ell} = \frac{N!}{\ell!(n-\ell)!}. \tag{1.9}$$

The random variable ℓ takes the values $0, 1, 2, \ldots, N-1, N$. The normalization is a direct consequence of the binomial expansion

$$(a+b)^N = \sum_{\ell=0}^{N} \binom{N}{\ell} a^\ell b^{N-\ell}. \tag{1.10}$$

Example 1.5 Negative binomial or Pascal distribution,

$$p_\ell = \frac{(N-1+\ell)!}{(N-1)!\ell!} a^\ell b^N, \tag{1.11}$$

valid for nonnegative integer ℓ, where $0 < a < 1$, $b = 1 - a$. The normalization comes from the negative binomial expansion

$$(1 - a)^{-N} = \sum_{\ell=0}^{\infty} \frac{(N - 1 + \ell)!}{(N - 1)! \ell!} a^{\ell}. \qquad (1.12)$$

1.3   Continuous Random Variable

A continuous random variable x can take any value on the real line. In this case we associate a probability to each interval on the line. The probability that the random variable x is in the interval $[a, b]$ is

$$\int_a^b \rho(x) dx, \qquad (1.13)$$

where $\rho(x)$ is the probability density, which must have the properties

$$\rho(x) \geq 0, \qquad (1.14)$$

$$\int_{-\infty}^{\infty} \rho(x) dx = 1. \qquad (1.15)$$

The cumulative probability distribution $F(x)$ is defined by

$$F(x) = \int_{-\infty}^{x} \rho(y) dy \qquad (1.16)$$

and is a monotonically increasing function. As $\rho(x)$ is normalized, then $F(x) \to 1$ when $x \to \infty$.

Example 1.6 Exponential distribution,

$$\rho(x) = \lambda e^{-\lambda x}, \qquad (1.17)$$

valid for $x \geq 0$. Integrating ρ from $x = 0$, one obtains the cumulative distribution

$$F(x) = 1 - e^{-\lambda x}. \qquad (1.18)$$

Example 1.7 Gaussian or normal distribution,

$$\rho(x) = \frac{1}{\sqrt{2\pi\sigma^2}} e^{-x^2/2\sigma^2}, \qquad (1.19)$$

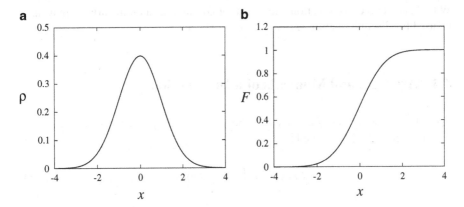

Fig. 1.1 (a) Gaussian probability distribution $\rho(x)$ of zero mean and variance equal to one. (b) Corresponding cumulative distribution $F(x)$

shown in Fig. 1.1. The normalization is a consequence of the identity

$$\int_{-\infty}^{\infty} e^{-\alpha x^2/2} dx = \sqrt{\frac{2\pi}{\alpha}}.$$ (1.20)

The cumulative distribution

$$F(x) = \frac{1}{\sqrt{2\pi\sigma^2}} \int_{-\infty}^{x} e^{-y^2/2\sigma^2} dy,$$ (1.21)

shown in Fig. 1.1, is known as the error function when $\sigma = 1$.

Example 1.8 Lorentz distribution,

$$\rho(x) = \frac{a}{\pi(a^2 + x^2)},$$ (1.22)

also known is Cauchy distribution. The cumulative distribution is given by

$$F(x) = \frac{1}{2} + \frac{1}{\pi} \arctan x,$$ (1.23)

obtained by integrating (1.22).

By resorting to the use of the Dirac delta function, a discrete probability distribution p_ℓ can be described by the following probability density,

$$\rho(x) = \sum_{\ell} p_\ell \delta(x - \ell).$$ (1.24)

With this resource, the notation employed for continuous random variable may also be used for discrete random variables, when convenient.

1.4   Averages and Moments of a Distribution

Consider a function $f(x)$ and let $\rho(x)$ be the probability density associated to x. The average $\langle f(x) \rangle$ is defined by

$$\langle f(x) \rangle = \int f(x)\rho(x)dx. \tag{1.25}$$

The moments μ_n are defined by

$$\mu_n = \langle x^n \rangle = \int x^n \rho(x)dx. \tag{1.26}$$

The first moment μ_1 is simply the average of x. The variance σ^2 is defined by

$$\sigma^2 = \langle (x - \langle x \rangle)^2 \rangle \tag{1.27}$$

and is always nonnegative. It is easy to see that

$$\sigma^2 = \langle x^2 \rangle - \langle x \rangle^2 = \mu_2 - \mu_1^2. \tag{1.28}$$

It suffices to use the following property of the average

$$\langle af(x) + bg(x) \rangle = a\langle f(x) \rangle + b\langle g(x) \rangle, \tag{1.29}$$

where a and b are constant. Indeed, starting from $(x - \mu_1)^2 = x^2 - 2\mu_1 x + \mu_1^2$, we see that $\langle (x - \mu_1)^2 \rangle = \langle x^2 \rangle - 2\mu_1 \langle x \rangle + \mu_1^2 = \mu_2 - \mu_1^2$ since $\langle x \rangle = \mu_1$.

Example 1.9 Moments of the Gaussian distribution. Consider the following identity

$$\int_{-\infty}^{\infty} e^{-\alpha x^2/2}dx = \sqrt{2\pi}\alpha^{-1/2}, \tag{1.30}$$

valid for $\alpha > 0$. Deriving both sides of this equation m times with respect to α, we get

$$\int_{-\infty}^{\infty} x^{2m} e^{-\alpha x^2/2}dx = 1 \cdot 3 \cdot 5 \cdot \ldots \cdot (2m - 1)\sqrt{2\pi}\alpha^{-1/2}\alpha^{-m}. \tag{1.31}$$

Dividing both sides by $\sqrt{2\pi}\alpha^{-1/2}$ and making the replacements $\alpha^{-1} = \sigma^2$ and $2m = n$, we obtain

$$\mu_n = \frac{1}{\sqrt{2\pi\sigma^2}} \int_{-\infty}^{\infty} x^n e^{-x^2/2\sigma^2} dx = 1 \cdot 3 \cdot 5 \cdot \ldots \cdot (n-1)\sigma^n, \qquad (1.32)$$

valid for n even. In particular,

$$\mu_2 = \sigma^2, \qquad \mu_4 = 3\sigma^4, \qquad \mu_6 = 15\sigma^6. \qquad (1.33)$$

The odd moments vanish.

Example 1.10 The Gaussian distribution

$$\rho(y) = \frac{1}{\sqrt{2\pi\sigma^2}} e^{-(y-\mu)^2/2\sigma^2} \qquad (1.34)$$

has average μ and variance σ^2. Indeed, defining the variable $x = y - \mu$, we see that $\langle x \rangle = 0$ and that $\langle x^2 \rangle = \sigma^2$ from which we get $\langle y \rangle = \mu$ and $\langle y^2 \rangle - \langle y \rangle^2 = \langle x^2 \rangle = \sigma^2$.

Example 1.11 Log-normal distribution. This distribution is given by

$$\rho(y) = \frac{1}{y\sqrt{2\pi\sigma^2}} e^{-(\ln y)^2/2\sigma^2}, \qquad (1.35)$$

valid for $y > 0$. The average and variance are, respectively,

$$\langle y \rangle = e^{\sigma^2/2}, \qquad \langle y^2 \rangle - \langle y \rangle^2 = e^{2\sigma^2} - e^{\sigma^2}. \qquad (1.36)$$

Example 1.12 Gamma distribution. This distribution is defined by

$$\rho(x) = \frac{x^{c-1} e^{-x/a}}{a^c \Gamma(c)}, \qquad (1.37)$$

valid for $x \geq 0$, where $a > 0$ and $c > 0$ are parameters and $\Gamma(c)$ is the Gamma function defined by

$$\Gamma(c) = \int_0^{\infty} z^{c-1} e^{-z} dz. \qquad (1.38)$$

The moments are obtained from

$$\mu_n = \frac{1}{a^c \Gamma(c)} \int_0^{\infty} x^{n+c-1} e^{-x/a} dx = \frac{\Gamma(n+c)}{\Gamma(c)} a^n, \qquad (1.39)$$

where we used the definition of the Gamma function.

1.5 Characteristic Function

The characteristic function $g(k)$ of a random variable x is defined as the Fourier transform of the probability density associated to x, that is,

$$g(k) = \int \rho(x)e^{ikx}dx = \langle e^{ikx} \rangle. \tag{1.40}$$

It has the following properties,

$$g(0) = 1, \qquad\qquad |g(k)| \leq 1. \tag{1.41}$$

The characteristic function is useful in obtaining the moments μ_n since the expansion of $g(k)$ in Taylor series, when it exists, give us

$$g(k) = 1 + \sum_{n=1}^{\infty} \frac{(ik)^n}{n!} \mu_n. \tag{1.42}$$

This expression is obtained directly from (1.40) through the expansion of e^{ikx} in powers of x. The characteristic function always exists. However, it is not always possible to expand it in Taylor series, which means that the probability distribution has no moments.

Example 1.13 Characteristic function of the Gaussian distribution. When all moments of a distribution are known, we may use them to get the corresponding characteristic function. This is what happens to the Gaussian distribution (1.19) whose moments are given by (1.32). Replacing μ_n, given by (1.32), into (1.42) and keeping in mind that the odd moments vanish, we get

$$g(k) = \langle e^{ikx} \rangle = \sum_{m=0}^{\infty} \frac{(ik)^{2m}}{(2m)!} 1 \cdot 3 \cdot 5 \cdot \ldots \cdot (2m-1)\sigma^{2m}. \tag{1.43}$$

Taking into account that

$$\frac{(2m)!}{1 \cdot 3 \cdot 5 \cdot \ldots \cdot (2m-1)} = 2^m m!, \tag{1.44}$$

we may write

$$\langle e^{ikx} \rangle = \sum_{m=0}^{\infty} \frac{(-k^2\sigma^2)^m}{2^m m!} = e^{-k^2\sigma^2/2}, \tag{1.45}$$

to that

$$g(k) = e^{-k^2\sigma^2/2}. \tag{1.46}$$

As we see, the characteristic function of the Gaussian distribution is also Gaussian.

The result (1.46) is valid for the distribution (1.19), which is a Gaussian distribution with zero mean. The characteristic function of the Gaussian (1.34) with nonzero mean is obtained from

$$g(k) = \langle e^{iky} \rangle, \tag{1.47}$$

where the average is performed with the use of (1.34). Carrying out the change of variable $y = \mu + x$, we may write

$$g(k) = e^{ik\mu} \langle e^{ikx} \rangle, \tag{1.48}$$

where the average is made by using the Gaussian distribution with zero mean, given by (1.19). Using (1.45), we reach the result

$$g(k) = e^{ik\mu - k^2\sigma^2/2}, \tag{1.49}$$

which is therefore the characteristic function of the Gaussian distribution (1.34) with mean μ and variance σ^2.

Example 1.14 Laplace distribution,

$$\rho(x) = \frac{1}{2\alpha} e^{-|x|/\alpha}. \tag{1.50}$$

It has the following characteristic function

$$g(k) = \frac{1}{1 + \alpha^2 k^2}. \tag{1.51}$$

Example 1.15 The Lorentz probability distribution (1.22) has the following characteristic function,

$$g(k) = e^{-a|k|}. \tag{1.52}$$

It is clear that $g(k)$ is not differentiable around $k = 0$ and therefore it has no expansion in Taylor series.

Example 1.16 Characteristic function of the Gamma distribution. Replacing μ_n, given by (1.39), into (1.42), we get

$$g(k) = \langle e^{ikx} \rangle = \sum_{n=0}^{\infty} \frac{\Gamma(n+c)}{\Gamma(c) \, n!} (ika)^n. \tag{1.53}$$

If $|k|a < 1$, the sum converges and the result gives

$$g(k) = (1 - ika)^{-c}. \tag{1.54}$$

However, by analytical continuation this result becomes valid for any value of k, and is therefore the characteristic function of the Gamma distribution.

The characteristic function is also useful in the generation of the cumulants κ_n which are defined through

$$g(k) = \exp\{\sum_{n=1}^{\infty} \frac{(ik)^n}{n!} \kappa_n\}. \tag{1.55}$$

Taking the logarithm of the right-hand side of (1.42), expanding it in a Taylor series and comparing it with the right-hand side of (1.55), we get the following relations between the cumulants and the moments,

$$\kappa_1 = \mu_1, \tag{1.56}$$

$$\kappa_2 = \mu_2 - \mu_1^2, \tag{1.57}$$

$$\kappa_3 = \mu_3 - 3\mu_2\mu_1 + 2\mu_1^3, \tag{1.58}$$

$$\kappa_4 = \mu_4 - 4\mu_3\mu_1 - 3\mu_2^2 + 12\mu_2\mu_1^2 - 6\mu_1^4, \tag{1.59}$$

etc. Comparing (1.49) and (1.55), we see that all cumulants of the Gaussian distribution, from the third on, vanish.

Two combinations of the cumulants are particularly relevant in the characterization of the probability distributions. One of them is called skewness and is defined by

$$\gamma_1 = \frac{\kappa_3}{\kappa_2^{3/2}}. \tag{1.60}$$

The other is called kurtosis and is defined by

$$\gamma_2 = \frac{\kappa_4}{\kappa_2^2}. \tag{1.61}$$

For symmetric distributions, whose odd moments vanish, the skewness vanishes and the kurtosis is reduces to

$$\gamma_2 = \frac{\mu_4 - 3\mu_2^2}{\mu_2^2}. \tag{1.62}$$

Consider the case of a discrete random variable which takes the values x_ℓ. Then

$$\rho(x) = \sum_\ell p_\ell \delta(x - x_\ell), \tag{1.63}$$

from which we conclude that the characteristic function of a discrete variable is given by

$$g(k) = \sum_\ell p_\ell e^{ikx_\ell}. \tag{1.64}$$

Example 1.17 A discrete random variable takes the values $+1$ and -1 with probabilities equal to $1/2$. Using the above notation, we have $x_0 = 1$ and $x_1 = -1$ and $p_0 = p_1 = 1/2$ so that the characteristic function is

$$g(k) = \cos k. \tag{1.65}$$

To obtain $\rho(x)$ from $g(k)$ we take the inverse Fourier transform, that is,

$$\rho(x) = \frac{1}{2\pi} \int g(k) e^{-ikx} dk. \tag{1.66}$$

1.6 Generating Function

For probability distributions corresponding to discrete variables that take the values $0, 1, 2, \ldots$, sometimes it is convenient to make use of the generating function $G(z)$ defined by

$$G(z) = \sum_{\ell=0}^{\infty} p_\ell z^\ell. \tag{1.67}$$

The series converges at least for $-1 \leq z \leq 1$. Deriving the generating function successively, we see that it has the following properties

$$G'(1) = \sum_{\ell=1}^{\infty} \ell p_\ell = \langle \ell \rangle, \tag{1.68}$$

$$G''(1) = \sum_{\ell=2}^{\infty} \ell(\ell - 1) p_\ell = \langle \ell^2 \rangle - \langle \ell \rangle. \tag{1.69}$$

Therefore, the moments can be calculated from the derivatives of the generating function determined at $z = 1$.

Example 1.18 The generating function of the binomial distribution, Eq. (1.8), is given by

$$G(z) = \sum_{\ell=0}^{N} \binom{N}{\ell} a^\ell b^{N-\ell} z^\ell = (az + b)^N. \tag{1.70}$$

After determining the derivatives $G'(z)$ and $G''(z)$ and using the formulas (1.68) and (1.69), we get the average

$$\langle \ell \rangle = Na, \tag{1.71}$$

and the variance

$$\langle \ell^2 \rangle - \langle \ell \rangle^2 = Nab, \tag{1.72}$$

of the binomial distribution.

Example 1.19 The generating function of the Pascal distribution, Eq. (1.11), is given by

$$G(z) = b^N \sum_{\ell=0}^{\infty} \frac{(N-1+\ell)!}{(N-1)!\ell!} (az)^\ell = \frac{(1-a)^N}{(1-az)^N}, \tag{1.73}$$

where we used the identity (1.12) and $b = 1 - a$. From the derivatives $G'(z)$ and $G''(z)$ and using (1.68) and (1.69), we obtain the average

$$\langle \ell \rangle = Nab^{-1}, \tag{1.74}$$

and the variance

$$\langle \ell^2 \rangle - \langle \ell \rangle^2 = 2Na^2 b^{-2}, \tag{1.75}$$

of the Pascal distribution.

1.7   Change of Variable

Consider two random variables x and y such that $y = f(x)$. Suppose that the probability density of the variable x is $\rho_1(x)$. How does one obtain the probability density $\rho_2(y)$ of the variable y? The answer is given by the formula

$$\rho_2(y) = \int \delta(y - f(x)) \rho_1(x) dx, \tag{1.76}$$

whose demonstration is given below,

Let $g_2(k)$ be the characteristic function corresponding to the variable y. Since x and y are tied by $y = f(x)$, then we may write

$$g_2(k) = \langle e^{iky} \rangle = \langle e^{ikf(x)} \rangle = \int e^{ikf(x)} \rho_1(x)dx. \tag{1.77}$$

On the other hand,

$$\rho_2(y) = \frac{1}{2\pi} \int e^{-iky} g_2(k)dk, \tag{1.78}$$

from which we get

$$\rho_2(y) = \frac{1}{2\pi} \int \int e^{-ik[y-f(x)]} \rho_1(x)dkdx. \tag{1.79}$$

Using the representation

$$\delta(z) = \frac{1}{2\pi} \int e^{-ikz} dk \tag{1.80}$$

of the Dirac delta function, we obtain the desired relation.

The integration of ρ_2, given by (1.76), in a certain interval of the variable y, give us

$$\int_{y_1}^{y_2} \rho_2(y)dy = \int_{x_1}^{x_2} \rho_1(x)dx + \int_{x_3}^{x_4} \rho_1(x)dx, \tag{1.81}$$

where $[x_1, x_2]$ and $[x_3, x_4]$ are the intervals of the variable x that are mapped into the interval $[y_1, y_2]$ of the variable y, as can be seen in Fig. 1.2. We are considering the case in which only two intervals of x are mapped into the same interval of y.

Fig. 1.2 Example of transformation $x \rightarrow y = f(x)$. The intervals $[x_1, x_2]$ and $[x_3, x_4]$ are mapped into the interval $[y_1, y_2]$ by the transformation $x \rightarrow y = f(x)$

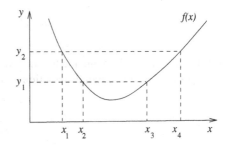

When the intervals become infinitesimal, the expression (1.81) is reduced to

$$\rho_2(y) = \rho_1(x') \left| \frac{dx'}{dy} \right| + \rho_1(x'') \left| \frac{dx''}{dy} \right|, \tag{1.82}$$

where x' and x'' denote the two branches of $f(x)$ that are mapped in y. If $f(x)$ is monotonic increasing, we may write

$$\rho_2(y)dy = \rho_1(x)dx. \tag{1.83}$$

If $f(x)$ is monotonic decreasing, we should introduce a minus sign in one side of the equation.

Example 1.20 Let $y = x^2$ and suppose that $\rho_1(x)$ is an even function. The two branches that we should consider here are $x' = -\sqrt{y}$ and $x = \sqrt{y}$. For both cases, $|dx'/dy| = |dx''/dy| = 1/2\sqrt{y}$ so that

$$\rho_2(y) = \frac{1}{2\sqrt{y}}\rho_1(-\sqrt{y}) + \frac{1}{2\sqrt{y}}\rho_1(\sqrt{y}) = \frac{1}{\sqrt{y}}\rho_1(\sqrt{y}), \tag{1.84}$$

where we have taken into account that $\rho_1(x)$ is an even function.

Alternatively, we may use (1.76) directly,

$$\rho_2(y) = \int_{-\infty}^{\infty} \delta(y - x^2)\rho_1(x)dx = 2\int_0^{\infty} \delta(y - x^2)\rho_1(x)dx, \tag{1.85}$$

so that

$$\rho_2(y) = \int_0^{\infty} \frac{1}{|x|}\delta(x - \sqrt{y})\rho_1(x)dx = \frac{1}{\sqrt{y}}\rho_1(\sqrt{y}). \tag{1.86}$$

If $\rho_1(x)$ is the probability distribution such that $\rho_1(x) = 1/2$ for $|x| \leq 1$ and zero otherwise, then

$$\rho_2(y) = \frac{1}{2\sqrt{y}}, \tag{1.87}$$

for $0 \leq y \leq 1$ and zero outside this interval.

If $\rho_1(x)$ is the Gaussian distribution (1.19), then

$$\rho_2(y) = \frac{1}{\sqrt{2y\pi\sigma^2}}e^{-y/2\sigma^2}, \tag{1.88}$$

valid for $y \geq 0$.

In numerical simulation of stochastic processes, the generation of random numbers with a certain probability distribution is an absolute necessity. The simplest and most used example is the generation of numbers with equal probability in the interval $[0, 1]$. Denoting by ξ a random variable with this property, the probability density $p(\xi)$ of ξ is $p(\xi) = 1$. However, in some situations, it is desirable to generate random numbers with other probability distributions, say with a probability density $\rho(x)$ defined in an interval $a \leq x \leq b$. If we manage to determine the relation $x = f(\xi)$ between x and ξ, then we can generate x from ξ by using this relation.

We will analyze here only the case in which $\rho(x)$ corresponds to a biunivocal function $f(\xi)$. In this case, using the expression (1.81), we get

$$\int_0^\xi p(\xi')d\xi' = \int_a^x \rho(x')dx'. \tag{1.89}$$

Taking into account that $p(\xi) = 1$, then

$$\xi = \int_a^x \rho(x')dx' = F(x), \tag{1.90}$$

where $F(x)$ is the cumulative probability distribution associated to the variable x. Therefore, $f(\xi)$ is the inverse function of $F(x)$, that is, $x = f(\xi) = F^{-1}(\xi)$.

Example 1.21 Suppose that $\rho(x) = 2x$ and $0 \leq x \leq 1$. Then $F(x) = x^2$ so that $x = f(\xi) = \sqrt{\xi}$.

The method described above is interesting only when $F(x)$ and its inverse can be obtained in closed forms. This is not the case, for example, of the Gaussian distribution. Later we will see how to avoid this problem for the Gaussian case.

1.8   Joint Distribution

Suppose that x and y are two random variables. The probability that x is in the interval $[a, b]$ and y in the interval $[c, d]$ is

$$\int_a^b \int_c^d \rho(x, y)dxdy, \tag{1.91}$$

where $\rho(x, y)$ is the joint probability density of x and y. It has the properties

$$\rho(x, y) \geq 0, \tag{1.92}$$

$$\int \int \rho(x, y)dxdy = 1. \tag{1.93}$$

From them we obtain the marginal probability densities $\rho_1(x)$ of x and $\rho_2(y)$ of y, given, respectively, by

$$\rho_1(x) = \int \rho(x, y)dy, \tag{1.94}$$

$$\rho_2(y) = \int \rho(x, y)dx. \tag{1.95}$$

The random variables x are y independent of each other if $\rho(x, y) = \rho_1(x)\rho_2(y)$. In this case, the average of the product of x and y equals the product of the averages, that is, $\langle xy \rangle = \langle x \rangle \langle y \rangle$.

A measure of the dependence between the variables x and y is given by the correlation or covariance C defined by

$$C = \langle (x - \langle x \rangle)(y - \langle y \rangle) \rangle, \tag{1.96}$$

which can be written in the form

$$C = \langle xy \rangle - \langle x \rangle \langle y \rangle. \tag{1.97}$$

If the variables x and y are independent, the covariance vanishes, $C = 0$.

Given $\rho(x, y)$, the probability distribution $\rho_3(z)$ of a third random variable z that depends on x and y through $z = f(x, y)$ can be obtained by means of the formula

$$\rho_3(z) = \int \int \delta(z - f(x, y))\rho(x, y)dxdy. \tag{1.98}$$

If two random variables u and v depend on x and y through the transformation $u = f_1(x, y)$ and $v = f_2(x, y)$, then the joint probability density $\rho_{12}(u, v)$ of the variables u and v is given by

$$\rho_{12}(u, v) = \int \int \delta(u - f_1(x, y))\delta(v - f_2(x, y))\rho(x, y)dxdy. \tag{1.99}$$

Both formulas (1.98) and (1.99) can be demonstrated by using a procedure analogous to that seen in the previous section.

The integration of ρ_{12}, given by expression (1.99), over a region $\mathscr{B}$ of the variables (u, v) give us

$$\int_{\mathscr{B}} \rho_{12}(u, v)dudv = \int_{\mathscr{A}} \rho(x, y)dxdy, \tag{1.100}$$

where $\mathscr{A}$ is the region of the variables (x, y) that is mapped into $\mathscr{B}$ by the transformation of variables. If the transformation is biunivocal, we may write

$$\rho_{12}(u, v)dudv = \rho(x, y)dxdy, \tag{1.101}$$

if the Jacobian of the transformation is positive. If the Jacobian is negative, we should introduce a minus sign in one side of the equation.

A very useful example in employing the transformation of variables is found in the following algorithm used to generate random numbers according to a Gaussian distribution from random numbers that are identically distributed in the interval $[0, 1]$. Let ξ and ζ be two independent random variable that are uniformly distributed in the interval $[0, 1]$ and consider two random variable r and θ defined by

$$r = \sqrt{\frac{2}{\alpha} |\ln(1 - \xi)|}, \qquad \theta = 2\pi\zeta, \qquad (1.102)$$

where α is a positive constant. Their probability densities are

$$\rho_1(r) = \alpha r e^{-\alpha r^2/2}, \qquad (1.103)$$

$$\rho_2(\theta) = \frac{1}{2\pi}, \qquad 0 \le \theta \le 2\pi. \qquad (1.104)$$

Next we define the variables x and y by

$$x = r \sin\theta, \qquad y = r \cos\theta. \qquad (1.105)$$

The joint probability distribution $\rho_c(x, y)$ of these variables are

$$\rho_c(x, y)dxdy = \rho_1(r)\rho_2(\theta)drd\theta. \qquad (1.106)$$

Since $dxdy = rdrd\theta$, then

$$\rho_c(x, y) = \frac{\alpha}{2\pi} e^{-\alpha(x^2+y^2)/2}, \qquad (1.107)$$

so that $\rho_c(x, y) = \rho(x)\rho(y)$, where

$$\rho(x) = \left(\frac{\alpha}{2\pi}\right)^{1/2} e^{-\alpha x^2/2} \qquad (1.108)$$

is the Gaussian distribution. Notice that x and y are independent random variables. Thus, from two random numbers ξ and ζ uniformly distributed in the interval $[0, 1]$, we can generate, using Eqs. (1.102) and (1.105), two independent random numbers x and y, each one distributed according to the Gaussian distribution (1.108).

Example 1.22   The Maxwell distribution of velocities is given by

$$\rho(x, y, z) = \left(\frac{\beta m}{2\pi}\right)^{3/2} e^{-\beta m(v_x^2+v_y^2+v_z^2)/2}, \qquad (1.109)$$

where v_x, v_y and v_z are the Cartesian components of the velocity of a molecule, m is the mass of a molecule and $\beta = 1/k_B T$, where k_B is the Boltzmann constant and T is the absolute temperature. We wish to determine the probability distribution $\rho_v(v)$ corresponding to the absolute value v of the velocity of a molecule, given by $v = (v_x^2 + v_y^2 + v_z^2)^{1/2}$. To this end, we determine first the joint probability density $\rho_3(v, \theta, \varphi)$, where θ and φ are the polar and azimuthal angle, respectively. From

$$\rho_3(v, \theta, \varphi) dv d\theta d\varphi = \rho(v_x, v_y, v_z) dv_x dv_y dv_z \tag{1.110}$$

and using the relation $dv_x dv_y dv_z = v^2 \sin\theta \, dv d\theta d\varphi$ between the Cartesian and spherical coordinates, we get

$$\rho_3(v, \theta, \varphi) = v^2 \sin\theta \left(\frac{\beta m}{2\pi} \right)^{3/2} e^{-\beta m v^2/2}. \tag{1.111}$$

Therefore,

$$\rho_v(v) = \int_0^\pi \int_0^{2\pi} \rho_3(v, \theta, \varphi) d\theta d\varphi = 4\pi v^2 \left(\frac{\beta m}{2\pi} \right)^{3/2} e^{-\beta m v^2/2}. \tag{1.112}$$

To obtain the probability density $\rho_e(E)$ of the kinetic energy $E = mv^2/2$ of a molecule, it suffices to use the relation $\rho_e(E) dE = \rho_v(v) dv$, from which we get

$$\rho_e(E) = 2\beta \left(\frac{\beta E}{\pi} \right)^{1/2} e^{-\beta E}. \tag{1.113}$$

Example 1.23 The Chi-squared distribution is defined as the probability distribution of the random variable $z = x_1^2 + x_2^2 + \ldots + x_N^2$, where x_i are independent random variables with the same probability distribution, a Gaussian with zero mean and variance σ^2. To find it, we determine first the probability distribution of the variable $r = (x_1^2 + x_2^2 + \ldots + x_N^2)^{1/2}$, called Chi distribution. According to the formula (1.98),

$$\rho(r) = \int \int \ldots \int \delta(r - (x_1^2 + x_2^2 + \ldots + x_N^2)^{1/2})$$
$$\times (2\pi\sigma^2)^{-N/2} e^{-(x_1^2+x_2^2+\ldots+x_N^2)/2\sigma^2} dx_1 dx_2 \ldots dx_N, \tag{1.114}$$

expression that can be written as

$$\rho(r) = \frac{e^{-r^2/2\sigma^2}}{(2\pi\sigma^2)^{N/2}} S_N, \tag{1.115}$$

where S_N is the integral

$$S_N = \int\int \cdots \int \delta(r - (x_1^2 + x_2^2 + \ldots + x_N^2)^{1/2}) dx_1 dx_2 \ldots dx_N, \qquad (1.116)$$

This N-dimensional integral is nothing but the area of an N-dimensional spherical surface of radius r and thus proportional to r^{N-1}, that is, $S_N = A_N r^{N-1}$. Therefore

$$\rho(r) = A_N \frac{r^{N-1} e^{-r^2/2\sigma^2}}{(2\pi\sigma^2)^{N/2}}. \qquad (1.117)$$

Next we should determine A_N. To this end, it suffices to use $\int_0^\infty \rho(r)dr = 1$, which follows directly from (1.114), from which we get

$$\frac{1}{A_N} = \int_0^\infty \frac{r^{N-1} e^{-r^2/2\sigma^2}}{(2\pi\sigma^2)^{N/2}} dr = \frac{1}{2\pi^{N/2}} \int_0^\infty z^{N/2-1} e^{-z} dz = \frac{\Gamma(N/2)}{2\pi^{N/2}}, \qquad (1.118)$$

where we performed the change of variables $z = r^2/2\sigma^2$ and used the definition of Gamma function. Thus, $A_N = 2\pi^{N/2}/\Gamma(N/2)$, from which we obtain

$$\rho(r) = \frac{2r^{N-1} e^{-r^2/2\sigma^2}}{(2\sigma^2)^{N/2} \Gamma(N/2)}, \qquad (1.119)$$

which is the Chi distribution, and also the area of the N-dimensional spherical surface of radius r,

$$S_N = \frac{2\pi^{N/2}}{\Gamma(N/2)} r^{N-1}. \qquad (1.120)$$

It is worth mentioning that the volume V_N of the N-dimensional sphere of radius r is related to S_N by $dV_N/dr = S_N$ so that

$$V_N = \frac{2\pi^{N/2}}{N\Gamma(N/2)} r^N. \qquad (1.121)$$

To get the distribution of the variable z, if suffices to remember that $z = r^2$. Performing the change of variables from r to z, we obtain the Chi-squared distribution

$$\rho_1(z) = \frac{z^{(N-2)/2} e^{-z/2\sigma^2}}{(2\sigma^2)^{N/2} \Gamma(N/2)}. \qquad (1.122)$$

When $N = 3$, the Chi distribution (1.119) becomes the distribution (1.112) of the absolute value of the velocity of a particle and the Chi-squared (1.122) becomes the distribution (1.113) of the kinetic energy of a particle. When $N = 2$, the Chi distribution (1.119) reduces to

$$\rho(r) = \frac{r}{\sigma^2} e^{-r^2/2\sigma^2}, \tag{1.123}$$

which is called Rayleigh distribution.

Example 1.24 The Student distribution is defined as the distribution of the variable $x = \sqrt{N} y/\sqrt{z}$, where y has a Gaussian distribution with zero mean and variance σ^2,

$$\rho_2(y) = \frac{1}{\sqrt{2\pi\sigma^2}} e^{-y^2/2\sigma^2}, \tag{1.124}$$

and z has the Chi-squared distribution (1.122). It is given by

$$\rho(x) = \int_{-\infty}^{\infty} \int_{0}^{\infty} \delta\left(x - \frac{\sqrt{N}y}{\sqrt{z}}\right) \rho_1(z) \rho_2(y) \, dz \, dy. \tag{1.125}$$

Replacing $\rho_2(y)$ and integrating in y,

$$\rho(x) = \frac{1}{\sqrt{2N\pi\sigma^2}} \int_{0}^{\infty} \sqrt{z}\, \rho_1(z) e^{-x^2 z/2N\sigma^2} \, dz. \tag{1.126}$$

Replacing the distribution (1.122) and performing the change of variable $z \to u$, where $u = z(1 + x^2/N)/2\sigma^2$, we obtain the expression

$$\rho(x) = \frac{1}{\sqrt{N\pi}} \frac{1}{\Gamma(N/2)} \left(1 + \frac{x^2}{N}\right)^{-(N+1)/2} \int_{0}^{\infty} u^{(N-1)/2} e^{-u} \, du. \tag{1.127}$$

After the integration in u, we reach the Student distribution

$$\rho(x) = \frac{1}{\sqrt{N\pi}} \frac{\Gamma((N+1)/2)}{\Gamma(N/2)} \left(1 + \frac{x^2}{N}\right)^{-(N+1)/2}. \tag{1.128}$$

When $N = 1$, it reduces to the Lorentz distribution $\rho(x) = 1/\pi(1 + x^2)$ and when $N \to \infty$, it reduces to the Gaussian distribution $\rho(x) = e^{-x^2/2}/\sqrt{2\pi}$.

Exercises

1. Obtain the Poisson distribution from the binomial distribution by taking the limits $N \to \infty$ and $a \to 0$ such that $aN = \alpha$, constant.
2. Obtain the cumulative distribution of the Laplace and Lorentz distribution.

3. Determine the mean and variance of Bernoulli, binomial, Poisson, geometric and Pascal distributions.

4. Find the mean and variance of the following probability distributions: (a) rectangular,

$$\rho(x) = \begin{cases} 0, & |x| > a, \\ (2a)^{-1}, & |x| \le a, \end{cases}$$

(b) triangular,

$$\rho(x) = \begin{cases} 0, & |x| > a, \\ a^{-2}(a - |x|), & |x| \le a, \end{cases}$$

(c) exponential, (d) Laplace, (e) log-normal and (f) Gamma.

5. Determine all the moments of the distributions of the previous exercise.

6. Obtain by integration the characteristic functions corresponding to the rectangular, triangular, exponential and Laplace distributions. Expand in Taylor series and get the moments. Compare with the results of the previous exercise.

7. From all the moments of the Gamma and Laplace distributions, find the respective characteristic functions.

8. Determine by integration the characteristic function of the Gaussian distribution of mean μ and variance σ^2, and also of the Lorentz and Gamma distribution.

9. Determine the characteristic function of the Poisson distribution. Show that all cumulants are equal.

10. Determine the probability distribution and the moments corresponding to the characteristic function $g(k) = a + b \cos k$, where $a + b = 1$.

11. Determine the generating function and from it the mean and the variance of the following probability distributions: (a) Poisson, (b) geometric and (c) Pascal.

12. Show that the log-normal distribution can be obtained from the normal distribution by the transformation $y = e^x$.

13. The probability density of the random variable x is given by $\rho_1(x) = 1$ for $0 \le x \le 1$ and $\rho_1(x) = 0$ otherwise. Obtain the probability density $\rho_2(y)$ of the variable $y = f(x)$ for the following cases: (a) $f(x) = -\cos(\pi x)$, (b) $f(x) = -\ln(1 - x)$.

14. A particle has equal probability of being found in any point of the surface of a disc of radius R whose center coincides with the origin of a system of polar coordinates (r, ϕ). Find the probability density of these variables. Determine the probability of finding the particle inside the sector defined by the angles $\phi = \phi_1$ and $\phi = \phi_2$.

15. A particle has equal probability of being found in any point of a spherical surface of radius R whose center coincides with the origin of a system of spherical coordinates (r, θ, ϕ). Find the probability density of these variables. Determine the probability of finding the particle between the two latitudes described by $\theta = \theta_1$ and $\theta = \theta_2$.

16. Determine an algorithm to generate random numbers that are distributed according to the exponential distribution from random numbers ξ that are equally distributed in the interval $[0, 1]$. Generate the numbers from these algorithm, make a histogram and compare it with the analytical expression. Do the same with the Laplace and Lorentz distributions.

17. Generate random numbers that are distributed according to the Gaussian distribution with width $\sigma = 1$. Make a histogram and compare with the analytical curve.

18. The numerical calculation of the area of a plane figure contained inside a square can be made as follows. Generate points equally distributed inside the square. An estimate of the ratio between the area of the figure and the area of the square is given by the ratio between the number of points inside the figure and the total number of points generated. Use this method to determine the area of a quarter of a circle and thus a numerical estimate of pi. The same method can be used to determined the volume of a solid figure contained inside a cube. Determine the volume of an eighth of a sphere of unit radius and again a numerical estimate of pi.

Chapter 2
Sequence of Independent Variables

2.1 Sum of Independent Variables

Many random phenomena are made up by a set or by a succession of independent trials and therefore described by independent random variables. One example is the random walk, which serves as a model for several random phenomena. At regular intervals of time, a walker takes a step forward or backward at random and independent of the previous steps. There are two basic theorems concerning the behavior of a sequence of independent random variables, valid when the number of them is very large: the law of large numbers and the central limit theorem.

Consider a random variable y which is a sum of two independent random variables x_1 and x_2, whose characteristic functions are $g_1(k)$ and $g_2(k)$, respectively. The characteristic function $G(k)$ corresponding to y is related to $g_1(k)$ and $g_2(k)$ through

$$G(k) = g_1(k)g_2(k). \tag{2.1}$$

That is, the characteristic function of a sum of independent random variables equals the product of the corresponding characteristic functions. To show this result it suffices to use the relation

$$\rho(y) = \int \delta(y - x_1 - x_2)\rho_1(x_1)\rho_2(x_2)dx_1dx_2, \tag{2.2}$$

obtained in Chap. 1, where $\rho_1(x_1)$, $\rho(x_2)$ and $\rho(y)$ are the probability densities corresponding to x_1, x_2 and y, respectively. Multiplying both sides by e^{iky} and integrating in y, we get the result (2.1). Alternatively, we may start from the definition of the characteristic function and write

$$G(k) = \int e^{iky}\rho(y)dy = \langle e^{iky} \rangle = \langle e^{ikx_1}e^{ikx_2} \rangle. \tag{2.3}$$

© Springer International Publishing Switzerland 2015
T. Tomé, M.J. de Oliveira, *Stochastic Dynamics
and Irreversibility*, Graduate Texts in Physics,
DOI 10.1007/978-3-319-11770-6_2

But the variables are independent, so that

$$\langle e^{ikx_1} e^{ikx_2}\rangle = \langle e^{ikx_1}\rangle \langle e^{ikx_2}\rangle, \tag{2.4}$$

from which we get the result (2.1) since

$$g_1(k) = \langle e^{ikx_1}\rangle = \int e^{ikx_1} \rho_1(x_1) dx_1, \tag{2.5}$$

$$g_2(k) = \langle e^{ikx_2}\rangle = \int e^{ikx_2} \rho_2(x_2) dx_2. \tag{2.6}$$

Suppose now that the variable y is a sum of N independent variables, that is,

$$y = x_1 + x_2 + x_3 + \ldots + x_N = \sum_{j=1}^{N} x_j. \tag{2.7}$$

Then the above result generalizes to

$$G(k) = g_1(k)g_2(k)g_3(k)\ldots g_N(k). \tag{2.8}$$

Denoting by κ_n the n-th cumulant of y and by $\kappa_n^{(j)}$ the n-th cumulant of x_j, then, from (2.8),

$$\kappa_n = \sum_{j=1}^{N} \kappa_n^{(j)}. \tag{2.9}$$

To get this result it suffices to take the logarithm of both sides of (2.8) and compare the coefficients of the n-th power of k. Two important cases of this general result correspond to $n = 1$ (mean) and $n = 2$ (variance). When $n = 1$,

$$\langle y\rangle = \sum_{j=1}^{N} \langle x_j\rangle, \tag{2.10}$$

and, when $n = 2$,

$$\langle y^2\rangle - \langle y\rangle^2 = \sum_{j=1}^{N} \{\langle x_j^2\rangle - \langle x_j\rangle^2\}. \tag{2.11}$$

These results may also be obtained in a direct way. Taking the average of both sides of (2.7), we get the relation (2.10) between the means. Taking the square of both sides of (2.7) and the average, we get

$$\langle y^2\rangle = \sum_{j=1}^{N} \langle x_j^2\rangle + 2\sum_{j<k} \langle x_j\rangle \langle x_k\rangle, \tag{2.12}$$

where we have taken into account that the random variables are independent. Taking the square of both sides of (2.10) and comparing with (2.12), we get the relation (2.11) between the variances.

If the N independent variables have the same probability distribution and therefore the same characteristic function $g(k)$, then

$$G(k) = [g(k)]^N \qquad (2.13)$$

so that

$$\kappa_n = N\kappa_n^{(j)}. \qquad (2.14)$$

In particular, when $n = 1$,

$$\langle y \rangle = N \langle x_j \rangle \qquad (2.15)$$

and, when $n = 2$,

$$\langle y^2 \rangle - \langle y \rangle^2 = N\{\langle x_j^2 \rangle - \langle x_j \rangle^2\}. \qquad (2.16)$$

That is, the mean and the variance of y are equal to N times the mean and variance of x_j, respectively.

Example 2.1 Bernoulli trials. Consider a sequence of N independent trials. In each trial, only two mutually exclusive events A and B may occur, which we call success and failure, respectively. Below we show examples of ten trials in which six successes occur

ABAABABAAB   AABABABBAA   AAABAABBBA

Let p be the probability of the occurrence of A and $q = 1 - p$ the probability of occurrence of B. We wish to determine the probability $P_N(\ell)$ of occurring A ℓ times in a sequence of N trials. For each trial define a random variable ξ_i ($i = 1, 2, \ldots, N$) that takes the value 1 when A occurs and the value 0 when B occurs. Therefore, the probability of $\xi_i = 1$ or 0 equals p or q, respectively. We want to determine the probability distribution $P_N(\ell)$ of the random variable

$$\ell = \xi_1 + \xi_2 + \ldots + \xi_N, \qquad (2.17)$$

which counts how many times the event A has occurred in a sequence of N trials.

The characteristic function $G(k)$ of the variable ℓ is, by definition, given by

$$G(k) = \sum_{\ell=0}^{N} P_N(\ell)e^{ik\ell}. \qquad (2.18)$$

On the other hand, since the variables ξ_i are independent and have the same probability distribution, then the characteristic function $G(k)$ of the random variable ℓ is given by

$$G(k) = [g(k)]^N, \tag{2.19}$$

where $g(k)$ is the characteristic function of each one of the random variables ξ_j and given by

$$g(k) = \langle e^{ik\xi_1} \rangle = pe^{ik} + q. \tag{2.20}$$

Thus

$$G(k) = (pe^{ik} + q)^N. \tag{2.21}$$

Using the binomial expansion, we get

$$G(k) = \sum_{\ell=0}^{N} \binom{N}{\ell} p^\ell e^{ik\ell} q^{N-\ell}, \tag{2.22}$$

which, when compared to the expression (2.18), gives us

$$P_N(\ell) = \binom{N}{\ell} p^\ell q^{N-\ell}, \tag{2.23}$$

which is the binomial probability distribution.

Example 2.2 Negative binomial or Pascal distribution. The distribution (2.23) gives the probability of occurrence of ℓ successes in a sequence of N trials. The probability of occurrence of ℓ successes in N trials such that a failure occurs in the N-th trial is therefore

$$P_{N-1}(\ell)q = \binom{N-1}{\ell} p^\ell q^{N-\ell}. \tag{2.24}$$

Since the number of failure is $r = N - \ell$, we interpret this expression as the probability $P(\ell)$ of occurrence of ℓ successes before the occurrence of r failures, and that we write in the form

$$P(\ell) = \binom{r+\ell-1}{\ell} p^\ell q^r. \tag{2.25}$$

It is called negative binomial distribution because of the identity

$$\sum_{\ell=0}^{\infty} \binom{r+\ell-1}{\ell} p^{\ell} = (1-p)^{-r}, \tag{2.26}$$

which guarantees the normalization of $P(\ell)$, recalling that $q = 1 - p$.

2.2 Law of Large Numbers

We analyze here a sequence of N independent random variables $\xi_1, \xi_2, \ldots, \xi_N$ that are identically distributed, that is, with the same probability distribution. The law of large numbers asserts that

$$\frac{1}{N} \sum_{j=1}^{N} \xi_j \to a, \qquad N \to \infty, \tag{2.27}$$

where $a = \langle \xi_j \rangle$ is the mean of the common distribution. The only condition for the validity of this theorem is that the mean exists. This theorem allows us to use the frequentist interpretation of probability of an event. Suppose that we are interested to know the frequency of an event A in a sequence of N trials. If we denote by ξ_j the variable that takes the value 1 if the event A occurs and the value 0 if it does not occur, then $\ell = (\xi_1 + \xi_2 + \ldots + \xi_N)$ will count the number of times the event A has occurred and the frequency will be thus ℓ/N. On the other hand, $a = \langle \xi_j \rangle = p$, where p is the probability of occurrence of A. Therefore, the law of large numbers guarantees that $\ell/N = p$ when $N \to \infty$.

The result (2.27) means that, in the limit $N \to \infty$, the only possible result of the variable y is a, which is equivalent to say that y takes the value a with probability one or yet that the probability density associated to y is

$$\rho(y) = \delta(y - a). \tag{2.28}$$

This result is shown as follows.

Let $G_y(k)$ be the characteristic function corresponding to the random variable y defined by

$$y = \frac{1}{N} \sum_{j=1}^{N} \xi_j \tag{2.29}$$

and $g(k)$ the characteristic function of each one of the variables ξ_j. Thus,

$$G_y(k) = \langle e^{iky} \rangle = \prod_{j=1}^{N} \langle e^{ik\xi_j/N} \rangle = [g(\frac{k}{N})]^N. \tag{2.30}$$

Since the mean exists, we may write

$$g(k) = 1 + ika + \mathbf{o}(k), \tag{2.31}$$

where $\mathbf{o}(x)$ means that $\mathbf{o}(x)/x \to 0$ when $x \to 0$. Thus

$$[g(\frac{k}{N})]^N = [1 + \frac{ika}{N} + \mathbf{o}(\frac{k}{N})]^N \to e^{ika}, \qquad N \to \infty, \tag{2.32}$$

so that

$$G_y(k) = e^{ika}. \tag{2.33}$$

The probability distribution $\rho(y)$ of y is obtained by the inverse Fourier transform, which is the density given by expression (2.28).

As mentioned above, the condition for the validity of the law of large numbers is that the mean exists. The *existence* of the mean means not only that integral $\int x\rho(x)dx$ should be finite but also that the integral $\int |x|\rho(x)dx$ should be finite. A counterexample is the Lorentz distribution for which this integral diverges.

2.3 Central Limit Theorem

The central limit theorem asserts that a random variable z, defined by

$$z = \frac{1}{\sqrt{Nb}}\{\sum_{j=1}^{N} \xi_j - Na\}, \tag{2.34}$$

where ξ_j are identically distributed random variables with mean a and variance b, has the Gaussian probability distribution

$$\frac{1}{\sqrt{2\pi}}e^{-z^2/2}, \tag{2.35}$$

in the limit $N \to \infty$. For the theorem to be valid, it suffices the existence of the mean a and the variance b. Notice that this is not the case, for example, of the Lorentz distribution.

Let $g(k)$ be the characteristic function of each one of the random variables ξ_j and consider the cumulant expansion. We should have

$$g(k) = \exp\{iak - \frac{1}{2}bk^2 + \mathbf{o}(k^2)\}. \tag{2.36}$$

The characteristic function $G_z(k)$ corresponding to the variable z is given by

$$G_z(k) = \langle e^{ikz} \rangle = \langle \exp\{i \frac{k}{\sqrt{Nb}} \sum_{j=1}^{N} (\xi_j - a)\} \rangle, \tag{2.37}$$

or by

$$G_z(k) = \prod_{j=1}^{N} \langle e^{iK\xi_j} \rangle e^{-iKa} = \{g(K)e^{-iKa}\}^N \tag{2.38}$$

since the variables are independent and have the same probability distribution, and the variable K is defined by $K = k/\sqrt{Nb}$. Using the cumulant expansion of $g(k)$, expression (2.36), we get

$$G_z(k) = \exp\{-\frac{1}{2}NbK^2 + N\mathbf{o}(K^2)\}. \tag{2.39}$$

Now $NbK^2 = k^2$ and being $\mathbf{o}(K^2) = \mathbf{o}(N^{-1})$, then $N\mathbf{o}(N^{-1}) \rightarrow 0$ when $N \rightarrow \infty$. Therefore,

$$G_z(k) = e^{-k^2/2}, \tag{2.40}$$

which corresponds to a Gaussian distribution

$$\rho(z) = \frac{1}{\sqrt{2\pi}} e^{-z^2/2}. \tag{2.41}$$

For N large enough, this result constitutes a good approximation so that, in terms of the variable $\ell = \xi_1 + \xi_2 + \ldots + \xi_N = \sqrt{Nb}z + Na$, it is written as

$$P_N(\ell) = \frac{1}{\sqrt{2\pi Nb}} e^{-(\ell-Na)^2/2Nb} \tag{2.42}$$

since $\rho(z)dz = P_N(\ell)d\ell$ and $d\ell = \sqrt{Nb}dz$.

Example 2.3 In the Bernoulli trials of Example 2.1, the probability of ξ_j taking the value 1 (occurrence of event A) is p and taking the value 0 (occurrence of event B) is $q = 1 - p$. We may use the central limit theorem to obtain the probability distribution $P_N(\ell)$ for a large number N of trials. To this end, it suffices to know the mean and the variance of ξ_j, given by $a = \langle \xi_j \rangle = p$ and $b = \langle \xi_j^2 \rangle - \langle \xi_j \rangle^2 = p - p^2 = pq$, respectively. According to the central limit theorem, the probability distribution of the variable ℓ, which counts how many times the event A has occurred, is

$$P_N(\ell) = \frac{1}{\sqrt{2\pi Npq}} e^{-(\ell-Np)^2/2Npq}, \tag{2.43}$$

which is a Gaussian of mean Np and variance Npq, the same, of course, of the original binomial distribution.

Example 2.4 Consider a paramagnetic crystal comprised by N magnetic ions. In the presence of an external magnetic field, the component of the magnetic dipole of each ion along the field, can be in two states: in the same direction of the field (state A) or in the opposite direction (state B). According to statistical mechanics, the probability of the occurrence of A is $p = e^{\beta H}/(e^{\beta H} + e^{-\beta H})$, and of B is $q = e^{-\beta H}/(e^{\beta H} + e^{-\beta H})$, where β is proportional to the inverse of the absolute temperature and H is proportional to the magnetic field.

Define the magnetization M as the number ℓ of ions in state A subtracted form the number $N - \ell$ of ions in state B, that is, $M = 2\ell - N$. If we use the variable σ_j which takes the values $+1$ or -1 according to whether the j-th ion is in state A or B, respectively, then $M = \sigma_1 + \sigma_2 + \ldots + \sigma_N$. These variables are independent and have the same probability distribution. The probability of $\sigma_j = +1$ is p and of $\sigma_j = -1$ is q so that the mean is

$$m = \langle \sigma_j \rangle = p - q = \tanh \beta H \tag{2.44}$$

and the variance is

$$\chi = \langle \sigma_j^2 \rangle - \langle \sigma_j \rangle^2 = 1 - m^2 = 1 - \tanh^2 \beta H. \tag{2.45}$$

For large N, the probability distribution $\mathscr{P}_N(M)$ of the variable M is thus

$$\mathscr{P}_N(M) = \frac{1}{\sqrt{2\pi N\chi}} e^{-(M-Nm)^2/2N\chi}. \tag{2.46}$$

It is clear that $\langle M \rangle = Nm$ and that $\langle M^2 \rangle - \langle M \rangle^2 = N\chi$, or, in other terms, not only the mean but also the variance of M are proportional to the number N of magnetic ions that comprises the crystal, that is, they are extensive quantities.

2.4 Random Walk in One Dimension

A walker moves along a straight line, starting from the origin. At each time interval τ, the walker takes a step of length h to the right with probability p or to the left with probability $q = 1 - p$. To describe the walk, we introduce independent random variables $\sigma_1, \sigma_2, \sigma_3, \ldots$ that take the values $+1$ or -1 according to whether the step is to the right or to the left, respectively. The variable σ_j indicates whether the j-th step is to the right or to the left and thus it takes the value $+1$ with probability p and the value -1 with probability q. The position of the walker after n steps, that is, at time $t = n\tau$ is $x = hm$, where $m = \sigma_1 + \sigma_2 + \ldots + \sigma_n$.

The mean and variance of σ_j are

$$a = \langle \sigma_j \rangle = p - q, \tag{2.47}$$

$$b = \langle \sigma_j^2 \rangle - \langle \sigma_j \rangle^2 = 1 - (p - q)^2 = 4pq, \tag{2.48}$$

respectively. The characteristic function $g(k)$ of the variable σ_j is

$$g(k) = \langle e^{ik\sigma_j} \rangle = pe^{ik} + qe^{-ik}. \tag{2.49}$$

To obtain the probability $P_n(m)$ of the walker being at the position $x = hm$ after n steps, that is, at time $t = n\tau$, we determine first the characteristic function

$$G_n(k) = [g(k)]^n = (pe^{ik} + qe^{-ik})^n. \tag{2.50}$$

Performing the binomial expansion

$$G_n(k) = \sum_{\ell=0}^{n} \binom{n}{\ell} p^\ell q^{n-\ell} e^{ik(2\ell-n)} \tag{2.51}$$

and comparing it with the definition of $G_n(k)$, given by

$$G_n(k) = \sum_{m=-n}^{n} P_n(m)e^{ikm}, \tag{2.52}$$

where m takes the values $-n, -n + 2, \ldots, n - 2$, and n, we see that

$$P_n(m) = \frac{n!}{(\frac{n+m}{2})!(\frac{n-m}{2})!} p^{(n+m)/2} q^{(n-m)/2}. \tag{2.53}$$

To accomplish the comparison it is convenient, first, to change variable, in the sum, passing from ℓ to $m = 2\ell - n$. The mean and variance of m are

$$\langle m \rangle = na = n(p - q), \tag{2.54}$$

$$\langle m^2 \rangle - \langle m \rangle^2 = nb = 4npq. \tag{2.55}$$

If we wish to obtain the probability distribution for $n \gg 1$, it suffices to use the central limit theorem since the variables $\sigma_1, \sigma_2, \sigma_3, \ldots$ are independent. From the result (2.42), we get

$$P_n(m) = \frac{1}{\sqrt{2\pi nb}} e^{-(m-na)^2/2nb}, \tag{2.56}$$

valid for $n \gg 1$.

The probability density $\rho(x,t) = P_n(m)/h$ of x at time t is

$$\rho(x,t) = \frac{1}{\sqrt{4\pi Dt}}e^{-(x-ct)^2/4Dt},\tag{2.57}$$

where

$$c = \frac{ha}{\tau} = \frac{h(p-q)}{\tau},\tag{2.58}$$

$$D = \frac{h^2b}{2\tau} = \frac{h^2 2pq}{\tau}.\tag{2.59}$$

We get in addition the results,

$$\langle x \rangle = ct,\tag{2.60}$$

$$\langle x^2 \rangle - \langle x \rangle^2 = 2Dt,\tag{2.61}$$

that allow us to say that c is the mean velocity of the walker. If the random walk is interpreted as the random motion of a particle in a certain environment, then D is called diffusion coefficient.

We consider next a generic one-dimensional random walk. Suppose that at each time step τ the walker displaces a value x_j from the present position. Assuming that the walker starts from the origin, then the position at time $t = \tau n$ is $x = x_1 + x_2 + \ldots + x_n$. Let $P(x_j)$ be the probability density of x_j and let $g(k)$ be the corresponding characteristic function, that is,

$$g(k) = \langle e^{ikx_j} \rangle = \int P(x_j)e^{ikx_j}\,dx_j.\tag{2.62}$$

The characteristic function $G(k)$ corresponding to the variable x is

$$G(k) = [g(k)]^n.\tag{2.63}$$

To get the probability density of the variable x for large n, we employ the same technique used to show the central limit theorem. This technique amounts to expand the characteristic function $g(k)$ in cumulants up to second order, that is,

$$g(k) = e^{iAk - Bk^2/2},\tag{2.64}$$

provided the mean A and the variance B of x_j exist. Therefore

$$G(k) = e^{inAk - nBk^2/2}.\tag{2.65}$$

Recalling that $t = n\tau$ and defining $c = A/\tau$ and $D = B/2\tau$, then

$$G(k) = e^{ictk - Dtk^2}, \qquad (2.66)$$

which is the characteristic function of a Gaussian distribution of mean ct and variance $2Dt$, so that the probability density $\rho(x, t)$ of x is

$$\rho(x, t) = \frac{1}{\sqrt{4\pi Dt}} e^{-(x - ct)^2/4Dt}. \qquad (2.67)$$

The result (2.67) can be understood more clearly if we examine a system composed of many particles performing independent random motion and if we take into account that the density of particles is proportional to the probability density ρ given above. Thus, if we imagine that at the initial time all particles are found around the origin, after some time they will be spread according to the distribution above. For large times, the density of particles is a Gaussian centered at $x = ct$ with width $\Delta = \sqrt{2Dt}$.

It is important to notice that $\rho(x, t)$ fulfills the differential equation

$$\frac{\partial \rho}{\partial t} = -c \frac{\partial \rho}{\partial x} + D \frac{\partial^2 \rho}{\partial x^2}. \qquad (2.68)$$

When $c = 0$, this equation reduces to the form

$$\frac{\partial \rho}{\partial t} = D \frac{\partial^2 \rho}{\partial x^2}, \qquad (2.69)$$

which is the diffusion equation. Equation (2.68) is a diffusion equation with drift. Both are particular cases of the Fokker-Planck that will be studied in Chaps. 4 and 5.

2.5   Two-Dimensional Random Walk

We consider now a random walk in a two-dimensional space. The walk in three or more dimensions can be treated similarly. At each time interval τ, the walker moves from the present position to a new position. At the j-th time step we denote the displacement by $\mathbf{r}_j = (x_j, y_j)$. Assuming that at the initial time the walker is at the origin of a coordinate system, the position at time $t = n\tau$ is $\mathbf{r} = \mathbf{r}_1 + \mathbf{r}_2 + \ldots + \mathbf{r}_n$. The variables $\mathbf{r}_1, \mathbf{r}_2, \ldots, \mathbf{r}_n$ are thus considered to be independent random variables, or rather independent random vectors, with a given probability distribution $P(\mathbf{r}_j) = P(x_j, y_j)$. The corresponding characteristic function $g(\mathbf{k}) = g(k_1, k_2)$ is given by

$$g(\mathbf{k}) = \langle e^{i\mathbf{k}\cdot\mathbf{r}_j} \rangle = \langle e^{i(k_1 x_j + k_2 y_j)} \rangle, \qquad (2.70)$$

or yet, by

$$g(\mathbf{k}) = \int \int e^{i\mathbf{k}\cdot\mathbf{r}_j} P(\mathbf{r}_j) dx_j \, dy_j. \tag{2.71}$$

Notice that x_j and y_j may not be independent.

The characteristic function $G(\mathbf{k})$ corresponding to the vector $\mathbf{r} = (x, y)$ is given by

$$G(\mathbf{k}) = \langle e^{i\mathbf{k}\cdot\mathbf{r}} \rangle = \langle e^{i\mathbf{k}\cdot(\mathbf{r}_1+\mathbf{r}_2+\dots+\mathbf{r}_n)} \rangle = \langle e^{i\mathbf{k}\cdot\mathbf{r}_j} \rangle^n = [g(\mathbf{k})]^n. \tag{2.72}$$

To get the probability density of the random vector $\mathbf{r}$ we employ the same technique used to show the central limit theorem. This technique amounts to use the expansion of $g(\mathbf{k})$ in cumulants up to order k^2, that is,

$$g(\mathbf{k}) = \exp\{i(a_1 k_1 + a_2 k_2) - \frac{1}{2}(b_{11}k_1^2 + 2b_{12}k_1 k_2 + b_{22}k_2^2)\}, \tag{2.73}$$

where

$$a_1 = \langle x_j \rangle \qquad \text{and} \qquad a_2 = \langle y_j \rangle \tag{2.74}$$

are the cumulants of first order and

$$b_{11} = \langle x_j^2 \rangle - \langle x_j \rangle^2, \tag{2.75}$$

$$b_{12} = \langle x_j y_j \rangle - \langle x_j \rangle \langle y_j \rangle, \tag{2.76}$$

$$b_{22} = \langle y_j^2 \rangle - \langle y_j \rangle^2 \tag{2.77}$$

are the cumulants of second order. Thus, for large $t = n\tau$ we get

$$G(\mathbf{k}) = \exp\{in(a_1 k_1 + a_2 k_2) - \frac{n}{2}(b_{11}k_1^2 + 2b_{12}k_1 k_2 + b_{22}k_2^2)\}. \tag{2.78}$$

The probability density $P_n(\mathbf{r}) = P_n(x, y)$ of the random vector $\mathbf{r} = (x, y)$ is obtained from

$$P_n(\mathbf{r}) = \frac{1}{(2\pi)^2} \int \int e^{-i\mathbf{k}\cdot\mathbf{r}} G(\mathbf{k}) dk_1 dk_2 \tag{2.79}$$

and is a two-dimensional Gaussian given by

$$P_n(x, y) = \frac{1}{2\pi \sqrt{n^2 2D}}$$

$$\times \exp\{-\frac{1}{4nD}[b_{22}(x - na_1)^2 + 2b_{12}(x - na_1)(y - na_2) + b_{11}(y - na_2)^2]\}, \tag{2.80}$$

where $D = (b_{11}b_{22} - b_{12}^2)/2$.

Example 2.5 Suppose that, at each time interval τ, a walker moves a distance h in directions $+x$, $-x$, $+y$ or $-y$ with equal probability. In this case

$$P(x_j, y_j) = \frac{1}{4}\delta(x_j - h)\delta(y_j) + \frac{1}{4}\delta(x_j + h)\delta(y_j)$$

$$+ \frac{1}{4}\delta(x_j)\delta(y_j - h) + \frac{1}{4}\delta(x_j)\delta(y_j + h). \tag{2.81}$$

The corresponding characteristic function is

$$g(k_1, k_2) = \langle e^{i(k_1 x_j + k_2 y_j)} \rangle = \frac{1}{4}(e^{ihk_1} + e^{-ihk_1} + e^{ihk_2} + e^{-ihk_2}), \tag{2.82}$$

or

$$g(k_1, k_2) = \frac{1}{2}(\cos hk_1 + \cos hk_2). \tag{2.83}$$

To get results valid for large n, we use the expansion in cumulants up to order $\mathbf{k}^2$, given by

$$g(k_1, k_2) = e^{-h^2(k_1^2 + k_2^2)/4}. \tag{2.84}$$

Therefore

$$G(k_1, k_2) = e^{-nh^2(k_1^2 + k_2^2)/4}. \tag{2.85}$$

so that, taking the inverse Fourier transform,

$$P_n(x, y) = \frac{1}{\pi n h^2} e^{-(x^2 + y^2)/nh^2}. \tag{2.86}$$

Defining the diffusion coefficient $D = h^2/4\tau$, then the probability density $\rho(x, y, t)$ of x and y at time t is

$$\rho(x, y, t) = \frac{1}{4\pi Dt} e^{-(x^2 + y^2)/4Dt}. \tag{2.87}$$

It is easy to see that $\langle x^2 \rangle = \langle y^2 \rangle = 2Dt$ so that $\langle \mathbf{r}^2 \rangle = \langle x^2 + y^2 \rangle = 4Dt$.

Example 2.6 Suppose that, at each time interval τ, a walker moves the same distance h in any direction with equal probability. Using polar coordinates, then

$$P(\mathbf{r}_j)dr_j d\theta_j = \frac{1}{2\pi}\delta(r_j - h)dr_j d\theta_j. \tag{2.88}$$

The corresponding characteristic function is

$$g(\mathbf{k}) = \langle e^{i(k_1 x_j + k_2 y_j)} \rangle = \langle e^{i(k_1 r_j \cos\theta_j + k_2 r_j \sin\theta_j)} \rangle, \tag{2.89}$$

or

$$g(\mathbf{k}) = \frac{1}{2\pi} \int_0^{2\pi} e^{ih(k_1 \cos\theta_j + k_2 \sin\theta_j)} d\theta_j. \tag{2.90}$$

Defining k and ϕ such that $k_1 = k\cos\phi$ and $k_2 = k\sin\phi$ are the Cartesian coordinates of $\mathbf{k}$, then

$$g(\mathbf{k}) = \frac{1}{2\pi} \int_0^{2\pi} e^{ihk \cos(\theta_j - \phi)} d\theta_j = \frac{1}{2\pi} \int_0^{2\pi} e^{ihk \cos\theta_j} d\theta_j \tag{2.91}$$

so that $g(\mathbf{k})$ depends only on the absolute value $k = (k_1^2 + k_2^2)^{1/2}$. To obtain the behavior of the probability density for large n, we need only the cumulants of first and second order. In this case $a_1 = a_2 = 0$,

$$b_{11} = \langle x_j^2 \rangle = \int_0^\infty \int_0^{2\pi} r^2 \cos^2\theta_j \, P(\mathbf{r}_j) dr_j d\theta_j = \frac{1}{2} h^2, \tag{2.92}$$

$b_{12} = 0$ and $b_{22} = b_{11}$. Thus

$$g(\mathbf{k}) = e^{-h^2(k_1^2 + k_2^2)/4}, \tag{2.93}$$

so that again we obtain the same probability distribution of the previous example.

2.6 Multidimensional Gaussian Distribution

We saw in the previous section that the random walk in two dimensions is described by a probability distribution which is an exponential of a quadratic form in two random variables. Here we examine probability distributions $\rho(x_1, x_2, \ldots, x_d)$ which are exponential of quadratic forms in several random variables, called multidimensional Gaussian distributions, and given by

$$\rho(x_1, x_2, \ldots, x_d) = \Omega \exp\{-\frac{1}{2} \sum_{ij} B_{ij} x_i x_j\}, \tag{2.94}$$

where Ω is a normalization factor, to be determined later, and B_{ij} are constants such that $B_{ji} = B_{ij}$. It is convenient to define the vector $x = (x_1, x_2, \ldots, x_d)$ and interpret B_{ij} as the elements of a square matrix B of dimension d. The multidimensional Gaussian distribution can then be written in the compact form

$$\rho(x) = \Omega e^{-x^\dagger Bx/2}, \tag{2.95}$$

where x in this equation is understood as a column matrix whose elements are x_i, and $x^\dagger$ is the row matrix whose elements are x_i or, in other terms, $x^\dagger$ is the transpose of x.

We now wish to determine not only Ω but also the characteristic function $g(k_1, \ldots, k_d)$, defined by

$$g(k) = \langle e^{ik^\dagger x} \rangle = \int e^{ik^\dagger x} \rho(x) dx, \tag{2.96}$$

where $k^\dagger$ is the matrix transpose of the column matrix k whose elements are k_i, and $dx = dx_1 dx_2 \ldots dx_d$. Replacing ρ,

$$g(k) = \Omega \int e^{ik^\dagger x - x^\dagger Bx/2} dx. \tag{2.97}$$

Let U be the matrix that diagonalize B, that is, the matrix such that

$$U^{-1} BU = \Lambda, \tag{2.98}$$

where Λ is the diagonal matrix whose elements are the eigenvalues λ_i of B, which we assume to be nonzero. Since B is symmetric, then $U^{-1} = U^\dagger$ and U is a unitary matrix. Defining new variables y_i, which comprise the vector $y = U^{-1}x$, then $x = Uy$ and $x^\dagger = y^\dagger U^{-1}$. These relations lead us to the equalities

$$x^\dagger Bx = y^\dagger U^{-1} BUy = y^\dagger \Lambda y = \sum_j \lambda_j y_j^2, \tag{2.99}$$

$$k^\dagger x = k^\dagger Uy = q^\dagger y = \sum_j q_j y_j, \tag{2.100}$$

where q is defined by $q = U^{-1}k$, so that $q^\dagger = k^\dagger U$.

Replacing these results into (2.97), $g(k)$ becomes a product of d Gaussian integrals,

$$g = \Omega \prod_j \int_{-\infty}^{\infty} e^{-\lambda_j y_j^2/2 - iq_j y_j} dq_j = \Omega \prod_j \sqrt{\frac{2\pi}{\lambda_i}} e^{-q_j^2/2\lambda_j}. \tag{2.101}$$

From this expression, we can determine the normalization factor Ω. To this end we recall that the normalization of the probability distribution is equivalent to impose $g = 1$ when $k = 0$, that is, when $q = 0$. Therefore,

$$\Omega = \prod_j \sqrt{\frac{\lambda_i}{2\pi}} = \sqrt{\frac{\det B}{(2\pi)^d}}, \tag{2.102}$$

where we have taken into account that the product of the eigenvalues of a matrix equals its determinant. We may conclude that

$$g = \exp\{-\sum_j q_j^2 / 2\lambda_j\}. \tag{2.103}$$

Taking into account that the matrix Λ^{-1}, the inverse of Λ, is diagonal,

$$\sum_j \frac{q_j^2}{\lambda_j} = q^\dagger \Lambda^{-1} q = k^\dagger U \Lambda^{-1} U^{-1} k = k^\dagger C k, \tag{2.104}$$

where $C = U\Lambda^{-1}U^{-1}$. We remark that the matrix $C = B^{-1}$ because, being $B = U^{-1}\Lambda U$, the product of C and B results in the identity matrix.

The characteristic function can thus be written in the form

$$g(k) = e^{-k^\dagger C k/2}, \qquad k^\dagger C k = \sum_{ij} C_{ij} k_i k_j. \tag{2.105}$$

Replacing the normalization factor Ω into (2.94), the multidimensional Gaussian probability distribution is written as

$$\rho(x) = \sqrt{\frac{\det B}{(2\pi)^d}} e^{-x^\dagger B x/2}, \qquad x^\dagger B x = \sum_{ij} B_{ij} x_i x_j. \tag{2.106}$$

The characteristic function (2.105) allows us to determine the moments of the variables x_i concerning the distribution (2.106). The derivative $\partial g / \partial k_i$ taken at $k = 0$ is identified with the average $\langle x_i \rangle$. The derivative $\partial^2 g / \partial k_i \partial k_j$ taken at $k = 0$ is identified with twice the average $\langle x_i x_j \rangle$. From this, we conclude that the distribution (2.106) has zero means and covariances equal to C_{ij},

$$C_{ij} = \langle x_i x_j \rangle. \tag{2.107}$$

It is still possible to construct Gaussian distributions with nonzero means. Performing the transformation $x_i \to x_i - a_i$, we get the following distribution

$$\rho(x) = \sqrt{\frac{\det B}{(2\pi)^d}} \, e^{-(x^\dagger - a^\dagger)B(x-a)/2}, \tag{2.108}$$

where a is a column matrix with elements a_i and $a^\dagger$ is the transpose of a. The respective characteristic function is given by

$$g(k) = e^{ia^\dagger k - k^\dagger Ck/2}. \tag{2.109}$$

By construction, the multidimensional Gaussian distribution (2.108) has means equal to a_i, that is,

$$a_i = \langle x_i \rangle. \tag{2.110}$$

We remark that the elements C_{ij} are identified as the covariances of x_i and not with the second moments,

$$C_{ij} = \langle x_i x_j \rangle - \langle x_i \rangle \langle x_j \rangle. \tag{2.111}$$

We see therefore that a and C define completely the multidimensional Gaussian distribution since B is the inverse of C.

2.7 Stable Distributions

According to the central limit theorem, the probability distribution of the variable

$$x = \frac{1}{N^{1/2}}(\xi_1 + \xi_2 + \ldots + \xi_N) \tag{2.112}$$

is a Gaussian if each probability distributions of the variables ξ_j has the second moment, and as long as N is large enough. Here we are assuming only distribution with zero mean. If the distributions of the variables ξ_j are already Gaussian it is not necessary that N to be large. For any value of N, the distribution of x will be Gaussian if the distribution of ξ_j are Gaussian. We say thus that the Gaussian distribution is a stable distribution with respect to the linear combination (2.112).

Other stable distribution can be obtained if we consider the following linear combination,

$$x = \frac{1}{N^{\beta}}(\xi_1 + \xi_2 + \ldots + \xi_N), \tag{2.113}$$

where $\beta > 0$, which reduces to the form (2.112) when $\beta = 1/2$. Given the distribution $\rho(\xi_j)$, which we consider to be the same for all variables ξ_j, we wish to obtain the distribution $\rho^*(x)$ of x. Denoting by $g^*(k)$ the characteristic function associated to the variable x, then

$$g^*(k) = \langle e^{ikx} \rangle = \prod_j \langle e^{ik\xi_j/N^{\beta}} \rangle = \langle e^{ik\xi_1/N^{\beta}} \rangle^N, \tag{2.114}$$

so that

$$g^*(k) = [g(kN^{-\beta})]^N, \tag{2.115}$$

where $g(k)$ is the characteristic function associated to the variables ξ_i.

Assuming that the resulting distribution $\rho^*(x)$ is the same distribution $\rho(x_i)$, then

$$g(k) = [g(kN^{-\beta})]^N, \tag{2.116}$$

which is the equation that gives $g(k)$. The corresponding probability distribution is generically called stable distribution. Taking the logarithm of both sides of this equation,

$$\ln g(k) = N \ln g(kN^{-\beta}), \tag{2.117}$$

so that $\ln g(k)$ is a homogeneous function in k. A possible solution of this equation is given by

$$\ln g(k) = -a|k|^\alpha, \tag{2.118}$$

where $\alpha = 1/\beta$ and $a > 0$, or yet

$$g(k) = e^{-a|k|^\alpha}. \tag{2.119}$$

When $\alpha = 2$, and therefore $\beta = 1/2$, the corresponding distribution is the Gaussian distribution. When $0 < \alpha < 2$ they are called Levy distributions. The Lorentz distribution is a particular Levy distribution corresponding to $\alpha = 1$, and therefore to $\beta = 1$.

The probability distribution $\rho(x)$ is obtained from the characteristic function by means of the inverse Fourier transform

$$\rho(x) = \frac{1}{2\pi} \int_{-\infty}^{\infty} e^{-ikx - a|k|^\alpha} dk. \tag{2.120}$$

Figure 2.1 shows these distributions for the cases $\alpha = 2$ (Gaussian), $\alpha = 1$ (Lorentz) and $\alpha = 0.5$. It is worth to note that, for $0 < \alpha < 2$, these distributions decay algebraically when $x \to \infty$. The behavior of $\rho(x)$ for sufficient large x is given by

$$\rho(x) = \frac{A_\alpha}{|x|^{1+\alpha}}, \tag{2.121}$$

where $A_\alpha = \sin(\pi\alpha/2)\Gamma(1 + \alpha)/\pi$. Taking the logarithm

$$\ln \rho(x) = \ln A_\alpha - (1 + \alpha)\ln|x|, \tag{2.122}$$

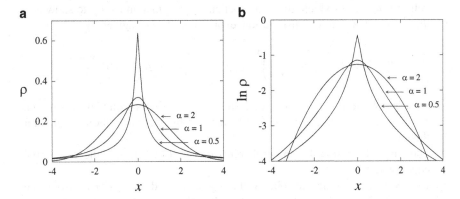

Fig. 2.1 Three stable distributions, defined by (2.120), corresponding to $\alpha = 2$ (Gaussian), $\alpha = 1$ (Lorentzian) and $\alpha = 0.5$. (a) Probability density ρ versus x. (b) Same plot, with ρ in logarithm scale

from which follows that the plot of $\ln \rho$ versus $\ln |x|$ has a linear behavior.

The algebraic behavior (2.121) leads us to an important result: the stable distributions with $0 < \alpha < 2$ have infinite variances. They do not obey the hypotheses of the central limit theorem.

Exercises

1. Generate a sequence of random numbers $\xi_1, \xi_2, \xi_3, \ldots$ that take the values -1 or $+1$ with equal probability. Plot the frequency $f_n = (\xi_1 + \xi_2 + \ldots + \xi_n)/n$ versus the number n. Confirm that $f_n \to 0$. Repeat the procedure for random number generated according to the Lorentz distribution given by $1/[\pi(1 + \xi^2)]$.

2. Generate a sequence of N random numbers $\xi_1, \xi_2, \ldots, \xi_N$ that take the values 0 or 1 and calculate $z = (\xi_1 + \xi_2 + \ldots + \xi_N - Na)/\sqrt{Nb}$, where $a = 1/2$ is the mean and $b = 1/4$ is the variance of the numbers generated. Repeat the procedure L times and make a histogram of the values of z. Compare it with the Gaussian distribution $(2\pi)^{-1/2} \exp\{-z^2/2\}$.

3. Consider a sequence of N independent and equally distributed random variables $\sigma_1, \sigma_2, \ldots, \sigma_N$, that take the values $+1$ or -1 with equal probability $1/2$, and let x be the variable $x = (\sigma_1 + \sigma_2 + \ldots + \sigma_N)/N$. Determine the averages $\langle x \rangle$, $\langle |x| \rangle$, and $\langle x^2 \rangle$ for large N.

4. Use Stirling formula

$$n! = n^n e^{-n} \sqrt{2\pi n},$$

valid for $n \gg 1$, which gives an excellent approximation for $n!$, to show that the binomial probability distribution, expression (2.23), is given by

$$P_N(\ell) = \left(\frac{N}{2\pi\ell(N-\ell)}\right)^{1/2} \exp\{-\ell \ln \frac{\ell}{Np} - (N-\ell) \ln \frac{N-\ell}{Nq}\},$$

for large N and ℓ. Expand the expression between braces around its maximum, $\ell = Np$, to reach the result (2.42), derived by means of the central limit theorem.

5. Show by substitution that the probability density (2.67) fulfills the differential equation (2.68).

6. Use Stirling formula to obtain the probability density (2.67) from the result (2.53), valid for the random walk. The appropriated definitions are $x = hm, t = n\tau, D = h^2b/2\tau, c = ha/\tau, a = p - q$ and $b = 4pq$.

7. Consider the one-dimensional random walk described by the sequence of independent random variables $\sigma_1, \sigma_2, \sigma_3, \ldots, \sigma_n$ that take the values $+1$ (step to the right), -1 (step to the left) and 0, (remaining in the same position). The probability of remaining in the same position, $\sigma_i = 0$, is p while the probability of taking a step to the right, $\sigma_i = +1$ is $q/2$ and to the left, $\sigma_i = -1$, is also $q/2$, where $q = 1 - p$. Determine the probability $P_n(m)$ of finding the walker at position m after n time steps. Find the probability distribution for large n.

8. Consider the one-dimensional random walk described by the sequence of independent random variables $\sigma_1, \sigma_2, \sigma_3, \ldots, \sigma_n$ that take the values $+1$ (step to the right) and -1 (step to the left). Suppose, however, that the probability of $\sigma_j = +1$ is p if j is odd and q if j is even, where $p + q = 1$. Consequently, the probability of $\sigma_j = -1$ is q if j is odd and p if j is even. Determine, for large n, the probability distribution of the position $x = h(\sigma_1 + \sigma_2 + \ldots + \sigma_n)$ of the walker, for large n.

9. A walker makes a two-dimensional random walk. At each time interval τ the possible displacements are $(\pm h, \pm h)$ all with the same probability. Determine, for large n, the probability of finding the walker at position $\mathbf{r} = (hm_1, hm_2)$ after n intervals of time, for large n.

10. A gas molecule moves a distance h between two collisions with equal probability in any direction. Denoting the j-th displacement by $\mathbf{r}_j = (x_j, y_j, z_j)$, then the total displacement from the origin after n collisions, is $\mathbf{r} = \mathbf{r}_1 + \mathbf{r}_2 + \ldots + \mathbf{r}_n$. Determine the average square displacement $\langle \mathbf{r}^2 \rangle$ of the molecule. Find also the characteristic function $G(\mathbf{k}) = \langle \exp(i\mathbf{k}\cdot\mathbf{r}) \rangle$ of the variable $\mathbf{r} = (x, y, z)$ and the probability distribution of $\mathbf{r}$. Find the expression of this probability for large n.

Chapter 3
Langevin Equation

3.1 Brownian Motion

According to Langevin, a particle performing a random movement, which we call Brownian motion, is subject to two forces. One dissipative, which we assume to be proportional to its velocity, and another of random character due to the impact of the particle with the molecules of the medium. Considering the simple case of a one-dimensional motion along a straight line, the equation of motion for a particle of mass m is given by

$$m\frac{dv}{dt} = -\alpha v + F_a(t), \tag{3.1}$$

where

$$v = \frac{dx}{dt} \tag{3.2}$$

is the velocity and x the position of the particle. The first term on the right-hand side of Eq. (3.1) is the friction force and hence of dissipative nature, where α is the friction coefficient, and $F_a(t)$ is the random force that has the following properties,

$$\langle F_a(t) \rangle = 0, \tag{3.3}$$

since in the average the force due to collisions with the molecules is zero and

$$\langle F_a(t) F_a(t') \rangle = B\delta(t - t') \tag{3.4}$$

because we assume that the impacts are independent. Equation (3.1), supplemented by the properties (3.3) and (3.4), is called Langevin equation.

© Springer International Publishing Switzerland 2015
T. Tomé, M.J. de Oliveira, *Stochastic Dynamics and Irreversibility*, Graduate Texts in Physics,
DOI 10.1007/978-3-319-11770-6_3

Dividing both sides of Eq. (3.1) by m, the Langevin takes the form

$$\frac{dv}{dt} = -\gamma v + \zeta(t),\tag{3.5}$$

where $\gamma = \alpha/m$ and $\zeta(t) = F_a(t)/m$. The noise $\zeta(t)$ is a stochastic variable, that is, a random variable that depends on time, with the properties,

$$\langle \zeta(t) \rangle = 0,\tag{3.6}$$

$$\langle \zeta(t)\zeta(t') \rangle = \Gamma\delta(t - t'),\tag{3.7}$$

where $\Gamma = B/m^2$.

Mean-square velocity The generic solution of the differential equation (3.5) is obtained as follows. We start by writing $v(t) = u(t)e^{-\gamma t}$, where $u(t)$ is a function of t to be determined. Replacing in (3.5), we see that it fulfills the equation

$$\frac{du}{dt} = e^{\gamma t}\zeta(t),\tag{3.8}$$

whose solution is

$$u = u_0 + \int_0^t e^{\gamma t'}\zeta(t')dt'.\tag{3.9}$$

Therefore

$$v = v_0 e^{-\gamma t} + e^{-\gamma t}\int_0^t e^{\gamma t'}\zeta(t')dt',\tag{3.10}$$

where v_0 is the velocity of the particle at time $t = 0$. This solution is valid for any time function $\zeta(t)$. Next, we use the specific properties of the noise to determine the mean and the variance of the velocity.

Using the property (3.6),

$$\langle v \rangle = v_0 e^{-\gamma t}.\tag{3.11}$$

Therefore

$$v - \langle v \rangle = e^{-\gamma t}\int_0^t e^{\gamma t'}\zeta(t')dt',\tag{3.12}$$

from which we get

$$(v - \langle v \rangle)^2 = e^{-2\gamma t} \int_0^t \int_0^t \zeta(t')\zeta(t'')e^{\gamma(t'+t'')}dt'dt'', \tag{3.13}$$

$$\langle (v - \langle v \rangle)^2 \rangle = e^{-2\gamma t} \int_0^t \Gamma e^{2\gamma t'} dt', \tag{3.14}$$

where we used the property (3.7). Performing the integral, we get the velocity variance

$$\langle v^2 \rangle - \langle v \rangle^2 = \frac{\Gamma}{2\gamma}(1 - e^{-2\gamma t}). \tag{3.15}$$

For large times, that is, in the stationary regime, $\langle v \rangle = 0$ and the mean-square velocity becomes

$$\langle v^2 \rangle = \frac{\Gamma}{2\gamma}. \tag{3.16}$$

From the kinetic theory, we know that

$$\frac{1}{2}m\langle v^2 \rangle = \frac{1}{2}k_B T, \tag{3.17}$$

where k_B is the Boltzmann constant and T is the absolute temperature. Comparing these two equations, we get the relation between the coefficient Γ and the temperature,

$$\Gamma = \frac{2\gamma k_B T}{m}. \tag{3.18}$$

Recalling that $B = \Gamma m^2$ and that $\alpha = \gamma m$, we reach the relation between B and the temperature,

$$B = 2\alpha k_B T. \tag{3.19}$$

Mean-square displacement Next we determine the mean-square displacement of the particle. To this end, we calculate first $x(t)$, given by

$$x = x_0 + \int_0^t v(t')dt', \tag{3.20}$$

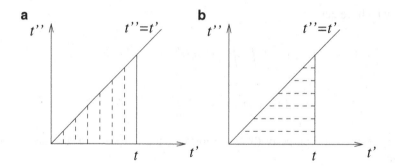

Fig. 3.1 Two equivalent ways of carrying out the integration over the hatched area. We may integrate first in t'' and then in t', as illustrated in (**a**), or we may integrate first in t' and then in t'', as illustrated in (**b**)

where x_0 is the position of the particle at time $t = 0$ or, in view of Eq. (3.10),

$$x = x_0 + v_0 \int_0^t e^{-\gamma t'} dt' + \int_0^t e^{-\gamma t'} \int_0^{t'} \zeta(t'') e^{\gamma t''} dt'' dt'. \tag{3.21}$$

Performing the first integral and changing the order of the integrals of the last term, according to the illustration in Fig. 3.1,

$$x = x_0 + v_0 \frac{1}{\gamma}(1 - e^{-\gamma t}) + \int_0^t \zeta(t'') e^{\gamma t''} \int_{t''}^t e^{-\gamma t'} dt' dt''. \tag{3.22}$$

Integrating in t', we get the result

$$x = x_0 + v_0 \frac{1}{\gamma}(1 - e^{-\gamma t}) + \frac{1}{\gamma} \int_0^t \zeta(t'')(1 - e^{\gamma(t''-t)}) dt'', \tag{3.23}$$

valid for any time function $\zeta(t)$. Using the property (3.6), we get the mean displacement of the particle

$$\langle x \rangle = x_0 + v_0 \frac{1}{\gamma}(1 - e^{-\gamma t}). \tag{3.24}$$

The mean-square displacement is calculated by first determining, from (3.23) and (3.24), the difference

$$x - \langle x \rangle = \frac{1}{\gamma} \int_0^t \zeta(t'')(1 - e^{\gamma(t''-t)}) dt'', \tag{3.25}$$

from which we get

$$(x - \langle x \rangle)^2 = \frac{1}{\gamma^2} \int_0^t \int_0^t \zeta(t')\zeta(t'')(1 - e^{\gamma(t'-t)})(1 - e^{\gamma(t''-t)}) dt' dt''. \qquad (3.26)$$

Next, using the property (3.7), it follows that

$$\langle (x - \langle x \rangle)^2 \rangle = \frac{\Gamma}{\gamma^2} \int_0^t (1 - e^{\gamma(t'-t)})^2 dt', \qquad (3.27)$$

from which we obtain, after performing the integral,

$$\langle x^2 \rangle - \langle x \rangle^2 = \frac{\Gamma}{\gamma^2} \{ t - \frac{2}{\gamma}(1 - e^{-\gamma t}) + \frac{1}{2\gamma}(1 - e^{-2\gamma t}) \}. \qquad (3.28)$$

For large times, the dominant term is the first one so that, in this case, the mean-square displacement is proportional to t, that is,

$$\langle x^2 \rangle - \langle x \rangle^2 = 2Dt, \qquad (3.29)$$

where $D = \Gamma/2\gamma^2 = B/2\alpha^2$ is the diffusion coefficient. Using relation (3.18), we can write the following relation between the diffusion coefficient and the temperature,

$$D = \frac{k_B T}{\alpha}. \qquad (3.30)$$

Although we have derived the relation (3.30) for the one-dimensional case, it is also valid in two and three dimensions. For spherical particles of radius a immersed in a liquid of viscosity coefficient μ, the constant α is given by the Stokes law,

$$\alpha = 6\pi\mu a, \qquad (3.31)$$

so that

$$D = \frac{k_B T}{6\pi\mu a}. \qquad (3.32)$$

which is the Sutherland-Einstein relation. The knowledge of the diffusion coefficient D, obtained through the measurement of the mean-square displacement $\langle x^2 \rangle - \langle x \rangle^2$, and of the quantities μ and a allows for the determination of the Boltzmann constant k_B. In fact, such an experiment was performed by Perrin examining the Brownian motion of particles suspended in a liquid.

The observation of the Brownian motion of a particle immersed in a liquid is possible because it is much larger than the molecules with which it collides. The molecules also perform Brownian motion, but according to Perrin (1913), "l'agitation moléculaire échappe à notre perception directe comme le mouvement de vagues de la mer à un observateur trop éloigné. Cependant, si quelque bateau se trouve alors en vue, le même observateur pourra voir un balancement que lui révélera l'agitation qu'il ne soupçonnait pas."

Energy The random impact of the molecules of the medium on the particle transfers kinetic energy from the molecules to the particles. The particle, in turn, dissipates energy due to the friction force and, at the same time, has its kinetic energy modified. The rate at which the energy is transferred to the particle is equal to the sum of the rate at which energy is dissipated and the variation of the kinetic energy of the particle. The energy balance is obtained as follows. We multiply both sides of the Langevin equation by v and use the equality $v\,dv/dt = (1/2)dv^2/dt$ to get

$$\frac{m}{2}\frac{d}{dt}v^2 = -\alpha v^2 + vF_a \tag{3.33}$$

or

$$\frac{d}{dt}\left(\frac{m}{2}v^2\right) + vF_{dis} = vF_a, \tag{3.34}$$

where $F_{dis} = \alpha v$ is the dissipative force.

Taking the average,

$$\frac{dE_c}{dt} + P_{dis} = P, \tag{3.35}$$

where $E_c = m\langle v^2\rangle/2$ is the average kinetic energy, $P_{dis} = \langle vF_{dis}\rangle = \langle \alpha v^2\rangle$ is the rate of dissipation of energy, or dissipated power, and $P = \langle vF_a\rangle$ is the rate at which energy is transferred to the particle, or transferred power.

Each term on the left side of Eq. (3.35) can be obtained in an explicit way as functions of time using the expression for $\langle v^2\rangle$ given by (3.15). Doing the calculations, we reach the result

$$P = \frac{m\Gamma}{2} = \frac{B}{2m} = \frac{\alpha k_B T}{m}, \tag{3.36}$$

that is, the transferred power is independent of time.

We remark that, although $P_{dis} \geq 0$, the variation of the kinetic energy dE_c/dt can be positive, zero or negative. This last case occurs when the initial kinetic energy of the particle, $mv_0^2/2$, is great than the average kinetic energy of the stationary state, $k_B T/2$. In the stationary regime, however, the variation in kinetic energy vanishes, $dE_c/dt = 0$, so that $P_{dis} = P$ and therefore all energy transferred to the particle is dissipated.

Langevin's solution The solution of Eq. (3.1) originally done by Langevin is as follows. Defining the variable $z = (d/dt)x^2$, its derivative with respect to time is

$$\frac{dz}{dt} = 2v^2 + 2x\frac{dv}{dt}. \tag{3.37}$$

Replacing (3.1) in this equation, and taking into account that $2xv = 2xdx/dt = z$, we get the result

$$m\frac{dz}{dt} = 2mv^2 - \alpha z + 2xF_a(t). \tag{3.38}$$

Next, Langevin assumes that $\langle xF_a \rangle = 0$, to get the equation

$$m\frac{d}{dt}\langle z \rangle = 2m\langle v^2 \rangle - \alpha \langle z \rangle. \tag{3.39}$$

In the stationary estate, we use the equipartition the energy, $m\langle v^2 \rangle = k_B T$, to get

$$\langle z \rangle = \frac{2k_B T}{\alpha} = 2D. \tag{3.40}$$

Since $z = (d/dt)x^2$, then

$$\frac{d}{dt}\langle x^2 \rangle = 2D, \tag{3.41}$$

whose integration gives the result

$$\langle x^2 \rangle - \langle x_0^2 \rangle = 2Dt, \tag{3.42}$$

valid for large times and obtained by Langevin in 1908.

3.2 Probability Distribution

Velocity distribution We have seen that the velocity $v(t)$ of a free particle in a viscous medium and subjected to random forces varies according to the Langevin equation (3.5) where $\zeta(t)$ is a stochastic variable, that is, a time-dependent random variable. Likewise, the velocity $v(t)$ is also a stochastic variable. The difference between them is that the probability distribution of $\zeta(t)$ is known in advance whereas that of $v(t)$ we wish to determine.

To find the probability distribution of $v(t)$, we start by discretizing time in intervals equal to τ, writing $t = n\tau$. Thus, Eq. (3.5) is written in the approximate form

$$v_{n+1} = av_n + \sqrt{\tau \Gamma} \xi_n, \tag{3.43}$$

where $a = (1 - \tau\gamma)$ and the random variables ξ_j have the properties

$$\langle \xi_j \rangle = 0, \qquad\qquad \langle \xi_j \xi_k \rangle = \delta_{jk}. \tag{3.44}$$

From (3.43), it follows the relation

$$v_n = \sum_{\ell=0}^{n-1} w_\ell, \tag{3.45}$$

where w_ℓ is defined by

$$w_\ell = a^\ell \sqrt{\tau \Gamma} \xi_{n-1-\ell} \tag{3.46}$$

and we consider the initial condition $v_0 = 0$. Thus v_n becomes a sum of independent random variables.

Denoting by $g_n(k)$ the characteristic function related to the variable v_n,

$$g_n(k) = \langle e^{ikv_n} \rangle = \prod_{\ell=0}^{n-1} \langle e^{ikw_\ell} \rangle \tag{3.47}$$

because the variables w_ℓ are independent. Assuming that the probability distribution of the variable ξ_ℓ is Gaussian of zero mean and variance 1, it follows that the probability distribution of the variable w_ℓ is also Gaussian of zero mean, but with variance $a^{2\ell}\tau\Gamma$. Therefore,

$$\langle e^{ikw_\ell} \rangle = e^{-a^{2\ell}\tau\Gamma k^2/2}, \tag{3.48}$$

from which we get

$$g_n(k) = e^{-b_n k^2/2}, \tag{3.49}$$

where

$$b_n = \tau\Gamma \sum_{\ell=0}^{n-1} a^{2\ell} = \frac{1 - a^{2n}}{1 - a^2} \tau\Gamma. \tag{3.50}$$

The probability density of the variable v_n is obtained by calculating the inverse Fourier transform of (3.49), that is,

$$P_n(v_n) = \frac{1}{\sqrt{2\pi b_n}} e^{-v_n^2/2b_n}. \tag{3.51}$$

Taking the limits $\tau \to 0$ and $n \to \infty$ with $n\tau = t$ fixed, the probability density $\rho(v, t)$ of the variable v at time t is

$$\rho(v, t) = \frac{1}{\sqrt{2\pi b(t)}} e^{-v^2/2b(t)}, \tag{3.52}$$

where $b(t)$ is the limit of b_n, given by

$$b(t) = \frac{\Gamma}{2\gamma}(1 - e^{-2\gamma t}) = \frac{k_B T}{m}(1 - e^{-2\gamma t}). \tag{3.53}$$

In the stationary regime, that is, when $t \to \infty$, $b(t) \to k_B T/m$, and we reach the result

$$\rho(v) = \sqrt{\frac{m}{2\pi k_B T}}\, e^{-mv^2/2k_B T}, \tag{3.54}$$

which is the Maxwell velocity distribution, for the one-dimensional case.

Position distribution To obtain the probability distribution of the position, we proceed in the same way as above. Using the same discretization we write

$$x_{n+1} = x_n + \tau v_n, \tag{3.55}$$

from which follows the relation

$$x_n = \tau \sum_{\ell=1}^{n-1} v_\ell, \tag{3.56}$$

in which we have set $x_0 = 0$ and $v_0 = 0$. Notice that the variables v_ℓ are not independent. However, if we use Eq. (3.45) we may write

$$x_n = \sum_{\ell=1}^{n-1} u_\ell, \tag{3.57}$$

where

$$u_\ell = \frac{1}{\gamma}(1 - a^\ell)\sqrt{\tau \Gamma}\xi_{n-1-\ell}, \tag{3.58}$$

to that now x_n is a sum of independent variables u_ℓ.

Denoting by $G_n(k)$ the characteristic function related to the random variable x_n,

$$G_n(k) = \langle e^{ikx_n} \rangle = \prod_{\ell=1}^{n-1} \langle e^{iku_\ell} \rangle. \tag{3.59}$$

Since the variable ξ_ℓ has a Gaussian distribution of zero mean and variance 1, then the variable u_ℓ also has the the same distribution, but with variance $(1-a^\ell)^2 \tau \Gamma / \gamma^2$. Hence

$$G_n(k) = e^{-d_n k^2/2}, \tag{3.60}$$

where

$$d_n = \frac{\tau \Gamma}{\gamma^2} \sum_{\ell=1}^{n-1}(1 - a^\ell)^2 = \frac{\tau \Gamma}{\gamma^2}\{n - 2\frac{1-a^n}{1-a} + \frac{1-a^{2n}}{1-a^2}\}. \tag{3.61}$$

In the limits $\tau \to 0$ and $n \to \infty$ with $n\tau = t$ fixed, we get the probability density $\rho_1(x,t)$ of the variables x in the Gaussian form

$$\rho_1(x,t) = \frac{1}{\sqrt{2\pi d(t)}} e^{-x^2/2d(t)}, \tag{3.62}$$

where $d(t)$ is the limit of d_n, given by

$$d(t) = \frac{\Gamma}{\gamma^2}\{t - \frac{2}{\gamma}(1 - e^{-\gamma t}) + \frac{1}{2\gamma}(1 - e^{-2\gamma t})\}, \tag{3.63}$$

which in turn is the variance of $x(t)$ given by Eq. (3.28). In the regime where t is very large $d(t) \to 2Dt$, where $D = \Gamma/2\gamma^2$, and thus

$$\rho_1(x,t) = \frac{1}{\sqrt{4\pi Dt}} e^{-x^2/4Dt}, \tag{3.64}$$

which coincides with the expression seen previously in Sect. 2.4.

3.3 Time Evolution of the Moments

In this section we study a generic Langevin equation in one variable, that we write in the form

$$\frac{dx}{dt} = f(x) + \zeta(t), \tag{3.65}$$

where $f(x)$ is a function of x only, which we merely call force, and $\zeta(t)$ is the stochastic variable, or noise, whose properties are known. It has zero mean

$$\langle \zeta(t) \rangle = 0 \tag{3.66}$$

and, in addition,

$$\langle \zeta(t)\zeta(t') \rangle = \Gamma \delta(t - t'), \tag{3.67}$$

that is, the variables $\zeta(t)$ and $\zeta(t')$ are independent for $t \neq t'$. The random variable that has this property is called white noise because the Fourier transform of $\langle \zeta(0)\zeta(t) \rangle$, given by

$$\int e^{i\omega t} \langle \zeta(0)\zeta(t) \rangle dt = \Gamma \tag{3.68}$$

is independent of the frequency ω.

Example 3.1 In the Brownian motion, seen in Sect. 3.1, a free particle moves in a viscous medium and is subject to random forces. Suppose that in addition to these forces it is subject to an external force $F(x)$. The equation of motion is

$$m\frac{d^2x}{dt^2} = F(x) - \alpha\frac{dx}{dt} + F_a(t). \tag{3.69}$$

For cases in which the mass of the particle is negligible, this equation reduces to

$$\alpha\frac{dx}{dt} = F(x) + F_a(t). \tag{3.70}$$

Dividing both sides by α, this equation becomes of the type (3.65) with $f(x) = F(x)/\alpha$, and $\zeta(t) = F_a(t)/\alpha$.

The Langevin equation may be called stochastic equation because each of the variables entering in this equation is a stochastic variable, which means that each has a time-dependent probability distribution. Therefore, to solve the Langevin equation means to determine the probability distribution $P(x,t)$ at each instant of time $t > 0$, given the probability distribution $P(x,0)$ at $t = 0$. If at the initial time the particle is located at the point x_0, then $P(x, 0) = \delta(x - x_0)$. Alternatively, we can determine all the moments $\mu_\ell(t) = \langle x^\ell \rangle$ as functions of time, given the moments at the initial time. Later, we will see how to derive from the Langevin equation, a differential equation for $P(x, t)$, called Fokker-Planck equation, which is in this sense equivalent to the Langevin equation. In this section we derive evolution equations for the moments of x.

We start by discretizing time in intervals equal to τ. The position at time $t = n\tau$ is x_n and the Langevin equation in the discretized form is

$$x_{n+1} = x_n + \tau f_n + \sqrt{\tau \Gamma} \xi_n, \tag{3.71}$$

where $f_n = f(x_n)$ and ξ_n has the properties

$$\langle \xi_n \rangle = 0, \qquad \langle \xi_n \xi_{n'} \rangle = \delta_{nn'}. \tag{3.72}$$

The discretized Langevin can be viewed as a recurrence equation. Notice that the random variable x_{n+1} is independent of ξ_{n+1}, although it depends on ξ_n, ξ_{n-1}, ξ_{n-2} etc.

Next, we see that

$$\langle x_{n+1} \rangle = \langle x_n \rangle + \tau \langle f_n \rangle. \tag{3.73}$$

Taking into account that $(\langle x_{n+1} \rangle - \langle x_n \rangle)/\tau$ is identified with the derivative $d\langle x \rangle/dt$ when $\tau \to 0$, then

$$\frac{d}{dt} \langle x \rangle = \langle f(x) \rangle, \tag{3.74}$$

which is the equation for the time evolution of the average $\langle x \rangle$.

Squaring both sides of Eq. (3.71), we get

$$x_{n+1}^2 = x_n^2 + 2\sqrt{\tau \Gamma} \xi_n x_n + \tau \Gamma \xi_n^2 + 2\tau x_n f_n + \tau \sqrt{\tau \Gamma} \xi_n f_n + \tau^2 f_n^2. \tag{3.75}$$

Using the property that x_n and ξ_n are independent, and that $\langle \xi_n \rangle = 0$ and $\langle \xi_n^2 \rangle = 1$, we reach the result

$$\langle x_{n+1}^2 \rangle = \langle x_n^2 \rangle + \tau \Gamma + 2\tau \langle x_n f_n \rangle + \tau^2 \langle f_n^2 \rangle. \tag{3.76}$$

Taking into account that $(\langle x_{n+1}^2 \rangle - \langle x_n^2 \rangle)/\tau \to d\langle x^2 \rangle/dt$ when $\tau \to 0$, then

$$\frac{d}{dt} \langle x^2 \rangle = \Gamma + 2\langle x f(x) \rangle, \tag{3.77}$$

which gives the time evolution of the second moment $\langle x^2 \rangle$.

To determine the equation for the time evolution of the ℓ-th moment of x, we raise both sides of Eq. (3.71) to the ℓ-th power

$$x_{n+1}^{\ell} = \{x_n + \tau f_n + \sqrt{\tau \Gamma} \xi_n\}^{\ell}. \tag{3.78}$$

Neglecting terms of order greater than τ, we obtain

$$x_{n+1}^{\ell} = x_n^{\ell} + \ell \sqrt{\tau \Gamma} \xi_n x_n^{\ell-1} + \ell \tau x_n^{\ell-1} f_n + \frac{1}{2} \ell (\ell - 1) \tau \Gamma \xi_n^2 x_n^{\ell-2}. \tag{3.79}$$

Taking into account that $(\langle x_{n+1}^{\ell} \rangle - \langle x_n^{\ell} \rangle)/\tau \to d\langle x^{\ell} \rangle/dt$ when $\tau \to 0$, we get

$$\langle x_{n+1}^{\ell} \rangle = \langle x_n^{\ell} \rangle + \ell \tau \langle x_n^{\ell-1} f_n \rangle + \frac{1}{2} \ell (\ell - 1) \tau \Gamma \langle x_n^{\ell-2} \rangle \tag{3.80}$$

so that

$$\frac{d}{dt} \langle x^{\ell} \rangle = \ell \langle x^{\ell-1} f(x) \rangle + \frac{1}{2} \ell (\ell - 1) \Gamma \langle x^{\ell-2} \rangle, \tag{3.81}$$

which gives the time evolution of the moment $\langle x^{\ell} \rangle$.

Equation (3.81) is actually a set of equations for the various moments of the random variable x. The equation for the first moment may depend on the second moment and thus cannot be solved alone. Thus, we need the equation for the second moment. However, the equation for the second moment may depend on the third moment so that we have to use the equation for the third moment. Therefore, the set of equations (3.81) constitutes a hierarchical set of equations. In some cases it may happen that the equation for a certain moment has only moments of lower order. In this case, we have a finite set of equations to solve.

Example 3.2 If $f(x) = c$,

$$\frac{d}{dt} \langle x \rangle = c, \qquad \frac{d}{dt} \langle x^2 \rangle = 2c \langle x \rangle + \Gamma, \tag{3.82}$$

whose solutions for the initial condition $\langle x \rangle = x_0$ and $\langle x^2 \rangle = x_0^2$ at $t = 0$ are

$$\langle x \rangle = x_0 + ct, \qquad \langle x^2 \rangle = x_0^2 + (\Gamma + 2cx_0)t + c^2 t^2. \tag{3.83}$$

From these two moments, we get the variance

$$\langle x^2 \rangle - \langle x \rangle^2 = \Gamma t. \tag{3.84}$$

This example is identified with the problem of the random walk seen in Sect. 2.4.

Example 3.3 Suppose that $f(x) = -\nu x$. In this case, the two first equations give us

$$\frac{d}{dt}\langle x \rangle = -\nu\langle x \rangle, \tag{3.85}$$

$$\frac{d}{dt}\langle x^2 \rangle = -2\nu\langle x^2 \rangle + \Gamma. \tag{3.86}$$

Therefore, by integration of these two equations we may get the two first moments. With the initial condition $\langle x \rangle = x_0$ and $\langle x^2 \rangle = x_0^2$ at $t = 0$ we get the following solution,

$$\langle x \rangle = x_0 e^{-\nu t}, \tag{3.87}$$

$$\langle x^2 \rangle = \frac{\Gamma}{2\nu} + (x_0^2 - \frac{\Gamma}{2\nu}) e^{-2\nu t}. \tag{3.88}$$

From them we get the variance

$$\langle x^2 \rangle - \langle x \rangle^2 = \frac{\Gamma}{2\nu}(1 - e^{-2\nu t}). \tag{3.89}$$

When $t \to \infty$, $\langle x \rangle \to 0$ and $\langle x^2 \rangle \to \Gamma/2\nu$. This example can be understood as the original Langevin equation seen in Sect. 3.1. To see this, it suffice to make the replacements $x \to v$, $\nu \to \gamma$.

3.4 Simulation of the Random Motion

The motion of a particle that obeys the Langevin equation

$$\frac{dx}{dt} = f(x) + \zeta(t), \tag{3.90}$$

where $\zeta(t)$ has the properties

$$\langle \zeta(t) \rangle = 0, \tag{3.91}$$

$$\langle \zeta(t)\zeta(t') \rangle = \Gamma\delta(t - t'), \tag{3.92}$$

can be simulated as follows. We discretize the time in intervals equal to τ and denote by x_n the position of the particle at time $t = n\tau$. The Langevin equation becomes thus,

$$x_{n+1} = x_n + \tau f_n + \sqrt{\tau\Gamma}\xi_n. \tag{3.93}$$

where $f_n = f(x_n)$ and $\xi_0, \xi_1, \xi_2, \ldots$ comprise a sequence of independent random variables such that

$$\langle \xi_n \rangle = 0, \qquad \langle \xi_n^2 \rangle = 1. \tag{3.94}$$

Thus, if a sequence of random numbers is generated $\xi_0, \xi_1, \xi_2, \ldots$ and if the initial position x_0 is given, we can generate the sequence of points $x_1, x_2, x_3, \ldots$, which constitutes the discretize trajectory of the particle. The variables ξ_i must be generated according to a a distribution with zero mean and variance equal to unity.

Suppose that we wish to know the average position of the particle as a function of time. We should generate several trajectories starting from the same point x_0. An estimate of the average $\langle x \rangle$ at time $t = n\tau$ is

$$\overline{x}_n = \frac{1}{L} \sum_{j=1}^{L} x_n^{(j)}, \tag{3.95}$$

where L is the number of trajectories and $x_n^{(j)}$ denotes the position of the particle at time $t = n\tau$ belonging to the j-th trajectory. In an analogous manner, we can obtain an estimate $\overline{x_n^2}$ of the second moment $\langle x^2 \rangle$. From the first and second moment we obtain an estimate of the variance, $\overline{x_n^2} - (\overline{x}_n)^2$.

The simulation of the L trajectories may be understand as being the trajectories of L noninteracting particles that move along the straight line x, all starting at the same point x_0 at time $t = 0$. At each time step τ, each particle moves to a new position according to the discretized Langevin equation. At each instant of time t we can also construct a histogram of x, which is proportional to the probability distribution $P(x, t)$. To this end, we make a partition of the x-axis in windows of size Δx and for each window $[x, x + \Delta x]$ we determine the number $N(x, t)$ of particles whose positions are inside the window.

In the stationary state, we may use just one trajectory because the probability distribution becomes invariant. Therefore, an estimate of the average $\langle x \rangle$ is

$$\overline{x} = \frac{1}{M} \sum_{n=1}^{M} x_n \tag{3.96}$$

where M is the numbers of values used to determine the average, along the trajectory, inside the stationary regime. In an analogous way, we can obtain an estimate $\overline{x^2}$ of the second moment $\langle x^2 \rangle$. From this single trajectory we can also determine an histogram of x, which is proportional to the stationary probability distribution $P(x)$.

3.5 Fokker-Planck Equation

Let $P_n(x_n)$ be the probability distribution of the variable x_n and $g_n(k)$ its corresponding characteristic function, given by

$$g_n(k) = \langle e^{ikx_n} \rangle = \int e^{ikx_n} P_n(x_n) dx_n. \qquad (3.97)$$

Using the discretized Langevin equation, given by (3.71),

$$g_{n+1}(k) = \langle e^{ikx_{n+1}} \rangle = \langle e^{ik[x_n + \tau f_n + \sqrt{\tau \Gamma} \xi_n]} \rangle, \qquad (3.98)$$

where $f_n = f(x_n)$, or, taking into account that x_n and ζ_n are independent,

$$g_{n+1}(k) = \langle e^{ik(x_n + \tau f_n)} \rangle \langle e^{ik\sqrt{\tau \Gamma} \xi_n} \rangle. \qquad (3.99)$$

Next, we perform the expansion of $g_{n+1}(k)$, up to terms of first order in τ. The first term of the product gives

$$\langle e^{ikx_n} \{1 + ik\tau f_n\} \rangle = \langle e^{ikx_n} \rangle + ik\tau \langle f_n e^{ikx_n} \rangle \qquad (3.100)$$

and the second,

$$1 + ik\sqrt{\tau \Gamma} \langle \xi_n \rangle - \frac{1}{2} k^2 \tau \Gamma \langle \xi_n^2 \rangle = 1 - \frac{1}{2} k^2 \tau \Gamma, \qquad (3.101)$$

where we used the properties $\langle \xi_n \rangle = 0$ and $\langle \xi_n^2 \rangle = 1$. Therefore,

$$g_{n+1}(k) = g_n(k) + ik\tau \langle f_n e^{ikx_n} \rangle - \frac{\tau \Gamma}{2} k^2 \langle e^{ikx_n} \rangle. \qquad (3.102)$$

Taking into account that $(g_{n+1}(k) - g_n(k))/\tau \rightarrow dg(k)/dt$ we get the equation

$$\frac{d}{dt} g(k) = ik\langle f(x) e^{ikx} \rangle - \frac{\Gamma}{2} k^2 \langle e^{ikx} \rangle. \qquad (3.103)$$

We use now the following properties,

$$ik\langle f(x) e^{ikx} \rangle = \langle f(x) \frac{d}{dx} e^{ikx} \rangle = -\int e^{ikx} \frac{\partial}{\partial x} [f(x) P(x)] dx \qquad (3.104)$$

and

$$-k^2 \langle e^{ikx} \rangle = \langle \frac{d^2}{dx^2} e^{ikx} \rangle = \int e^{ikx} \frac{\partial^2}{\partial x^2} P(x) dx \qquad (3.105)$$

which together with

$$g(k) = \int P(x)e^{ikx}dx, \tag{3.106}$$

allows us to reach the equation

$$\frac{\partial}{\partial t}P(x,t) = -\frac{\partial}{\partial x}[f(x)P(x,t)] + \frac{\Gamma}{2}\frac{\partial^2}{\partial x^2}P(x,t), \tag{3.107}$$

which is the time evolution equation for the probability distribution $P(x,t)$. This equation is called Fokker-Planck equation.

Up to linear order in τ, Eq. (3.102) can be written in the equivalent form

$$g_{n+1}(k) = \langle e^{ikx_n + \tau[ikf(x_n) - \Gamma k^2/2]} \rangle. \tag{3.108}$$

Suppose that at time $n = 0$, the particle is at position x'. Then at $n = 0$, the probability is $P_0(x) = \delta(x - x')$ so that

$$g_1(k) = e^{ikx' + \tau[ikf(x') - \Gamma k^2/2]}. \tag{3.109}$$

which must be understood as the characteristic function related to the probability distribution $P_1(x)$ at time τ, that is, as the conditional probability distribution $K(x, \tau; x', 0)$ of finding the particle in position x at time $t = \tau$ given that it was in position x' at time $t = 0$. But (3.109) can be understood as the Fourier transform of a Gaussian distribution of mean $x' + \tau f(x')$ and variance Γ so that

$$K(x, \tau; x', 0) = \frac{1}{\sqrt{2\tau\pi\Gamma}}e^{[x - x' - \tau f(x')]^2/2\tau\Gamma}. \tag{3.110}$$

This expression can also be understood as the solution of the Fokker-Planck equation for small values of time and initial condition such that the particle is in the position x'. Notice that when $\tau \to 0$, in fact $K(x, \tau; x', 0) \to \delta(x - x')$.

3.6 Set of Langevin Equations

In the previous sections we studied the Langevin equations in one variable. Here we analyze a system described by a set of several variables. We consider a system described by N stochastic variables $x_1, x_2, x_3, \ldots, x_N$ whose time evolution is given by the set of equations

$$\frac{dx_i}{dt} = f_i(x_1, x_2, \ldots, x_N) + \zeta_i(t), \tag{3.111}$$

for $i = 1, 2, \ldots, N$, where the stochastic variables $\zeta_1(t), \zeta_2(t), \ldots, \zeta_N(t)$ have the properties

$$\langle \zeta_i(t) \rangle = 0 \tag{3.112}$$

and

$$\langle \zeta_i(t) \zeta_j(t') \rangle = \Gamma_i \delta_{ij} \delta(t - t'), \tag{3.113}$$

where $\Gamma_1, \Gamma_2, \ldots, \Gamma_N$ are constant.

The Langevin equations (3.111) constitute, in general, a set of coupled equations. The simplest example of a coupled set of equations occur when the functions f_i are linear, that is, when

$$f_i = \sum_{j=1}^{N} A_{ij} x_j. \tag{3.114}$$

In this case the Langevin equations are given by

$$\frac{d}{dt} x_i = \sum_{j=1}^{N} A_{ij} x_j + \zeta_i \tag{3.115}$$

and can be written in the matrix form

$$\frac{d}{dt} X = AX + Z, \tag{3.116}$$

where X and Z are column matrices with elements x_i and ζ_i, respectively, and A is the square matrix whose elements are A_{ij}.

To solve the matrix equation (3.116) we determine, first, the matrix M that diagonalize A, that is, we determine M such that

$$M^{-1} A M = \Lambda, \tag{3.117}$$

where Λ is the diagonal matrix whose elements λ_i are the eigenvalues of A. The matrix M is constructed from the right eigenvector of A and the matrix M^{-1}, the inverse of M, is constructed from the left eigenvector of A. Next, we set up the square matrix $R(t)$, defined by

$$R(t) = M D(t) M^{-1}, \tag{3.118}$$

where $D(t)$ is the diagonal matrix whose elements are $D_k(t) = e^{\lambda_k t}$. Notice that at time $t = 0$ the matrix $R(t)$ reduces to the identity matrix, $R(0) = I$, because at $t = 0$ the matrix $D(t)$ is the identity matrix, $D(0) = I$.

Deriving both sides of (3.118) with respect to time, we get

$$\frac{d}{dt}R(t) = M\Lambda D(t)M^{-1}. \tag{3.119}$$

But, given that $M\Lambda = AM$, the right-hand side of this equation becomes equal to $AR(t)$ so that $R(t)$ fulfills the equation

$$\frac{d}{dt}R(t) = AR(t). \tag{3.120}$$

The general solution of (3.116) is obtained with the help of $R(t)$. For the initial condition $X(0) = X_0$, the solution is given by

$$X(t) = R(t)X_0 + \int_0^t R(t-t')Z(t')dt', \tag{3.121}$$

which can be checked by direct substitution into (3.116), using the property (3.120) and bearing in mind that $R(0) = I$. From this solution we can obtain the various moments of x_i.

Define the square matrix C whose elements are given by

$$C_{ij} = \langle (x_i - \langle x_i \rangle)(x_j - \langle x_j \rangle) \rangle, \tag{3.122}$$

which can be written as

$$C_{ij} = \langle x_i x_j \rangle - \langle x_i \rangle \langle x_j \rangle. \tag{3.123}$$

The element C_{ij} is called the covariance between the stochastic variables x_i and x_j and the matrix C is called covariance matrix. From the definition of the column matrix X, whose elements are the stochastic variables x_i and defining $X^\dagger$ as the transpose of X, that is, the row matrix whose elements are x_i, we see that

$$C = (X - \langle X \rangle)(X^\dagger - \langle X^\dagger \rangle). \tag{3.124}$$

Using the general solution (3.121), we observe that

$$\langle X \rangle = R(t)X_0, \tag{3.125}$$

where we have taken into account that $\langle Z \rangle = 0$ so that

$$X - \langle X \rangle = \int_0^t R(t-t')Z(t')dt'. \tag{3.126}$$

Multiplying by its transpose $X^\dagger - \langle X^\dagger \rangle$, we get the following result for the covariance matrix, after taking the average

$$C(t) = \int_0^t R(t') \Gamma R^\dagger(t') dt',\qquad(3.127)$$

where we used (3.112) and the result $\langle Z(t') Z^\dagger(t'') \rangle = \Gamma \delta(t' - t'')$, coming from (3.113) and here Γ is the diagonal matrix whose elements are $\Gamma_i \delta_{ij}$.

3.7 Harmonic Oscillator

In the first sections of this chapter we have analyzed the Brownian motion of a free particle. Here we study the Brownian motion of a confined particle, that is, subject to forces that restrict its movement in a given region of space. Thus, we consider a particle of mass m that moves along the x-axis and subject to a force $F(x)$ that depends only on the position x. The motion equation is

$$m \frac{dv}{dt} = F(x) - \alpha v + F_a(t),\qquad(3.128)$$

where v is the particle velocity,

$$v = \frac{dx}{dt},\qquad(3.129)$$

α is the friction coefficient and $F_a(t)$ is the random force that has the property given by Eqs. (3.3) and (3.4). Dividing both sides of Eq. (3.128) by m, it is written in the form

$$\frac{dv}{dt} = f(x) - \gamma v + \zeta(t),\qquad(3.130)$$

where $\gamma = \alpha/m$, $f(x) = F(x)/m$ and $\zeta(t) = F_a(t)/m$. The noise $\zeta(t)$ has the properties given by Eqs. (3.6) and (3.7).

The simplest type of force that we may imagine is the elastic force derived from the harmonic potential, that is, $F(x) = -Kx$ where K is the elastic constant. In this case $f(x) = -kx$, where $k = K/m$ and Eq. (3.130) reduces to

$$\frac{dv}{dt} = -kx - \gamma v + \zeta(t).\qquad(3.131)$$

With the purpose of solving the set of Eqs. (3.129) and (3.131), we start by writing them in matrix form

$$\begin{pmatrix} dx/dt \\ dv/dt \end{pmatrix} = \begin{pmatrix} 0 & 1 \\ -k & -\gamma \end{pmatrix} \begin{pmatrix} x \\ v \end{pmatrix} + \begin{pmatrix} 0 \\ \zeta \end{pmatrix}.\qquad(3.132)$$

Defining the matrices

$$A = \begin{pmatrix} 0 & 1 \\ -k & -\gamma \end{pmatrix}, \qquad X = \begin{pmatrix} x \\ v \end{pmatrix}, \qquad Z = \begin{pmatrix} 0 \\ \zeta \end{pmatrix}, \qquad (3.133)$$

Eq. (3.132) reduces to the compact form

$$\frac{d}{dt}X = AX + Z, \qquad (3.134)$$

which is identified with Eq. (3.116) so that we may use the method presented in the previous section.

We start by determining the eigenvalues of the matrix A, which are the roots of equation

$$\lambda^2 + \gamma\lambda + k = 0 \qquad (3.135)$$

and are given by

$$\lambda_1 = \frac{1}{2}(-\gamma + \sqrt{\gamma^2 - 4k}), \qquad \lambda_2 = \frac{1}{2}(-\gamma - \sqrt{\gamma^2 - 4k}). \qquad (3.136)$$

Notice that $\lambda_1\lambda_2 = k$ and $\lambda_1 + \lambda_2 = -\gamma$.

The matrix M is constructed from the right eigenvector of A and its inverse M^{-1} is obtained from the left eigenvectors of A and are given by

$$M = \begin{pmatrix} 1 & 1 \\ \lambda_1 & \lambda_2 \end{pmatrix}, \qquad M^{-1} = \frac{1}{\lambda_1 - \lambda_2}\begin{pmatrix} -\lambda_2 & 1 \\ \lambda_1 & -1 \end{pmatrix}. \qquad (3.137)$$

Using the expression for M and M^{-1}, we obtain $R = MDM^{-1}$, keeping in mind that D is the diagonal matrix whose elements are $e^{\lambda_1 t}$ and $e^{\lambda_2 t}$. Performing the product,

$$R(t) = \frac{1}{\lambda_1 - \lambda_2}\begin{pmatrix} \lambda_1 e^{\lambda_2 t} - \lambda_2 e^{\lambda_1 t} & e^{\lambda_1 t} - e^{\lambda_2 t} \\ \lambda_1\lambda_2(e^{\lambda_2 t} - e^{\lambda_1 t}) & \lambda_1 e^{\lambda_1 t} - \lambda_2 e^{\lambda_2 t} \end{pmatrix}. \qquad (3.138)$$

Using Eq. (3.121), and considering the initial conditions such that the position and velocity are zero, $x(0) = 0$ and $v(0) = 0$, recalling that the elements of $Z(t)$ are 0 and $\zeta(t)$, we get the following expressions for the position and velocity

$$x(t) = \frac{1}{\lambda_1 - \lambda_2}\int_0^t (e^{\lambda_1(t-t')} - e^{\lambda_2(t-t')})\zeta(t')dt', \qquad (3.139)$$

$$v(t) = \frac{1}{\lambda_1 - \lambda_2}\int_0^t (\lambda_1 e^{\lambda_1(t-t')} - \lambda_2 e^{\lambda_2(t-t')})\zeta(t')dt'. \qquad (3.140)$$

From these expressions we can determine the moments $\langle x^2 \rangle$, $\langle xv \rangle$ and $\langle v^2 \rangle$. To this end, we use the property (3.7) to get the following results

$$\langle x^2 \rangle = \frac{\Gamma}{2(\lambda_1 - \lambda_2)^2} \left(\frac{e^{2\lambda_1 t} - 1}{\lambda_1} + 4\frac{e^{-\gamma t} - 1}{\gamma} + \frac{e^{2\lambda_2 t} - 1}{\lambda_2} \right), \tag{3.141}$$

$$\langle xv \rangle = \frac{\Gamma}{2(\lambda_1 - \lambda_2)^2} (e^{\lambda_1 t} - e^{\lambda_2 t})^2, \tag{3.142}$$

$$\langle v^2 \rangle = \frac{\Gamma}{2(\lambda_1 - \lambda_2)^2} \left(\lambda_1 (e^{2\lambda_1 t} - 1) + 4\frac{k}{\gamma}(e^{-\gamma t} - 1) + \lambda_2 (e^{2\lambda_2 t} - 1) \right), \tag{3.143}$$

where we used the equalities $\lambda_1 + \lambda_2 = -\gamma$ and $\lambda_1 \lambda_2 = k$. Alternatively, these results can be obtained directly from (3.127).

To get the results for large times, we take the limit $t \to \infty$ and use the equality $\lambda_1 - \lambda_2 = \sqrt{\gamma^2 - 4k}$,

$$\langle x^2 \rangle = \frac{\Gamma}{2k\gamma} = \frac{k_B T}{K}, \tag{3.144}$$

$$\langle v^2 \rangle = \frac{\Gamma}{2\gamma} = \frac{k_B T}{m}, \tag{3.145}$$

and $\langle xv \rangle = 0$. These results lead us to the equipartition of energy

$$\frac{m}{2}\langle v^2 \rangle = \frac{K}{2}\langle x^2 \rangle = \frac{1}{2}k_B T, \tag{3.146}$$

valid in equilibrium.

3.8 Linear System

We analyze again the system described by the set of Langevin equations given by (3.111). The time evolution of the moments of the variables x_i can be obtained by means of the discretization method used in Sect. 3.3. This method leads us to the following equation for first moments

$$\frac{d}{dt}\langle x_i \rangle = \langle f_i \rangle \tag{3.147}$$

and to the following equation for the second moments

$$\frac{d}{dt}\langle x_i x_j \rangle = \langle x_i f_j \rangle + \langle x_j f_i \rangle + \Gamma_i \delta_{ij}. \tag{3.148}$$

Equations for the time evolution for the moments of higher order can also be obtained. The equations for the various moments of x_i constitute a set of hierarchic coupled equations which are equivalent to the set of Langevin equations (3.111).

Again, we consider the simplest example, namely, that for which the forces f_i are linear,

$$f_i = \sum_{j=1}^{N} A_{ij} x_j. \tag{3.149}$$

In this case, Eqs. (3.147) and (3.148) reduce to

$$\frac{d}{dt}\langle x_i \rangle = \sum_{j=1}^{N} A_{ij}\langle x_j \rangle \tag{3.150}$$

and

$$\frac{d}{dt}\langle x_i x_j \rangle = \sum_{k=1}^{N} A_{jk}\langle x_i x_k \rangle + \sum_{k=1}^{N} A_{ik}\langle x_j x_k \rangle + \Gamma_i \delta_{ij}. \tag{3.151}$$

Equations (3.149) and (3.150) constitute two sets of linear differential equations of first order that can be solved. They are particularly interesting for determining the stationary properties, that is, the properties that one obtains in the limit $t \to \infty$.

From the definition of the covariance C_{ij}, given by (3.123), and using Eqs. (3.150) and (3.151), we reach the following equation for C_{ij},

$$\frac{d}{dt}C_{ij} = \sum_{k=1}^{N} C_{ik} A_{jk} + \sum_{k=1}^{N} A_{ik} C_{kj} + \Gamma_i \delta_{ij}, \tag{3.152}$$

which can be written in the compact form

$$\frac{d}{dt}C = CA^{\dagger} + AC + \Gamma, \tag{3.153}$$

where $A^{\dagger}$ is the transpose of A and in this equation Γ is the diagonal matrix whose elements are $\Gamma_i \delta_{ij}$. The time-dependent solution $C(t)$ of this equation together with the solution for $\langle x_i \rangle$ give us therefore the solution for $\langle x_i x_j \rangle$. But $C(t)$ has already been found and is given by Eq. (3.127). That the expression in the right-hand side of Eq. (3.127) is in fact the solution of (3.153) can be checked by its direct substitution into (3.153). To this end, is suffices to recall that $dR/dt = AR$ and that $R(0) = I$, the identity matrix.

Example 3.4 The equations of motion of a particle performing a Brownian motion are given by (3.2) and (3.5). Using the result (3.148), we obtain the following equations for the second moments of x and v,

$$\frac{d}{dt}\langle x^2 \rangle = 2\langle xv \rangle, \tag{3.154}$$

$$\frac{d}{dt}\langle xv \rangle = \langle v^2 \rangle - \gamma \langle xv \rangle, \tag{3.155}$$

$$\frac{d}{dt}\langle v^2 \rangle = -2\gamma \langle v^2 \rangle + \Gamma. \tag{3.156}$$

Example 3.5 From Eqs. (3.129) and (3.131), valid for a particle performing Brownian motion and subject to an elastic force, we get, using the result (3.148), the following equations for the second moments of x and v,

$$\frac{d}{dt}\langle x^2 \rangle = 2\langle xv \rangle, \tag{3.157}$$

$$\frac{d}{dt}\langle xv \rangle = \langle v^2 \rangle - k\langle x^2 \rangle - \gamma \langle xv \rangle, \tag{3.158}$$

$$\frac{d}{dt}\langle v^2 \rangle = -2k\langle xv \rangle - 2\gamma \langle v^2 \rangle + \Gamma. \tag{3.159}$$

In the stationary state, the left-hand side of these equations vanish, from which we conclude that in this regime $\langle xv \rangle = 0$ and that $\langle v^2 \rangle = \Gamma/2\gamma$ and $\langle x^2 \rangle = \Gamma/2\gamma k$, results obtained previously, as shown by Eqs. (3.144) and (3.145).

3.9   Electric Circuit

Consider an electric circuit comprising a resistor of resistance R, an inductor of inductance L and a capacitor of capacitance C, where we assume the occurrence of fluctuations both in tension and in electrical current. The equations for the charge Q of the capacitor and the current I in the circuit are given by

$$\frac{d}{dt}Q = -\gamma Q + I + I_r(t), \tag{3.160}$$

$$L\frac{d}{dt}I = -\frac{1}{C}Q - RI + V_r(t), \tag{3.161}$$

where $I_r(t)$ and $V_r(t)$ are the random current and random tension, respectively, considered to be independent, and called Johnson-Nyquist noise. These two stochastic variables have zero means and the following time correlations

$$\langle I_r(t)I_r(t')\rangle = A\delta(t - t'), \tag{3.162}$$

$$\langle V_r(t)V_r(t')\rangle = B\delta(t - t') \tag{3.163}$$

Equations (3.160) and (3.161) are written in the form

$$\frac{d}{dt}Q = -\gamma Q + I + \zeta_1(t), \tag{3.164}$$

$$\frac{d}{dt}I = -\frac{1}{LC}Q - \frac{R}{L}I + \zeta_2(t), \tag{3.165}$$

where the noises $\zeta_1(t) = I_r(t)$ and $\zeta_2(t) = V_r(t)/L$ have the following properties

$$\langle \zeta_i(t)\rangle = 0, \tag{3.166}$$

$$\langle \zeta_i(t)\zeta_j(t')\rangle = \Gamma_i\delta_{ij}\delta(t - t'), \tag{3.167}$$

for $i, j = 1, 2$. The coefficients Γ_i are related to A and B by $\Gamma_1 = A$ and $\Gamma_2 = B/L^2$.

From these equations, and using the results (3.148), we get the following equations for the second moments of Q and I,

$$\frac{d}{dt}\langle Q^2\rangle = -2\gamma\langle Q^2\rangle + 2\langle QI\rangle + \Gamma_1, \tag{3.168}$$

$$\frac{d}{dt}\langle QI\rangle = -\frac{1}{LC}\langle Q^2\rangle - \left(\frac{R}{L} + \gamma\right)\langle QI\rangle + \langle I^2\rangle, \tag{3.169}$$

$$\frac{d}{dt}\langle I^2\rangle = -2\frac{1}{LC}\langle QI\rangle - 2\frac{R}{L}\langle I^2\rangle + \Gamma_2. \tag{3.170}$$

Next, we consider only the stationary solution and assume that in this regime the average energy stored in the capacitor and inductor are equal to one another, that is, we assume the equipartition of energy in the form

$$\frac{1}{2C}\langle Q^2\rangle = \frac{L}{2}\langle I^2\rangle \tag{3.171}$$

This relation together with the equations valid in the stationary regime make up four equations in three variables, $\langle Q^2\rangle$, $\langle QI\rangle$ and $\langle I^2\rangle$. To find a solution, we assume that

Γ_1 and Γ_2 are not independent constants. Using Eq. (3.171), we see that a solution for the stationary state is

$$\langle QI \rangle = 0, \qquad \langle Q^2 \rangle = \frac{\Gamma_1}{2\gamma}, \qquad \langle I^2 \rangle = \frac{L\Gamma_2}{2R}, \qquad (3.172)$$

and that Γ_1 and Γ_2 are related to one another by

$$\frac{\Gamma_1}{\Gamma_2} = \frac{\gamma C L^2}{R}. \qquad (3.173)$$

Taking into account that the energy stored in the capacitor and in the inductor are $(1/2)k_B T$, we get the following relations between Γ_1 and Γ_2 and the temperature

$$\Gamma_1 = 2C\gamma k_B T, \qquad (3.174)$$

$$\Gamma_2 = \frac{2Rk_B T}{L^2}. \qquad (3.175)$$

Since $\Gamma_2 = B/L^2$, then we reach the following relation between the coefficient B and the temperature

$$B = 2Rk_B T, \qquad (3.176)$$

which is the result due to Nyquist.

3.10 Kramers Equation

We saw in Sect. 3.5 that the Langevin equation (3.65) is associated with an equation for the time evolution of the probability distribution, called Fokker-Planck equation. A set of equations such as that given by Eq. (3.111) may also be associated to a Fokker-Planck equation in several variables. In this section, we present a derivation of the Fokker-Planck equation for the set of Eqs. (3.129) and (3.130). To this end, we discretize these two equations in the form

$$x_{n+1} = x_n + \tau v_n, \qquad (3.177)$$

$$v_{n+1} = v_n + \tau f_n - \tau \gamma v_n + \sqrt{\tau \Gamma} \xi_n, \qquad (3.178)$$

where $f_n = f(x_n)$.

Proceeding in a manner analogous to what we did in Sect. 3.5, we denote by $P_n(x_n, v_n)$ the probability distribution of the variables x_n and v_n and by $g_n(k, q)$ the corresponding characteristic function

$$g_n(k, q) = \langle e^{ikx_n + iqv_n} \rangle = \int e^{ikx_n + iqv_n} P_n(x_n, v_n) dx_n dv_n. \tag{3.179}$$

Using the discretized Langevin equations, given by (3.177) and (3.178), we write

$$g_{n+1}(k, q) = \langle e^{ik(x_n + \tau v_n) + iq(v_n + \tau f_n - \tau \gamma v_n + \sqrt{\tau \Gamma} \xi_n)} \rangle \tag{3.180}$$

or, given that ζ_n is independent of x_n and v_n,

$$g_{n+1}(k, q) = \langle e^{ik(x_n + \tau v_n) + iq(v_n + \tau f_n - \tau \gamma v_n)} \rangle \langle e^{iq\sqrt{\tau \Gamma} \xi_n} \rangle. \tag{3.181}$$

Next, we perform the expansion of $g_{n+1}(k, q)$ up to terms of first order in τ, to get

$$g_{n+1}(k, q) = g_n(k, q) + \langle e^{ikx_n + iqv_n} (ik\tau v_n + iq\tau f_n - iq\tau \gamma v_n - \frac{q^2}{2} \tau \Gamma) \rangle. \tag{3.182}$$

Taking into account that $(g_{n+1}(k, q) - g_n(k, q))/\tau \;\rightarrow\; dg(k, q)/dt$, we get the equation

$$\frac{d}{dt} g(k, q) = \langle e^{ikx + iqv} (ikv + iqf(x) - iq\gamma v - \frac{q^2}{2} \Gamma) \rangle. \tag{3.183}$$

Now we use a procedure similar to that the one that led us to Eq. (3.107) from Eq. (3.103), to find the following equation

$$\frac{\partial}{\partial t} P(x, v, t) = -\frac{\partial}{\partial x}[vP(x, v, t)] - \frac{\partial}{\partial v}[(f(x) - \gamma v)P(x, v, t)] + \frac{\Gamma}{2} \frac{\partial^2}{\partial v^2} P(x, v, t), \tag{3.184}$$

which is the time evolution equation for the probability distribution $P(x, v, t)$. This equation is called Kramers equation.

Up to linear order in τ, Eq. (3.102) can be written in the equivalent form

$$g_{n+1}(k, q) = \langle e^{ikx_n + iqv_n + ik\tau v_n + iq\tau f_n - iq\tau \gamma v_n - \tau \Gamma q^2/2} \rangle. \tag{3.185}$$

Suppose that at time $n = 0$, the particle is in position x' with velocity v'. Then, at $n = 0$, the probability distribution is $P_0(x, v) = \delta(x - x')\delta(v - v')$ so that

$$g_1(k, q) = e^{ik(x' + \tau v') + iq(v' + \tau f(x') - \tau \gamma v') - \tau \Gamma q^2/2}. \tag{3.186}$$

which should be understood as the characteristic function of the probability distribution $P_1(x, v)$ at time τ, or, as the conditional probability distribution $K(x, v, \tau; x', v', 0)$ of finding the particle in position x with velocity v at time $t = \tau$ given that it was in position x' with velocity v' at time $t = 0$. The inverse Fourier transform of (3.186) leads us to the result

$$K(x, v, \tau; x', v', 0) = \delta(x - x' - \tau v') \frac{1}{\sqrt{2\tau\pi\Gamma}} e^{[v - v' - \tau f(x') + \tau\gamma v']^2 / 2\tau\Gamma}. \qquad (3.187)$$

This expression can also be understood as the solution of the Fokker-Planck equation for small time and with the initial condition that the particle is in position x' with velocity v'.

Exercises

1. The original Langevin's solution for the Brownian motion of a particle is based on the result $\langle xF_a \rangle = 0$, where x is the position and F_a is the random force. Prove this result.
2. Show that the time correlation of velocities of a free particle performing Brownian motion whose equation of motion is the Langevin equation (3.5), is given by

$$\langle v(t_0)v(t_0 + t) \rangle = \langle v^2(t^*) \rangle e^{-\gamma |t|},$$

where $t^* = t_0$ if $t \geq 0$ or $t^* = t_0 + t$ if $t < 0$. Determine, from it, the time autocorrelation in equilibrium $K(t)$ defined by

$$K(t) = \lim_{t_0 \to \infty} \langle v(t_0)v(t_0 + t) \rangle$$

and the Fourier transform $\hat{K}(\omega) = \int e^{i\omega t} K(t)dt$. Show that the diffusion coefficient D is given by

$$D = \int_0^\infty K(t)dt.$$

3. Determine explicitly, as functions of time, the variation of the kinetic energy dE_c/dt and the average dissipated power P_{dis} for a free particle performing a Brownian motion. Show that the sum $dE_c/dt + P_{dis} = P$ where P is the transferred power is a constant. Make a plot of these three quantities versus t. For what values of the initial velocity the average kinetic energy E_c decreases?
4. For the ordinary Brownian motion, the evolution equations for the second moments are given by (3.154)–(3.156). Solve these equations to get $\langle x^2 \rangle$, $\langle xv \rangle$

and $\langle v^2 \rangle$ as functions of time. Suppose that at time $t = 0$ the position and velocity are x_0 and v_0, respectively.

5. Consider a particle that performs a Brownian motion along the x-axis, subject to an elastic force, whose equations for the second moments are given by Eqs. (3.157)–(3.159). Solve directly these equation and show that the solutions are given by (3.141)–(3.143). Suppose that at time $t = 0$, $x(0) = 0$ and $v(0) = 0$.

6. Consider the set of Langevin equations,

$$\frac{dx_1}{dt} = -cx_1 - ax_2 + \zeta_1(t),$$

$$\frac{dx_2}{dt} = -cx_2 - bx_1 + \zeta_2(t),$$

where $\langle \zeta_1(t) \rangle = \langle \zeta_2(t) \rangle = 0$ and $\langle \zeta_i(t)\zeta_j(t') \rangle = \Gamma \delta_{ij}\delta(t - t')$. Determine $\langle x_1^2 \rangle$, $\langle x_2^2 \rangle$ e $\langle x_1 x_2 \rangle$ as functions of time. Solve for: (i) $b = a$, $c > a > 0$ and (ii) $b = -a$, $a > 0$, $c > 0$.

7. Simulate the random motion defined by the Langevin equation (3.5)–(3.7), discretizing time $t = n\tau$ in intervals equal to τ. For large times, make the histogram of the velocities. Determine also the time correlation $\langle v(t_0)v(t_0 + t) \rangle$ of the velocity. This can be done from the time series v_n, generated from the simulation as follows. Fix a certain value of $t = n\tau$ and calculate the arithmetic mean of the product $v_m v_{m+n}$ obtained from various values of m along the series. Repeat the procedure for other values of n. Make a plot of the correlation versus t.

8. Simulate the random motion of a particle that obeys the Langevin equation defined by (3.65)–(3.67), for: (a) $f(x) = c > 0$ for $x < 0$, $f(x) = 0$ for $x = 0$, and $f(x) = -c$ for $x > 0$; (b) $f(x) = -vx$, $v > 0$; and (c) $f(x) = ax - bx^3$, $b > 0$ and a any value. Make histograms of x in the stationary state, for several values of the parameters.

Chapter 4
Fokker-Planck Equation I

4.1 Equation in One Variable

We saw in Sect. 3.5 that the Langevin equation in one variable

$$\frac{dx}{dt} = f(x) + \zeta(t), \tag{4.1}$$

where the noise $\zeta(t)$ has the properties

$$\langle \zeta(t) \rangle = 0, \tag{4.2}$$

$$\langle \zeta(t)\zeta(t') \rangle = \Gamma \delta(t - t'), \tag{4.3}$$

is associated to the Fokker-Planck equation in one variable, or Smoluchowski equation,

$$\frac{\partial}{\partial t} P(x,t) = -\frac{\partial}{\partial x}[f(x)P(x,t)] + \frac{\Gamma}{2}\frac{\partial^2}{\partial x^2}P(x,t), \tag{4.4}$$

which gives the time evolution of the probability density $P(x,t)$. This association means that the probability distribution of the stochastic variable x obtained by means of the Langevin equation is identified with the solution of the Fokker-Planck equation.

Example 4.1 The Langevin equation above can be interpreted as the equation of motion of a particle of negligible mass that moves in a highly dissipative medium and subject to an external force. Indeed, the equation of motion of such a particle is

$$m\frac{d^2x}{dt^2} = -\alpha\frac{dx}{dt} + F(x) + F_{\mathrm{a}}(t), \tag{4.5}$$

© Springer International Publishing Switzerland 2015
T. Tomé, M.J. de Oliveira, *Stochastic Dynamics
and Irreversibility*, Graduate Texts in Physics,
DOI 10.1007/978-3-319-11770-6_4

where the first term in the right-hand side is the friction force, proportional to the velocity; the second is an external force and the third is a random force. When the mass is very small and the friction is very large, we can neglect the term at left and write

$$\alpha \frac{dx}{dt} = F(x) + F_a(t). \tag{4.6}$$

Dividing both sides of these equation by the friction coefficient α, we get Eq. (4.1). Thus, the quantity $f(x)$ in Eq. (4.1) is interpreted as the ratio between the external force and the friction coefficient, and the noise $\zeta(t)$ is interpreted as the ration between the random force and friction coefficient.

Example 4.2 When $f(x) = c$, a constant, the Langevin equation (4.1) describes the diffusion of particles subject to a constant force. The Langevin and Fokker-Planck equations are

$$\frac{dx}{dt} = c + \zeta(t), \tag{4.7}$$

$$\frac{\partial P}{\partial t} = -c \frac{\partial P}{\partial x} + \frac{\Gamma}{2} \frac{\partial^2 P}{\partial x^2}, \tag{4.8}$$

respectively. The characteristic function $G(k, t)$ obeys the equation

$$\frac{\partial G}{\partial t} = ikcG - \frac{\Gamma}{2} k^2 G, \tag{4.9}$$

which is obtained from the Fokker-Planck equation by integration by parts and assuming that P and $\partial P / \partial x$ vanish when $|x| \to \infty$. The solution of this equation is

$$G(k, t) = G(k, 0) e^{ikct - \Gamma t k^2 / 2}. \tag{4.10}$$

Using the initial condition that the particle is in $x = x_0$ at time $t = 0$, $P(x, 0) = \delta(x - x_0)$ and $G(k, 0) = e^{ikx_0}$. The solution becomes

$$G(k, t) = e^{ik(x_0 + ct) - \Gamma t k^2 / 2}. \tag{4.11}$$

We see that $G(k, t)$ is the characteristic function of a Gaussian distribution of mean $x_0 + ct$ and variance Γt, so that

$$P(x, t) = \frac{1}{\sqrt{2\pi \Gamma t}} e^{-(x - x_0 - ct)^2 / 2\Gamma t}, \tag{4.12}$$

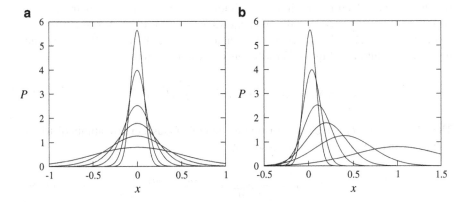

Fig. 4.1 Probability distribution $P(x,t)$ versus x for various instants of times, for the Brownian motion, (**a**) symmetric ($c = 0$) and (**b**) asymmetric with drift to the right ($c > 0$). Initially, the particle is at $x = 0$

as seen in Fig. 4.1.

Example 4.3 When $f(x) = -vx$, the Langevin equation describes the diffusion of particles subject to an elastic force. In this case, the Langevin and Fokker-Planck equations are

$$\frac{dx}{dt} = -vx + \zeta(t), \tag{4.13}$$

$$\frac{\partial P}{\partial t} = v\frac{\partial(xP)}{\partial x} + \frac{\Gamma}{2}\frac{\partial^2 P}{\partial x^2}, \tag{4.14}$$

respectively. This last equation is also known as Smoluchowski equation. The characteristic function $G(k,t)$ obeys the equation

$$\frac{\partial G}{\partial t} = -vk\frac{\partial G}{\partial k} - \frac{\Gamma}{2}k^2 G, \tag{4.15}$$

which is obtained from the Fokker-Planck equation by means of integration by parts and assuming that P and $\partial P/\partial x$ vanish when $x \to \pm\infty$. Supposing a solution of the type

$$G(k,t) = e^{ia(t)k - b(t)k^2/2}, \tag{4.16}$$

we see that $a(t)$ and $b(t)$ should fulfill the equations

$$\frac{da}{dt} = -va, \qquad \frac{db}{dt} = -2vb + \Gamma, \tag{4.17}$$

which must be solved for the initial condition such that the particle is in $x = x_0$ at time $t = 0$. This initial condition is equivalent to $P(x,0) = \delta(x - x_0)$ so that $G(k,0) = e^{ikx_0}$. This characteristic function tell us that at time $t = 0$, $a = x_0$ and $b = 0$. The solution of Eqs. (4.17) with this initial condition is

$$a(t) = x_0 e^{-\nu t}, \qquad\qquad b(t) = \frac{\Gamma}{2\nu}(1 - e^{-2\nu t}). \qquad (4.18)$$

The function $G(k,t)$ is a characteristic function of a Gaussian probability distribution of mean $a(t)$ and variance $b(t)$ so that

$$P(x,t) = \frac{1}{\sqrt{2\pi b(t)}} e^{-[x-a(t)]^2/2b(t)}. \qquad (4.19)$$

It is worth noticing that, unlike what happens in the previous example, in this, $P(x,t)$ reaches a stationary distribution when $t \to \infty$ because the variance $b(t) \to \Gamma/2\nu$ in this limit. The stationary distribution, which we denote by $P(x)$, is given by

$$P(x) = \sqrt{\frac{\nu}{\pi\Gamma}} e^{-\nu x^2/\Gamma}. \qquad (4.20)$$

4.2 Stationary Solution

Next, let us see how one can find the stationary solution of the Fokker-Planck equation (4.4), in the general case. To this end, we write the equation in the form

$$\frac{\partial}{\partial t} P(x,t) = -\frac{\partial}{\partial x} J(x,t), \qquad (4.21)$$

where $J(x,t)$ is given by

$$J(x,t) = f(x)P(x,t) - \frac{\Gamma}{2}\frac{\partial}{\partial x}P(x,t). \qquad (4.22)$$

In the form (4.21) the Fokker-Planck equation becomes a continuity equation, and $J(x,t)$ is the probability current. We suppose that the variable x takes values in the interval $[a,b]$.

If we integrate both sides of Eq. (4.21) in x, we get

$$\frac{d}{dt} \int_a^b P(x,t)dx = J(a,t) - J(b,t). \qquad (4.23)$$

Since the probability density $P(x,t)$ must be normalized at any time, that is,

$$\int_a^b P(x,t)dx = 1, \tag{4.24}$$

then the left-hand side of Eq. (4.23) should vanish, from which we may conclude that the boundary conditions are such that $J(a,t) = J(b,t)$. Thus, the conservation of total probability (4.24) is not only consequence of the Fokker-Planck equation but also of the boundary conditions. We will treat in this section only the case in which the probability current vanishes at the ends, $x = a$ and $x = b$, at any instant t, that is, $J(a,t) = J(b,t) = 0$. The boundary condition such that the probability current vanishes is called reflecting.

There are other boundary conditions for the Fokker-Planck equation, which we choose according to the problem we wish to solve. The periodic boundary conditions are such that $P(a,t) = P(b,t)$ and $J(a,t) = J(b,t)$ so that in this case the total probability is also conserved. However, unlike the reflecting boundary condition, the probability current at the boundaries is, in general, nonzero. The boundary condition, called absorbing, is such that only P vanishes at the boundary. The derivative $\partial P/\partial x$ as well as the probability current are nonzero. If the currents are distinct at the ends, the condition (4.24) is not fulfilled.

In the stationary state, the probability density is independent of t so that, taking into account Eq. (4.22), the probability current is also independent of t. Since the left-hand side of Eq. (4.21) vanishes, hence the probability current is also independent of x. Therefore, it must have the same value for any x. But since it vanishes at the ends of the interval $[a,b]$, it must be zero in the whole interval, that is,

$$J(x) = 0, \tag{4.25}$$

which is to be understood as the microscopic reversibility condition. Therefore, the stationary distribution $P(x)$ obeys the equation

$$f(x)P(x) - \frac{\Gamma}{2}\frac{d}{dx}P(x) = 0, \tag{4.26}$$

or yet

$$\frac{d}{dx}\ln P(x) = \frac{2}{\Gamma}f(x). \tag{4.27}$$

Denoting by $V(x)$ the potential corresponding to the force $f(x)$, that is,

$$f(x) = -\frac{d}{dx}V(x), \tag{4.28}$$

then

$$\ln P(x) = -\frac{2}{\Gamma} V(x) + \text{const,} \qquad (4.29)$$

from which we get

$$P(x) = Ae^{-2V(x)/\Gamma}, \qquad (4.30)$$

where A is a normalization constant.

Example 4.4 If in Example 4.1 we denote by $U(x)$ the potential energy associated to $F(x)$, we see that $U(x) = \alpha V(x)$. According to statistical mechanics, the equilibrium distribution is

$$P(x) = \frac{1}{Z} e^{-U(x)/k_BT}, \qquad (4.31)$$

where k_B is the Boltzmann constant, T is the absolute temperature and Z is a normalization constant. Comparing with (4.30), this expression gives the following relation between the intensity of the noise Γ and the temperature, $\Gamma = 2k_BT/\alpha$.

4.3 Evolution Operator

We saw in Sect. 4.2 how to obtain the stationary solution of the Fokker-Planck equation in one variable. Here we show that the solution $P(x,t)$ approaches the stationary solution $P(x)$ when $t \to \infty$. We also study the behavior of $P(x,t)$ for large times.

The Fokker-Planck equation in one variable

$$\frac{\partial}{\partial t} P(x,t) = -\frac{\partial}{\partial x}[f(x)P(x,t)] + \frac{\Gamma}{2}\frac{\partial^2}{\partial x^2} P(x,t), \qquad (4.32)$$

where $f(x)$ is a real function, can be written in the form

$$\frac{\partial}{\partial t} P(x,t) = \mathcal{W} P(x,t), \qquad (4.33)$$

where $\mathcal{W}$ is the evolution operator, that acts on functions $\phi(x)$, defined by

$$\mathcal{W}\phi(x) = -\frac{\partial}{\partial x}[f(x)\phi(x)] + \frac{\Gamma}{2}\frac{\partial^2}{\partial x^2}\phi(x). \qquad (4.34)$$

The stationary probability distribution $P(x)$ fulfills the equation

$$\mathcal{W} P(x) = 0. \tag{4.35}$$

The class of functions over which the operator $\mathcal{W}$ acts comprises the functions $\phi(x)$ such that $-f(x)\phi(x) + (\Gamma/2)\phi'(x)$ has the same value at the boundaries $x = a$ and $x = b$. That is, such that

$$- f(a)\phi(a) + \frac{\Gamma}{2}\phi'(a) = -f(b)\phi(b) + \frac{\Gamma}{2}\phi'(b). \tag{4.36}$$

For these functions, the following property is valid

$$\int_a^b \mathcal{W} \phi(x)dx = 0. \tag{4.37}$$

This is a fundamental property because from it we conclude that the probability distribution $P(x,t)$, which fulfills the Fokker-Planck equation (4.33), is normalized at any time $t > 0$, once it is normalized at $t = 0$. To perceive this, it suffices to integrate both sides of (4.33) to find

$$\frac{d}{dt} \int_a^b P(x,t)dx = 0. \tag{4.38}$$

The introduction of the evolution operator $\mathcal{W}$ allows us to write the solution of the Fokker-Planck equation in the form

$$P(x,t) = e^{t\mathcal{W}} P(x,0), \tag{4.39}$$

where $e^{t\mathcal{W}}$ is the operator defined by

$$e^{t\mathcal{W}} = 1 + t\mathcal{W} + \frac{t^2}{2!}\mathcal{W}^2 + \frac{t^3}{3!}\mathcal{W}^3 + \dots \tag{4.40}$$

Indeed, deriving both sides of Eq. (4.39) with respect to time and using the definition (4.40) we see that (4.33) is fulfilled. Suppose next that $\mathcal{W}$ has a discrete spectrum, that is, assume that

$$\mathcal{W} \phi_\ell(x) = \Lambda_\ell \phi_\ell(x), \tag{4.41}$$

for $\ell = 0, 1, 2, \dots$, where $\phi_\ell(x)$ are the eigenfunctions and Λ_ℓ are the eigenvalues of $\mathcal{W}$, and also that $P(x,0)$ has the expansion

$$P(x,0) = \sum_{\ell=0}^{\infty} a_\ell \phi_\ell(x). \tag{4.42}$$

Then, Eq. (4.39) gives

$$P(x,t) = \sum_{\ell=0}^{\infty} a_\ell e^{t\Lambda_\ell} \phi_\ell(x). \tag{4.43}$$

The eigenfunctions must meet the boundary conditions given by Eq. (4.36) and fulfill the following equation

$$\Lambda_\ell \int_a^b \phi_\ell(x)dx = 0, \tag{4.44}$$

which is obtained by using the property (4.37) in Eq. (4.41).

It is easy to see that one of the eigenfunctions of $\mathscr{W}$ must be $P(x)$, the stationary probability distribution. Indeed, examining Eq. (4.35), we see that $P(x)$ is an eigenfunction with a zero eigenvalue, which we set to be ϕ_0, that is, $\phi_0(x) = P(x)$ so that $\Lambda_0 = 0$. Thus, we may write

$$P(x,t) = P(x) + \sum_{\ell=1}^{\infty} a_\ell e^{t\Lambda_\ell} \phi_\ell(x), \tag{4.45}$$

where we have taken into account that $a_0 = 1$, which can be shown by integrating both sides of Eq. (4.42) and using the result (4.44).

We will see later, in Sect. 4.5, that the other eigenvalues are strictly negative so that all terms in the summation vanish when $t \to \infty$. Therefore, in this limit $P(x,t) \to P(x)$. The behavior of $P(x,t)$ for large times is thus exponential and characterized by the second dominant eigenvalue Λ_1, that is,

$$P(x,t) = P(x) + a_1\phi_1(x)e^{-t|\Lambda_1|}, \tag{4.46}$$

as long as $a_1 \neq 0$, otherwise it suffices to consider the next term whose coefficient a_ℓ is nonzero. The quantity $\tau = 1/|\Lambda_1|$ is then the relaxation time to the stationary solution. In many situations, such as that of Example 4.5 below, the probability $P(x,t)$ vanishes when $t \to \infty$, although the integral is always finite, and therefore the stationary distribution, in this sense, does not exist. It may also happen that $\tau \to \infty$, as can be seen in the Example 4.5 below, case in which the relaxation is no longer exponential and becomes algebraic.

Example 4.5 Consider a particle confined in the interval $-L/2 \leq x \leq L/2$, in the absence of external forces. In this case, $f(x) = 0$ and therefore we should solve the eigenvalue equation

$$\frac{\Gamma}{2}\frac{d^2}{dx^2}\phi(x) = \Lambda\phi(x), \tag{4.47}$$

with the boundary conditions $\phi'(-L/2) = \phi'(L/2) = 0$. The solutions that obey
these conditions are

$$\phi_\ell(x) = \begin{cases} L^{-1}\cos(kx), & \ell = 0, 2, 4, \ldots \\ L^{-1}\sin(kx), & \ell = 1, 3, 5, \ldots \end{cases} \qquad \Lambda_\ell = -\frac{\Gamma}{2}k^2, \qquad (4.48)$$

where $k = \pi\ell/L$ and we choose the normalization such that $\phi_0(x) = P(x)$. If the
particle is at the origin at $t = 0$, that is, if $P(x, 0) = \delta(x)$, then the solution of the
Fokker-Planck equation is

$$P(x, t) = \frac{1}{L} + \frac{2}{L}\sum_{\ell=2(\text{even})}^{\infty} e^{-t\Gamma k^2/2}\cos(kx). \qquad (4.49)$$

As long as L is finite the relaxation time is $\tau = 1/|\Lambda_2| = (2/\Gamma)(L/2\pi)^2$. This
time diverges when $L \to \infty$. In this case, however,

$$P(x, t) = \frac{1}{\pi}\int_0^{\infty} e^{-t\Gamma k^2/2}\cos kx\, dk = \frac{1}{\sqrt{2\pi\Gamma t}}e^{-x^2/2\Gamma t} \qquad (4.50)$$

and therefore the decay is algebraic, $P(x, t) \sim t^{-1/2}$.

4.4 Adjoint Equation

To the Fokker-Planck in one variable,

$$\frac{\partial}{\partial t}P(x, t) = \mathscr{W}\,P(x, t), \qquad (4.51)$$

we associate the adjoint equation

$$\frac{\partial}{\partial t}Q(x, t) = \mathscr{W}^\dagger Q(x, t), \qquad (4.52)$$

where $\mathscr{W}^\dagger$ is the adjoint operator $\mathscr{W}$, defined by

$$\int_a^b \phi^*(\mathscr{W}^\dagger\chi)dx = \int_a^b \chi(\mathscr{W}\phi)^* dx, \qquad (4.53)$$

for functions $\phi(x)$ that obey the boundary conditions (4.36) and for function $\chi(x)$
that obey the boundary conditions

$$\chi'(a) = \chi'(b). \qquad (4.54)$$

In correspondence with the theory of matrices, Eq. (4.53) tell us that the adjoint operator is analogous to the adjoint matrix, which is a conjugate transpose matrix.

From the definitions of $\mathscr{W}$ and $\mathscr{W}^\dagger$ we conclude that

$$\mathscr{W}^\dagger \chi(x) = f(x)\frac{\partial}{\partial x}\chi(x) + \frac{\Gamma}{2}\frac{\partial^2}{\partial x^2}\chi(x), \tag{4.55}$$

a result that is obtained by replacing (4.34) into (4.53), integrating by parts the right hand side of (4.53) and using the boundary conditions (4.36) for $\phi(x)$ and the boundary conditions (4.54) for $\chi(x)$. Comparing (4.34) with (4.55), we see that in general $\mathscr{W} \neq \mathscr{W}^\dagger$, that is, $\mathscr{W}$ is not Hermitian (self-adjoint), except when $f = 0$. Explicitly, the adjoint Fokker-Planck equation in one variable is given by

$$\frac{\partial}{\partial t}Q(x,t) = f(x)\frac{\partial}{\partial x}Q(x,t) + \frac{\Gamma}{2}\frac{\partial^2}{\partial x^2}Q(x,t). \tag{4.56}$$

Denoting by $\chi_\ell(x)$ the eigenfunctions of $\mathscr{W}^\dagger$, we may write

$$\mathscr{W}^\dagger \chi_\ell = \Lambda_\ell \chi_\ell \tag{4.57}$$

since the operator $\mathscr{W}^\dagger$, given by (4.55), must have the same eigenvalues of $\mathscr{W}$, given by (4.34), what may be understood by noting first that $\mathscr{W}$ is a real operator, that is, it has the following property, $(\mathscr{W}\phi)^* = \mathscr{W}\phi^*$. Second, using this property in (4.53), we find

$$\int_a^b \phi(\mathscr{W}^\dagger \chi)dx = \int_a^b \chi(\mathscr{W}\phi)dx. \tag{4.58}$$

In correspondence with the theory of matrices, this equation tell us that the operator $\mathscr{W}^\dagger$ is analogous to a transpose matrix, which has the same eigenvalues of the original matrix.

We have seen that $\phi_0(x) = P(x)$ is the eigenfunction with eigenvalue $\Lambda_0 = 0$. To this eigenfunction we associate the adjoint eigenfunction $\chi_0 = 1$. That $\chi_0 = 1$ is an eigenfunction with eigenvalue zero can be checked replacing it into Eq. (4.55), giving $\mathscr{W}^\dagger \chi_0 = 0$.

Example 4.6 For $f(x) = -\nu x$, the eigenfunctions are related to the Hermite polynomials $H_\ell(x)$, which fulfills the relation

$$H_\ell''(x) - 2xH_\ell'(x) = -2\ell H_\ell(x). \tag{4.59}$$

Comparing with Eq. (4.57), we may conclude that $\chi_\ell(x) = a_\ell H_\ell(x\sqrt{\nu/\Gamma})$ and that $\Lambda_\ell = -\ell\nu$.

We assume that the sets $\{\phi_\ell\}$ and $\{\chi_\ell\}$ of the eigenfunctions of the operators $\mathscr{W}$ and $\mathscr{W}^\dagger$ comprise a bi-orthonormal set, having the following properties,

$$\int_a^b \chi_{\ell'}(x)\phi_\ell(x)dx = \delta_{\ell'\ell}, \tag{4.60}$$

$$\sum_\ell \phi_\ell(x)\chi_\ell(x') = \delta(x - x'). \tag{4.61}$$

We have seen in Sect. 4.3 that the expansion of the probability distribution $P(x,t)$ in the eigenfunctions $\phi_\ell(x)$ is given by (4.43), where the constants a_ℓ are the coefficients of the expansion (4.42) of the initial distribution $P(x,0)$. Multiplying both sides of (4.42) by χ_ℓ and integrating in x, we find the following formula for the coefficients a_ℓ,

$$a_\ell = \int_a^b \chi_\ell(x)P(x,0)dx, \tag{4.62}$$

where we used the orthogonality relation (4.60). Setting this result into (4.42), we reach the following expression for the probability distribution at any time t,

$$P(x,t) = \int_a^b K(x,t,x',0)P(x',0)dx', \tag{4.63}$$

where

$$K(x,t,x',0) = \sum_\ell e^{t\Lambda_\ell}\phi_\ell(x)\chi_\ell(x'). \tag{4.64}$$

If $P(x,0) = \delta(x - x_0)$, then $P(x,t)$ reduces to the expression

$$P(x,t) = K(x,t,x_0,0) = \sum_\ell e^{t\Lambda_\ell}\phi_\ell(x)\chi_\ell(x_0). \tag{4.65}$$

4.5 Hermitian Operator

We have seen in the previous section that, in general, $\mathscr{W}$ is not Hermitian. However, when microscopic reversibility, expressed by relation (4.26), occurs, it is possible to perform a transformation on $\mathscr{W}$ to obtain a Hermitian operator, which we denote by $\mathscr{K}$, that holds the same eigenvalues of $\mathscr{W}$.

We define the operator $\mathscr{K}$ by

$$\mathscr{K}\phi(x) = [\psi_0(x)]^{-1}\mathscr{W}[\psi_0(x)\phi(x)], \tag{4.66}$$

where $\psi_0(x) = \sqrt{P(x)}$ and $P(x)$ is the stationary probability, that fulfills the relation (4.26). The eigenfunctions of $\mathcal{K}$ are $\psi_\ell(x) = [\psi_0(x)]^{-1}\phi_\ell(x)$ because

$$\mathcal{K}\psi_\ell = \psi_0^{-1}\mathcal{W}\phi_\ell = \psi_0^{-1}\Lambda_\ell\phi_\ell = \Lambda_\ell\psi_\ell, \tag{4.67}$$

from which we conclude that the eigenvalues are Λ_ℓ, the same ones of $\mathcal{W}$. To obtain an explicit form of $\mathcal{K}$, we apply the operator on any function $\psi(x)$ and use the definition of $\mathcal{W}$, that is,

$$\mathcal{K}\psi = \psi_0^{-1}\mathcal{W}(\psi_0\psi) = \psi_0^{-1}\{-\frac{\partial}{\partial x}(f\psi_0\psi) + \frac{\Gamma}{2}\frac{\partial^2}{\partial x^2}(\psi_0\psi)\}. \tag{4.68}$$

Afterwards we use the equality

$$\frac{\partial}{\partial x}\ln\psi_0 = \frac{1}{2}\frac{\partial}{\partial x}\ln P(x) = \frac{1}{\Gamma}f(x), \tag{4.69}$$

to obtain the desired form

$$\mathcal{K}\psi = -\frac{1}{2}\{\frac{\partial f}{\partial x} + \frac{1}{\Gamma}f^2\}\psi + \frac{\Gamma}{2}\frac{\partial^2\psi}{\partial x^2}. \tag{4.70}$$

Equation (4.70) reveals that the operator $\mathcal{K}$ can be formally written as proportional to the Hamiltonian operator $\mathcal{H}$, given by

$$\mathcal{H}\psi = -\frac{\hbar^2}{2m}\frac{\partial^2\psi}{\partial x^2} + V(x)\psi \tag{4.71}$$

that describes the motion of a particle of mass m subject to a potential $V(x)$. Choosing the proportionality constant in such a way that $\mathcal{H} = -m\Gamma\mathcal{K}$, we see that the constant Γ must be proportional to the Planck constant, $\Gamma = \hbar/m$, and that, moreover, the potential is given by

$$V(x) = \frac{m}{2}\{\Gamma\frac{d}{dx}f(x) + [f(x)]^2\}. \tag{4.72}$$

Example 4.7 For $f(x) = -vx$, we see that

$$\mathcal{K} = \frac{1}{2}(v - \frac{1}{\Gamma}v^2x^2) + \frac{\Gamma}{2}\frac{\partial^2}{\partial x^2}, \tag{4.73}$$

and $V(x) = m(-\Gamma v + v^2x^2)/2$. On the other hand, quantum mechanics tell us that

$$-\frac{\hbar^2}{2m}\frac{\partial^2}{\partial x^2}\psi_\ell + \frac{1}{2}m\omega^2x^2\psi_\ell = \hbar\omega(\ell + \frac{1}{2})\psi_\ell, \tag{4.74}$$

for $\ell = 0, 1, 2, \ldots$, where ψ_ℓ are the eigenfunctions of the harmonic oscillator. Performing the replacements $\hbar/m = \Gamma$, and $\omega = \nu$ and therefore $m\omega^2/\hbar = \nu^2/\Gamma$, we get

$$\frac{\Gamma}{2}\frac{\partial^2}{\partial x^2}\psi_\ell + \nu\frac{1}{2}\psi_\ell - \frac{\nu^2}{2\Gamma}x^2\psi_\ell = -\nu\ell\psi_\ell, \tag{4.75}$$

from which we conclude that

$$\mathscr{H}\psi_\ell = -\nu\ell\psi_\ell. \tag{4.76}$$

Therefore $\Lambda_\ell = -\nu\ell$, and the eigenvalues are negative, except $\Lambda_0 = 0$.

When microscopic reversibility, given by relation (4.26), occurs, there is yet a simple relationship between the eigenfunctions of $\mathscr{W}$ and of $\mathscr{W}^\dagger$, given by $\phi_\ell(x) = P(x)\chi_\ell(x)$. Indeed, replacing this expression in $\mathscr{W}\phi_\ell = \Lambda_\ell\phi_\ell$, using the definition of $\mathscr{W}$, given by (4.34), and the equality $\mathscr{W}P(x) = 0$, we get

$$-Pf\frac{\partial}{\partial x}\chi_\ell + \frac{\Gamma}{2}P\frac{\partial^2}{\partial x^2}\chi_\ell + \Gamma\frac{\partial P}{\partial x}\frac{\partial\chi_\ell}{\partial x} = \Gamma\Lambda_\ell P\chi_\ell. \tag{4.77}$$

Using the relation (4.26), that is, $2fP = \Gamma\partial P/\partial x$, we obtain

$$f\frac{\partial}{\partial x}\chi_\ell + \frac{\Gamma}{2}\frac{\partial^2}{\partial x^2}\chi_\ell = \Lambda_\ell\chi_\ell, \tag{4.78}$$

which is the eigenvalue equation for the adjoint operator $\mathscr{W}^\dagger$.

It is worthwhile to notice that the expression (4.70) tell us that $\mathscr{H}$ is in fact Hermitian. The Hermitian operators hold the following properties: the eigenvalues are real, the eigenfunctions are orthogonal and can form a complete set, as is the case of the second order differential operator defined by (4.70). The last property justify the expansion made in (4.42). Using the correspondence with the theory of stochastic matrices, we may state in addition that the dominant eigenvector, which corresponds to the largest eigenvalue, is strictly positive and nondegenerate. Therefore, we may identify it with the stationary probability density $P(x)$. The corresponding eigenvalue being zero and nondegenerate, all the other eigenvalue must be strictly negative.

Let us show, explicitly, that the eigenvalues are negative or zero. To this end it suffices to show that

$$\int_a^b \psi^*(x)\mathscr{H}\psi(x)dx \leq 0, \tag{4.79}$$

for any function $\psi(x)$. We start by writing Eq. (4.68) in the form

$$\mathscr{H}\psi = \frac{\Gamma}{2}\left(\psi'' - \frac{\psi}{\psi_0}\psi_0''\right) = \frac{\Gamma}{2}\frac{1}{\psi_0}\left(\psi_0^2(\frac{\psi}{\psi_0})'\right)', \tag{4.80}$$

where we use Eq. (4.69) to get rid of f. Thus

$$\int_a^b \psi^* \mathcal{K} \psi dx = \frac{\Gamma}{2} \int_a^b (\frac{\psi^*}{\psi_0}) \left(\psi_0^2 (\frac{\psi}{\psi_0})' \right)' dx \qquad (4.81)$$

and, performing an integration by parts, we find

$$\int_a^b \psi^* \mathcal{K} \psi dx = -\frac{\Gamma}{2} \int_a^b \left| (\frac{\psi}{\psi_0})' \right|^2 \psi_0^2 dx \leq 0, \qquad (4.82)$$

where the integrated part vanishes by taking into account the boundary conditions, that is, the condition $J(a) = J(b) = 0$.

4.6   Absorbing Wall

In this section we analyze the movement of a particle that performs a Brownian motion along the positive semi-axis x and is under the action of a constant force $f(x) = c$, which can be positive, negative or zero. At $x = 0$ there is an absorbing wall so that, when the particle reaches the point $x = 0$, it remains forever in this point. The Fokker-Planck equation that describes the evolution of the probability distribution $P(x, t)$ of x at time t is given by

$$\frac{\partial P}{\partial t} = -c \frac{\partial P}{\partial x} + \frac{\Gamma}{2} \frac{\partial^2 P}{\partial x^2}, \qquad (4.83)$$

which can be written in the form

$$\frac{\partial P}{\partial t} = -\frac{\partial J}{\partial x}, \qquad (4.84)$$

where $J(x, t)$ is the probability current, given by

$$J = cP - \frac{\Gamma}{2} \frac{\partial P}{\partial x}. \qquad (4.85)$$

A time $t = 0$, we are assuming that the particle is found at $x = x_0$ so that the initial condition is $P(x, 0) = \delta(x - x_0)$.

Since the wall is absorbing, the probability distribution at $x = 0$ must vanish at any time, $P(0, t) = 0$, which constitutes the boundary condition for solving the Fokker-Planck equation. The absorption of the particle by the wall is described by a nonzero probability current at $x = 0$, which means that the probability distribution $P(x, t)$ is nonconserving, that is, the integral

$$\int_0^\infty P(x, t) dx = \mathscr{P}(t) \qquad (4.86)$$

is distinct from unit and is understood as the permanence probability of the particle up to time t. When performing this integral, we assume that the current vanishes when $x \to \infty$. The integration of Eq. (4.84) in x leads us to the following relation between $\mathscr{P}(t)$ and the probability current at $x = 0$,

$$\frac{d}{dt}\mathscr{P}(t) = J(0,t). \qquad (4.87)$$

The absorbing probability of the particle up to time t is simply equal to $1 - \mathscr{P}(t)$. The absorbing probability of the particle between the time t and $t + \Delta t$ is therefore $-\mathscr{P}(t+\Delta) + \mathscr{P}(t)$. Defining the quantity $\Phi(t)$ such that $\Phi(t)\Delta t$ is this probability, then $\Phi(t) = -d\,\mathscr{P}(t)/dt$ so that

$$\Phi(t) = -J(0,t). \qquad (4.88)$$

In other terms, $\Phi(t)$ is the absorption probability of the particle at time t per unit time.

Here we determine $P(x,t)$ through the calculation of the eigenfunctions of the evolution operator $\mathscr{W}$ and by using the result (4.65). The eigenvalue equation is given by

$$\mathscr{W}\psi = -c\frac{\partial\psi}{\partial x} + \frac{\Gamma}{2}\frac{\partial^2\psi}{\partial x^2} = \lambda\psi, \qquad (4.89)$$

and should be solved with the absorbing condition $\psi(0) = 0$. The eigenvalue equation related to the adjoint operator is given by

$$\mathscr{W}^\dagger\phi = c\frac{\partial\phi}{\partial x} + \frac{\Gamma}{2}\frac{\partial^2\phi}{\partial x^2} = \lambda\phi, \qquad (4.90)$$

As can be seen, the evolution operator $\mathscr{W}$ is not Hermitian. The determination of the eigenfunctions becomes easier if the evolution operator is transformed into a Hermitian operator. In fact, this can be done by means of the transformation $\psi(x) = \varphi(x)e^{cx/\Gamma}$ or of the transformation $\phi(x) = \varphi(x)e^{-cx/\Gamma}$. Both transformations lead us to the following eigenvalue equation

$$-\frac{c^2}{2\Gamma}\varphi + \frac{\Gamma}{2}\frac{\partial^2\varphi}{\partial x^2} = \lambda\varphi, \qquad (4.91)$$

which is to be solved with the absorbing boundary condition $\varphi(0) = 0$. To this end we assume solutions of the type

$$\varphi_k(x) = \frac{2}{\sqrt{2\pi}}\sin kx, \qquad (4.92)$$

which substituted in the eigenvalue equation give us the eigenvalues

$$\lambda_k = -\frac{c^2}{2\Gamma} - \frac{\Gamma}{2}k^2. \tag{4.93}$$

The eigenfunctions φ_k are orthonormalized according to

$$\int_0^\infty \varphi_k(x)\varphi_{k'}(x)dx = \delta(k - k'), \tag{4.94}$$

and obey the relation

$$\int_0^\infty \varphi_k(x)\varphi_k(x')dk = \delta(x - x'). \tag{4.95}$$

The eigenfunctions $\psi_k(x)$ of $\mathscr{W}$ and the eigenfunctions $\phi_k(x)$ of the adjoint operator $\mathscr{W}^\dagger$ are related to the eigenfunctions $\varphi_k(x)$ through

$$\psi_k(x) = e^{cx/\Gamma}\varphi_k(x), \qquad \phi_k(x) = e^{-cx/\Gamma}\varphi_k(x). \tag{4.96}$$

For the initial condition $P(x, 0) = \delta(x - x_0)$, and according to (4.65),

$$P(x, t) = \int_0^\infty e^{t\lambda_k}\psi_k(x)\phi_k(x_0)dk. \tag{4.97}$$

Substituting the expression for the eigenvalues and eigenfunctions,

$$P(x, t) = \frac{2}{\pi}e^{c(x-x_0)/\Gamma}e^{-tc^2/2\Gamma}\int_0^\infty e^{-t\Gamma k^2/2}\sin kx \sin kx_0\, dk. \tag{4.98}$$

Performing the integral, we reach the following expression for the probability distribution

$$P(x, t) = e^{c(x-x_0)/\Gamma}e^{-tc^2/2\Gamma}\frac{1}{\sqrt{2\pi\Gamma t}}\left(e^{-(x-x_0)^2/2\Gamma t} - e^{-(x+x_0)^2/2\Gamma t}\right), \tag{4.99}$$

valid for $x \geq 0$.

From $P(x, t)$ we obtain the probability current by means of the definition (4.85). In particular, we obtain $J(0, t)$ and $\Phi(t) = -J(0, t)$, the probability of absorption at time t per unit time,

$$\Phi(t) = \frac{x_0}{t\sqrt{2\pi\Gamma t}}e^{-(x_0+ct)^2/2\Gamma t}, \tag{4.100}$$

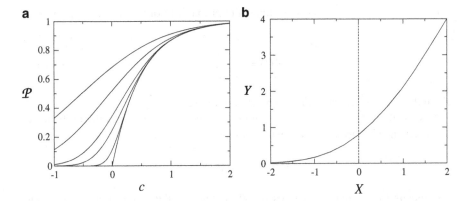

Fig. 4.2 Brownian motion with absorbing wall. (**a**) Probability of permanence $\mathscr{P}$ up to time t as a function of c, for several values of t. When $t \to \infty$, $\mathscr{P}$ behaves according to (4.102) and vanishes for $c \leq 0$. (**b**) Plot of $Y = \mathscr{P} t^{1/2}$ versus $X = c t^{1/2}$ for small values of $|c|$ and large values of t. The curves collapse into a universal function $Y = \mathscr{F}(X)$

from which, and using the result $\Phi(t) = -d\,\mathscr{P}(t)/dt$, we obtain the probability of permanence up to time t,

$$\mathscr{P}(t) = 1 - \int_0^t \frac{x_0}{t'\sqrt{2\pi\Gamma t'}} e^{-(x_0+ct')^2/2\Gamma t'}\,dt'. \tag{4.101}$$

In Fig. 4.2 we show the probability of permanence $\mathscr{P}(t)$ as a function of c for several values of time t. The total probability of permanence $\mathscr{P}^* = \lim_{t \to \infty} \mathscr{P}(t)$ is

$$\mathscr{P}^* = \begin{cases} 1 - e^{-2cx_0/\Gamma}, & c > 0, \\ 0, & c \leq 0, \end{cases} \tag{4.102}$$

result obtained by the use of the equality

$$\int_0^\infty \xi^{-3/2} e^{-(1+a\xi)^2/2\xi}\,d\xi = \sqrt{2\pi}\,e^{-(a+|a|)}. \tag{4.103}$$

We see therefore that a particle that performs a Brownian motion with negative or zero bias will be absorbed by the wall no matter what the initial position is. In other terms, whatever may be the initial position the particle will reach the point $x = 0$ and will be absorbed. On the other hand, if the bias is positive, the probability of the particle being absorbed is smaller that one and therefore there is a probability that the particle will never reach the wall.

When $c = 0$, we perform a transformation of integration variable to find

$$\mathscr{P}(t) = \frac{2}{\sqrt{\pi}} \int_0^{\xi_0} e^{-\xi^2}\,d\xi, \tag{4.104}$$

where $\xi_0 = x_0/\sqrt{2\Gamma t}$. For large times, the probability of permanence up to time t is

$$\mathscr{P}(t) = \frac{2x_0}{\sqrt{2\pi\Gamma t}}, \tag{4.105}$$

which decays algebraically to its zero asymptotic value.

It is worthwhile to notice that around $c = 0$ and for large times the probability of permanence $\mathscr{P}$ holds the following scaling form

$$\mathscr{P} = t^{-1/2}\mathscr{F}(ct^{1/2}). \tag{4.106}$$

This means to say that the curves of $\mathscr{P}$ versus c collapse into one curve when plotted in a plane in which the horizontal axis is $X = ct^{1/2}$ and the vertical axis is $Y = \mathscr{P}t^{1/2}$, as shown in Fig. 4.2. The curve $Y = \mathscr{F}(X)$ is called universal function. At $X = 0$, $\mathscr{F}(0) = 2x_0/\sqrt{2\pi\Gamma}$. When $X \to \infty$ and $c > 0$, $\mathscr{F}(X) \to X$ so that $\mathscr{P} = c$ when $t \to \infty$. In the opposite limit, $X \to -\infty$, $\mathscr{F}(X) \to 0$, so that $\mathscr{P} \to 0$ when $t \to \infty$ and $c < 0$.

4.7 Reflecting Wall

Here we study again the movement of a particle that performs a Brownian motion along the positive semi-axis x and is under the action of a constant force $f(x) = c$, that may be positive, negative or zero. Now there is a reflecting wall at $x = 0$. The Fokker-Planck equation that describes the evolution of the probability distribution $P(x, t)$ of x at time t is given by (4.83) and must be solved with the boundary condition such that the probability current $J(x, t)$, given by (4.85), vanish at $x = 0$, $J(0, t) = 0$. This condition and assuming that the current vanishes when $x \to \infty$ lead us to the conservation of probability $P(x, t)$ at any time.

In the stationary state we assume that the probability current vanishes at all points, what leads us to the following equation for the stationary distribution $P_e(x)$

$$cP_e - \frac{\Gamma}{2}\frac{\partial P_e}{\partial x} = 0, \tag{4.107}$$

from which we determined the stationary distribution

$$P_e(x) = \frac{\Gamma}{2|c|}e^{2cx/\Gamma}, \tag{4.108}$$

valid for $c < 0$. When $c \geq 0$ there is no nonzero solution for Eq. (4.107).

Here again we determine $P(x, t)$ through the calculation of the eigenfunctions ψ and ϕ of the evolution operator $\mathscr{W}$ and its adjoint operator and by the use of the result (4.65). The eigenvalue equations are given by (4.89) and (4.90). Using the transformation $\psi(x) = \varphi(x)e^{cx/\Gamma}$ or the transformation $\phi(x) = \varphi(x)e^{-cx/\Gamma}$ we reach the eigenvalue equation

$$-\frac{c^2}{2\Gamma}\varphi + \frac{\Gamma}{2}\frac{\partial^2\varphi}{\partial x^2} = \lambda\varphi, \tag{4.109}$$

which must be solved with the reflecting boundary condition, that is, in such a way that $c\varphi - (\Gamma/2)\partial\varphi/\partial x$ vanishes at $x = 0$.

Assuming solutions of the form

$$\varphi(x) = Ae^{ikx} + Be^{-ikx}, \tag{4.110}$$

we find

$$\lambda_k = -\frac{c^2}{2\Gamma} - \frac{\Gamma}{2}k^2. \tag{4.111}$$

The reflecting boundary condition leads to the following relation between A and B

$$(c - ik)A = -(c + ik)B. \tag{4.112}$$

There is only one constant to be determined, what is done by the normalization of φ, leading to the result

$$\varphi_k(x) = \frac{1}{\sqrt{\pi|\lambda_k|\Gamma}}(c \sin kx + \Gamma k \cos kx), \tag{4.113}$$

valid for $k \neq 0$. When $c < 0$, in addition to these eigenfunctions, there is an eigenfunction corresponding to the stationary state, $\varphi_0(x) = a^{1/2}e^{cx/\Gamma}$ related to zero eigenvalue.

The eigenfunctions are orthonormalized according to (4.94) and obey the relation (4.95). The eigenfunctions $\psi_k(x)$ of $\mathcal{W}$ and the eigenfunctions $\phi_k(x)$ of the adjoint operator $\mathcal{W}^\dagger$ are related to the eigenfunctions $\varphi_k(x)$ through (4.96). For the initial condition $P(x, 0) = \delta(x)$, particle placed on $x = 0$ at time $t = 0$, and according to (4.65),

$$P(x, t) = P_e(x) + \int_0^\infty e^{t\lambda_k}\psi_k(x)\phi_k(0)dk, \tag{4.114}$$

where $P_e(x)$ is given by (4.108) when $c < 0$ and vanishes when $c \geq 0$. Substituting the expression for the eigenvalues,

$$P(x, t) = P_e(x) + e^{cx/\Gamma}\int_0^\infty \frac{e^{t\lambda_k}}{|\lambda_k|}(ck \sin kx + \Gamma k^2 \cos kx)\frac{dk}{\pi}, \tag{4.115}$$

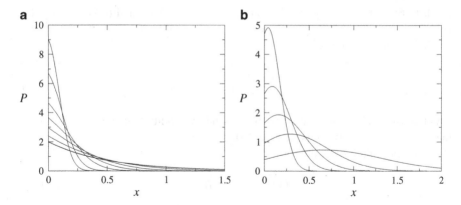

Fig. 4.3 Brownian motion with a reflecting wall at $x = 0$ for: (**a**) bias to the left, $c < 0$, and (**b**) bias to the right, $c > 0$. Probability distribution $P(x,t)$ versus x for several instants of time. Initially, the particle is found at $x = 0$. When $c < 0$, $P(x,t)$ approaches the stationary distribution $P_e(x)$ given by (4.85). When $c > 0$, $P(x,t)$ develops a bump that moves to the right

Using its definition (4.85), the probability current can be determined from $P(x,t)$,

$$J(x,t) = e^{cx/\Gamma} \int_0^\infty k \sin kx \, dk. \qquad (4.116)$$

Substituting the eigenvalue λ_k, the integral can be carried out with the result

$$J(x,t) = \frac{x}{t} \frac{1}{\sqrt{2\pi \Gamma t}} e^{-(x-ct)^2/2\Gamma t}. \qquad (4.117)$$

It is clear that it vanishes at $x = 0$ and in the limit $x \to \infty$.

Another representation of $P(x,t)$ can be obtained by replacing the result (4.117) into (4.85) and integrating (4.85). We get

$$P(x,t) = \frac{2e^{2cx/\Gamma}}{\Gamma t \sqrt{2\pi \Gamma t}} \int_x^\infty x' e^{-(x'+ct)^2/2\Gamma t} dx', \qquad (4.118)$$

where we have taken into account that $P(x,t)$ must vanish in the limit $t \to \infty$. Figure 4.3 shows the probability distribution $P(x,t)$ as a function of x for several values of t. When $c < 0$, $P(x,t)$ approaches its stationary value, $P_e(x)$, given by (4.108). When $c > 0$, $P(x,t)$ develops a bump which moves to the right with an increasing width, as can be seen in Fig. 4.3. For large times the bump acquires a constant velocity equal to c and a width equal to $\sqrt{\Gamma t}$. Therefore, we are faced with two very distinct regimes. When $c < 0$, the particle is bound to the wall, moving next to it. When $c > 0$, it detaches from the wall, moving away from it permanently. We notice, in addition, that in the regime where the particle is bound, the eigenvalue spectrum is composed by an single eigenvalue, equal to zero and corresponding to the stationary state, and a continuous band. The difference between the single eigenvalue and the top of the continuous band equals $c^2/2\Gamma$ and vanishes when $c = 0$, that is, at the threshold of the regime in which the particle detaches from the wall.

When $c = 0$, we may integrate (4.118) to get

$$P(x,t) = \frac{2}{\sqrt{2\pi\Gamma t}} e^{-x^2/2\Gamma t}. \tag{4.119}$$

In this case, the bump does not move, but its width increases with time.

The mean particle velocity $v = (d/dt)\langle x\rangle$ is determined from

$$v = \int_0^\infty x \frac{\partial P}{\partial t} dx. \tag{4.120}$$

Using the Fokker-Planck equation (4.83) and integrating by parts

$$v(t) = c + \frac{\Gamma}{2} P(0,t). \tag{4.121}$$

We see therefore that the mean velocity is related to the probability density at $x = 0$. When $t \to \infty$, $P(0,t) \to P_e(0) = 2|c|/\Gamma$ for $c < 0$ and the terminal velocity $v^* = v(\infty)$ vanishes. For $c > 0$, $P(0,t) \to 0$ and $v^* = c$.

Using (4.84), Eq. (4.120) is written as

$$v = -\int_0^\infty x \frac{\partial J}{\partial x} dx = \int_0^\infty J dx. \tag{4.122}$$

Substituting J, given by (4.117), in this expression, we reach the result

$$v(t) = \frac{1}{t\sqrt{2\pi\Gamma t}} \int_0^\infty x e^{-(x-ct)^2/2\Gamma t} dx. \tag{4.123}$$

Figure 4.4 shows the velocity v as a function of c for several instants of time t.

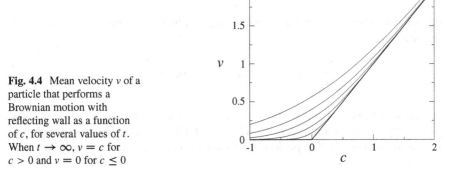

Fig. 4.4 Mean velocity v of a particle that performs a Brownian motion with reflecting wall as a function of c, for several values of t. When $t \to \infty$, $v = c$ for $c > 0$ and $v = 0$ for $c \le 0$

Notice that, when $c = 0$ the integral in (4.123) can be done, with the result

$$v(t) = \frac{\Gamma}{\sqrt{2\pi\Gamma t}}, \tag{4.124}$$

and the velocity decays algebraically with time. Integrating in time we see that

$$\langle x \rangle = \frac{2\Gamma}{\sqrt{2\pi\Gamma}} t^{1/2}, \tag{4.125}$$

which characterizes the diffusive motion.

4.8   Transition Probability

We have seen in Sect. 4.3 that the Fokker-Planck equation can be written in the form

$$\frac{\partial}{\partial t} P(x,t) = \mathscr{W} P(x,t), \tag{4.126}$$

where $\mathscr{W}$ is the evolution operator, that acts on functions $\phi(x)$, and is defined by (4.34). If the distribution $P(x,t')$ is known at time t', then the distribution $P(x,t)$ at time $t \geq t'$ can be obtained by

$$P(x,t) = e^{(t-t')\mathscr{W}} P(x,t'), \tag{4.127}$$

which is the solution of the Fokekr-Planck equation in the form (4.126). Introducing a Dirac delta function, this solution can be written in the form

$$P(x,t) = \int e^{(t-t')\mathscr{W}} \delta(x - x') P(x',t') dx'. \tag{4.128}$$

Defining the transition probability $K(x,t;x',t')$ by

$$K(x,t;x',t') = e^{(t-t')\mathscr{W}} \delta(x - x'), \tag{4.129}$$

the solution of the Fokker-Planck equation, given by (4.128), can be written in the integral form

$$P(x,t) = \int K(x,t;x',t') P(x',t') dx'. \tag{4.130}$$

The transition probability $K(x,t;x',t')$ must be understood as a conditional probability density. More precisely, $K(x,t;x',t')\Delta x$ is the conditional probability of finding the particle that performs the random motion in the interval between x

and $x + \Delta x$ at time t, given that it was in position x' at time t'. Integrating both sides of Eq. (4.129) in x, we conclude that the transition probability is normalized,

$$\int K(x,t;x',t')dx = 1, \tag{4.131}$$

at any time, as we should expect for a conditional probability density. From (4.129) we conclude that

$$\lim_{t \to t'} K(x,t;x',t') = \delta(x - x'). \tag{4.132}$$

Deriving both sides of Eq. (4.129) with respect to time t, we see that the transition probability also fulfills the Fokker-Planck equation

$$\frac{\partial}{\partial t} K(x,t;x',t') = \mathscr{W} K(x,t;x',t'). \tag{4.133}$$

In other terms, the transition probability is the solution of the Fokker-Planck equation for $t \geq t'$ complied with the condition given by (4.132).

It is worth mentioning that the transition probability $K(x,t;x',t')$ also fulfills the adjoint Fokker-Planck equation, but with respect to the second variable x'. To show this result it suffices to rewrite Eq. (4.129) in the equivalent form

$$K(x,t;x',t') = e^{(t-t')\mathscr{W}^\dagger} \delta(x - x'), \tag{4.134}$$

where here $\mathscr{W}^\dagger$ acts on functions of x'. This result is a consequence of the equality

$$\mathscr{W}^\dagger \delta(x - x') = \mathscr{W} \delta(x - x'), \tag{4.135}$$

where here we should understand that $\mathscr{W}^\dagger$ acts on functions of x' and $\mathscr{W}$ on functions of x. This equality follows directly from the definition of adjoint operator, given by Eq. (4.53). Deriving both sides of Eq. (4.134) with respect to time, we see that $K(x,t;x',t')$ fulfills the equation

$$\frac{\partial}{\partial t} K(x,t;x',t') = \mathscr{W}^\dagger K(x,t;x',t'), \tag{4.136}$$

which is the adjoint Fokker-Planck equation. We remark that in this equation the adjoint operator $\mathscr{W}^\dagger$ acts only on functions of the variable x'.

Using the following representation of the delta function,

$$\delta(x - x') = \int e^{-ik(x-x')} \frac{dk}{2\pi}, \tag{4.137}$$

in the definition of the transition probability, Eq. (4.129), we see that it has the following representation

$$K(x,t;x',t') = \int e^{ikx'} e^{(t-t')\mathscr{W}} e^{-ikx} \frac{dk}{2\pi}, \qquad (4.138)$$

We should notice that $\mathscr{W}$ acts on functions of variable x. If the representation of the delta function (4.137) is used in (4.134), we see that the transition probability can also be written in the form

$$K(x,t;x',t') = \int e^{-ikx} e^{(t-t')\mathscr{W}^\dagger} e^{ikx'} \frac{dk}{2\pi}, \qquad (4.139)$$

bearing in mind that in this equation the adjoint operator acts on functions of the variable x'.

Another representation of the transition probability is obtained by using the representation of the delta function in terms of the eigenfunctions of $\mathscr{W}$, given by (4.61). Introducing this representation in the definition of the transition probability, Eq. (4.129), we obtain the representation

$$K(x,t;x',t') = \sum_\ell e^{(t-t')\Lambda_\ell} \phi_\ell(x) \chi_\ell(x'). \qquad (4.140)$$

Notice in addition that the definition (4.129) of the transition probability shows us that it depends on t and t' only through the difference $t - t'$. From its definition we see that

$$K(x,t;x',t') = K(x,t-t';x',0), \qquad (4.141)$$

so that $K(x,t;x',t')$ can be obtained from $K(x,t;x',0)$ by replacing t by $t - t'$.

For the initial condition $P(x,0) = \delta(x-x')$, the probability distribution $P(x,t)$ is simply

$$P(x,t) = K(x,t;x',0) = \sum_\ell e^{t\Lambda_\ell} \phi_\ell(x) \chi_\ell(x'). \qquad (4.142)$$

Example 4.8 For a free particle that performs a Brownian motion, the operator $\mathscr{W}$ is given by

$$\mathscr{W} = \frac{\Gamma}{2} \frac{\partial^2}{\partial x^2}, \qquad (4.143)$$

so that

$$\mathscr{W} e^{-ikx} = -\frac{\Gamma}{2} k^2 e^{-ikx}. \qquad (4.144)$$

Inserting into (4.138), we find

$$K(x,t;x',0) = \int e^{ik(x'-x)} e^{-t\Gamma k^2/2} \frac{dk}{2\pi}. \tag{4.145}$$

Just integrate in k to get

$$K(x,t;x',0) = \frac{1}{\sqrt{2\pi t\Gamma}} e^{-(x-x')^2/2t\Gamma}. \tag{4.146}$$

Example 4.9 Consider a particle that performs a Brownian motion and is subject to an elastic force $f(x) = -\nu x$. The transition probability fulfills the Fokker-Planck equation

$$\frac{\partial}{\partial t} K = \nu \frac{\partial}{\partial x}(xK) + \frac{\Gamma}{2} \frac{\partial^2}{\partial x^2} K, \tag{4.147}$$

and must be solved with the initial condition $K = \delta(x - x')$. Using the solution presented in the Example 4.3, we get

$$K(x,t;x',0) = \frac{1}{\sqrt{2\pi b(t)}} e^{-[x-x'c(t)]^2/2b(t)}, \tag{4.148}$$

where c and b depend on t according to

$$c(t) = e^{-\nu t}, \qquad b(t) = \frac{\Gamma}{2\nu}(1 - e^{-\nu t}). \tag{4.149}$$

4.9 Chapman-Kolmogorov Equation

Using the identity $e^{(t-t')\mathcal{W}} = e^{(t-t'')\mathcal{W}} e^{(t''-t')\mathcal{W}}$ in Eq. (4.129) we find

$$K(x,t;x',t') = e^{(t-t'')\mathcal{W}} K(x,t'';x',t'), \tag{4.150}$$

which is the solution of the Fokker-Planck equation. Next, we insert a Dirac delta function $\delta(x - x'')$, in a manner analogous to that made when deriving (4.130) from (4.127), to reach the following relation

$$K(x,t;x',t') = \int K(x,t;x'',t'') K(x'',t'';x',t') dx'', \tag{4.151}$$

which is the Chapman-Kolmogorov equation for Markovian stochastic processes.

The iterative application of relation (4.151) leads us to the multiple integral

$$K(x_N, t_N; x_0, t_0) = \int K(x_N, t_N; x_{N-1}, t_{N-1}) \dots K(x_2, t_2; x_1, t_1)$$

$$\times K(x_1, t_1; x_0, t_0) dx_{N-1} \dots dx_2 dx_1, \tag{4.152}$$

valid for $t_0 \leq t_1 \leq t_2 \leq \dots \leq t_{N-1} \leq t_N$. Inserting (4.146) into this equation results in the Wiener integral.

The multiple integral (4.152), which relates the transition probability corresponding to two instants of time $t_0 < t_N$ with the transition probability calculated at several intermediate instants of time $t_0 \leq t_1 \leq t_2 \leq \dots \leq t_N$ can be used to determine the transition probability itself. To perceive how this can be done, we divide the interval $t_N - t_0$ in N equal parts Δt. In addition we set $t_0 = 0$ and $t_N = t$ so that $t = N\Delta t$. Using (4.152), the transition probability $K(x, t; x_0, 0)$ reads

$$K(x_N, t; x_0, 0) = \int K(x_N, \Delta t; x_{N-1}, 0) \dots K(x_2, \Delta t; x_1, 0)$$

$$\times K(x_1, \Delta t; x_0, 0) dx_{N-1} \dots dx_2 dx_1. \tag{4.153}$$

Next, we take N to be a very large number so that Δt will be very small. Once the transition probability $K(x, \Delta t; x', 0)$ is known for small values of the interval Δt, we can determine $K(x, t; x_0, 0)$ by performing the integral (4.153). This approach requires thus the knowledge of the transition probability for small values of the time interval, which can be obtained either from the Fokker-Planck equation, as we will see below, or from the corresponding Langevin equation, as seen in Sect. 3.5.

To determine the transition probability for small values of $\Delta t = t - t'$, we use the result (4.139)

$$K(x, \Delta t; x', 0) = \int e^{-ikx} e^{\Delta t \, \mathcal{W}^\dagger} e^{ikx'} \frac{dk}{2\pi}, \tag{4.154}$$

where $\mathcal{W}^\dagger$ acts on functions of x'. For small values of Δt,

$$e^{\Delta t \, \mathcal{W}^\dagger} e^{ikx'} = e^{ik[x' + \Delta t f(x')] - \Delta t \Gamma k^2/2}, \tag{4.155}$$

where we used (4.55) and therefore

$$K(x, \Delta t; x', 0) = \int e^{ik[-x+x'+\Delta t f(x')] - \Delta t \Gamma k^2/2} \frac{dk}{2\pi}. \tag{4.156}$$

Performing this integral

$$K(x, \Delta t; x', 0) = \frac{1}{\sqrt{2\pi \Delta t \Gamma}} \exp\{-\frac{\Delta t}{2\Gamma}[\frac{x - x'}{\Delta t} - f(x')]^2\}, \tag{4.157}$$

which inserted into (4.153) results in the Kac integral,

$$K(x_N,t;x_0,0) = \frac{1}{(2\pi\Delta t\Gamma)^{N/2}} \int e^{-A(x_N,\dots,x_2,x_1,x_0)} dx_{N-1}\dots dx_2 dx_1, \qquad (4.158)$$

where A is given by

$$A = \frac{\Delta t}{2\Gamma}\left[\left(\frac{x_N-x_{N-1}}{\Delta t} - f_{N-1}\right)^2 + \dots + \left(\frac{x_2-x_1}{\Delta t} - f_1\right)^2 + \left(\frac{x_1-x_0}{\Delta t} - f_0\right)^2\right],$$

$$(4.159)$$

where we use the notation $f_n = f(x_n)$.

Fourier space The Chapman-Kolmogorov equation can be written in the alternative form in the Fourier space. To this end we define the Fourier transform of the transition probability

$$G(k,t;k',t') = \int e^{-ikx} K(x,t;x',t') e^{ik'x'} dx dx'. \qquad (4.160)$$

The inverse transformation is

$$K(x,t;x',t') = \int e^{ikx} G(k,t;k',t') e^{-ik'x'} \frac{dk}{2\pi}\frac{dk'}{2\pi}. \qquad (4.161)$$

Using this relation in (4.151), we obtain the Chapman-Kolmogorov equation in the Fourier space,

$$G(k,t;k',t') = \int G(k,t;k'',t'') G(k'',t'';k',t') \frac{dk''}{2\pi}. \qquad (4.162)$$

The iterative application of relation (4.162) leads us to the multiple integral

$$G(k_N,t_N;k_0,t_0) = \int G(k_N,t_N;k_{N-1},t_{N-1})\dots G(k_2,t_2;k_1,t_1)$$

$$\times G(k_1,t_1;k_0,t_0)\frac{dk_{N-1}}{2\pi}\dots\frac{dk_2}{2\pi}\frac{dk_1}{2\pi}, \qquad (4.163)$$

valid for $t_0 \leq t_1 \leq t_2 \leq \dots \leq t_{N-1} \leq t_N$, from which we get the result

$$G(k_N,t;k_0,0) = \int G(k_N,\Delta t;k_{N-1},0)\dots G(k_2,\Delta t;k_1,0)$$

$$\times G(k_1,\Delta t;k_0,0)\frac{dk_{N-1}}{2\pi}\dots\frac{dk_2}{2\pi}\frac{dk_1}{2\pi}. \qquad (4.164)$$

To determine the Fourier transform of the transition probability for small values of $\Delta t = t - t'$, we use the result (4.139) or (4.154) to write

$$G(k, \Delta t; k', 0) = \int e^{ik'x'} e^{\Delta t \, \mathscr{W}^\dagger} e^{-ikx'} dx, \qquad (4.165)$$

where $\mathscr{W}^\dagger$ acts on functions of x'. For small values of Δt, we use the result (4.155) to get

$$G(k, \Delta t; k', 0) = e^{-\Delta t \Gamma k^2/2} \int e^{ik'x' - ik[x' + \Delta t f(x')]} dx'. \qquad (4.166)$$

Once integrated in x', the result is inserted in (4.164).

Harmonic oscillator We apply the above result to determine the transition probability corresponding to the harmonic oscillator, for which $f(x) = -\nu x$. Although the result has already been presented before, as in the Example 4.9, here we will do a derivation of that result. For $f(x) = -\nu x$,

$$G(k, \Delta t; k', 0) = e^{-\Delta t \Gamma k^2/2} \int e^{i(k' - ak)x'} dx' = 2\pi\delta(k' - ak)e^{-\Delta t \Gamma k^2/2}, \qquad (4.167)$$

where $a = 1 - \nu\Delta t$. Replacing this result into (4.164) and performing the integration in $k_1, k_2, \ldots, k_{N-1}$, we get

$$G(k_N, t; k_0, 0) = 2\pi\delta(k_0 - ck_N)e^{-bk_N^2/2}, \qquad (4.168)$$

where

$$b = \Delta t \Gamma(1 + a^2 + a^4 \ldots + a^{2(N-1)}) = \Delta t \Gamma \frac{1 - a^{2N}}{1 - a^2}, \qquad (4.169)$$

$$c = a^N. \qquad (4.170)$$

Bearing in mind that $\Delta t = t/N$, then in the limit $N \to \infty$

$$b = \frac{\Gamma}{2\nu}(1 - e^{-2\nu t}), \qquad c = e^{-\nu t}, \qquad (4.171)$$

what leads us to the following result

$$G(k, t; k_0, 0) = 2\pi\delta(k_0 - ck)e^{-bk^2/2}, \qquad (4.172)$$

from which we obtain the transition probability, using (4.161),

$$K(x,t;x_0,0) = \int e^{ik(x-x_0 c)} e^{-bk^2/2} \frac{dk}{2\pi}. \tag{4.173}$$

Finally, integrating in k,

$$K(x,t;x_0,0) = \frac{1}{\sqrt{2\pi b}} e^{-(x-x_0 c)^2/2b}, \tag{4.174}$$

which is the result presented before in Example 4.9.

4.10 Escape from a Region

In this section we will give an interpretation for the solution $Q(y,t)$ of the adjoint Fokker-Planck equation,

$$\frac{\partial}{\partial y} Q(y,t) = f(y) \frac{\partial}{\partial y} Q(y,t) + \frac{\Gamma}{2} \frac{\partial^2}{\partial y^2} Q(y,t), \tag{4.175}$$

studied in Sect. 4.4. To this end, we notice, as seen in Sect. 4.8, that the transition probability $K(x,t|y,0)$ fulfills the adjoint Fokker-Planck equation in the second variable y. Therefore, the integral of $K(x,t|y,0)$ in a certain interval of the variable x, say, in the interval $[x_1, x_2]$, also fulfills the same equation and can thus be identified with $Q(y,t)$, that is,

$$Q(y,t) = \int_{x_1}^{x_2} K(x,t|y,0)dx, \tag{4.176}$$

Bearing in mind the meaning of the transition probability $K(x,t|y,0)$, we may interpreted $Q(y,t)$, given by (4.176), as the probability that, at time t, the particle that performs the Brownian motion is in the interval $[x_1, x_2]$, given that it was in position y at the initial time $t = 0$.

We examine here the problem of determining the time a particle takes to escape from a given region $[x_1, x_2]$ assuming that it is in position y at time $t = 0$. It is clear that y must belong to this interval otherwise the particle is already out of the region. To solve this problem we should determine the probability $Q(y,t)$ that the particle is inside this region at time t, given that it was at position y at time $t = 0$. This can be done by solving the adjoint Fokker-Planck (4.175). However, we should know which initial condition and which boundary condition we should use.

Since $K(x,0|y,0) = \delta(x - y)$ then at time $t = 0$, we should have $Q(y,0) = 1$ for y inside the interval $[x_1, x_2]$ and $Q(y,0) = 0$ for y outside it, a result that follows directly from Eq. (4.176). This is therefore the initial condition. Now, suppose that, at initial time, the particle is at the boundary or outside the region $[x_1, x_2]$. We wish that the particle remains outside this region forever, what may be done by setting $Q(x_1, t) = Q(x_2, t) = 0$ at any time $t > 0$, which are the desired boundary conditions. The boundary condition for which $Q(y, t)$ vanishes corresponds to an absorbing boundary. When $\partial Q(y, t)/\partial y$ vanishes at the boundary, it is reflecting.

The function $Q(y, t)$ can also be understood as the probability that the time of permanence of the particle inside the region $[x_1, x_2]$ is larger than t. Then, the probability that the time of permanence in this region is between t_1 and t_2 is

$$Q(y, t_1) - Q(y, t_2) = -\int_{t_1}^{t_2} \frac{\partial}{\partial t} Q(y, t) dt. \tag{4.177}$$

We see thus that $-\partial Q/\partial t$ is the probability density corresponding to the time of permanence in the region $[x_1, x_2]$. We assume that $Q(y, t) \to 0$ when $t \to \infty$, that is, that the particle will leave the region for times large enough.

The mean time of permanence in the interval $[x_1, x_2]$ of a particle that was at the position y at time $t = 0$ is then

$$\Theta(y) = -\int_0^\infty t \frac{\partial Q(y, t)}{\partial t} dt = \int_0^\infty Q(y, t) dt, \tag{4.178}$$

where we performed an integration by parts and use the property $Q(y, \infty) = 0$. Integrating both sides of Eq. (4.175) in t, from zero to infinity, and taking into account that $Q(y, \infty) - Q(y, 0) = -1$, we reach the following equation for $\Theta(y)$

$$f(y) \frac{\partial}{\partial y} \Theta(y) + \frac{\Gamma}{2} \frac{\partial^2}{\partial y^2} \Theta(y) = -1, \tag{4.179}$$

valid in the interval $x_1 < y < x_2$. At the boundaries of the interval we should have $\Theta(x_1) = \Theta(x_2) = 0$.

Example 4.10 Suppose that a particle performs a Brownian motion along the x-axis and that it was in position y at time $t = 0$. We wish to determine the mean time $\Theta(y)$ it takes to escape from the region $[x_1, x_2]$. To this end, it suffices to solve the equation

$$\frac{\Gamma}{2} \frac{\partial^2}{\partial y^2} \Theta(y) = -1. \tag{4.180}$$

The solution is

$$\Theta(y) = \frac{2}{\Gamma}(y - x_1)(x_2 - y). \tag{4.181}$$

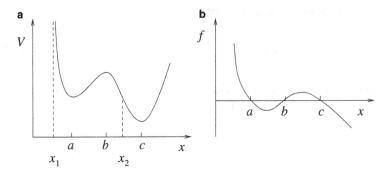

Fig. 4.5 (a) Double well potential V, one shallow and the other deep. The minima occur in $x = a$ and $x = c$ and the maximum in $x = b$. (b) Corresponding force, $f = -dV/dx$

It is clear that the maximum time of permanence will occur when, at the initial time, the particle is in the middle point of the interval, that is, when $y = (x_1 + x_2)/2$.

Escape from a metastable state Consider a particle subject to a force $f(x)$ associated with a bistable potential $V(x)$, as shown in Fig. 4.5, with two wells, one shallow and the other deep. The shallow well represents a metastable while the deep well represents a stable state. At the initial time the particle is at the bottom of the shallow well. We wish to know the mean time that the particle takes to leave the shallow well and reach the other well. To this end, we focus on the region $[x_1, x_2]$, shown in Fig. 4.5. The point x_1 is such that $V(x) \to \infty$ when $x \to x_1$ so that the physical region corresponds to $x > x_1$. The point x_2 must be between b and c. To be precise, we choose x_2 such that $V(x_2) = V(a)$. The mean time of permanence $\Theta(x)$ in this region, of a particle which initially was in the point x inside the region, obeys the equation

$$f(x)\Theta'(x) + \frac{2}{\Gamma}\Theta''(x) = -1, \tag{4.182}$$

where $f(x) = -V'(x)$. One boundary condition is $\Theta(x_2) = 0$. Since the potential diverges in $x = x_1$, the particle will not escape through this end. At this end, which corresponds to a reflecting boundary, we should impose the condition $\Theta'(x_1) = 0$.

The solution of (4.182) is obtained by means of the auxiliary function $\Phi(x)$ defined by

$$\Theta'(x) = -\Phi(x)e^{2V(x)/\Gamma}. \tag{4.183}$$

inserting this in the differential equation, we obtain the result

$$\Phi'(x) = \frac{2}{\Gamma}e^{-2V(x)/\Gamma}, \tag{4.184}$$

from which it follows, by integration, that

$$\Phi(x) = \frac{2}{\Gamma} \int_{x_1}^{x} e^{-2V(x')/\Gamma} dx', \tag{4.185}$$

where we chose the integration constant as to fulfill the boundary condition in $x = x_1$. Therefore, replacing this result in

$$\Theta(x) = \int_{x}^{x_2} \Phi(x') e^{2V(x')/\Gamma} dx', \tag{4.186}$$

where we choose the integration constant as to fulfill the boundary condition in $x = x_2$, we obtain

$$\Theta(x) = \frac{2}{\Gamma} \int_{x}^{x_2} \int_{x_1}^{x'} e^{2[V(x')-V(x'')]/\Gamma} dx'' dx', \tag{4.187}$$

An approximation for $\Theta(a)$, the mean time the particle takes to leave the metastable state starting at the bottom of the shallow well, can be done as follows. First, we note that the absolute maximum of the function $[V(x') - V(x'')]$ in the integration region, which is defined by $x_1 \leq x'' \leq x'$ and $a \leq x' \leq x_2$, occurs when $x' = b$ and $x'' = a$. Next, we observe that the relevant contribution for the integral comes from a small region around the maximum of the function. Thus, we expand this function around the maximum of the function up to quadratic terms. The integral becomes equivalent to the product of two Gaussian integrals which can thus be integrated leading to the result

$$\Theta(a) = \frac{4\pi}{\Gamma \sqrt{|V''(b)||V''(a)|}} e^{2[V(b)-V(a)]/\Gamma}, \tag{4.188}$$

which is the Arrhenius formula. Since Γ is proportional to the temperature we see that $\Theta \sim e^{\Delta E/k_B T}$, where ΔE is the difference between the maximum of the potential energy and the metastable minimum of the potential energy.

4.11 Time Correlation

The values of a stochastic variable x calculated at a certain time t may be correlated with the values of the same variable at an earlier time t'. The time correlation of the variable x between two instants of time t and t' such that $t > t'$ is denoted by

$$\langle x(t)x(t')\rangle, \tag{4.189}$$

where t' is called waiting time and t, observation time. The transition probability $K(x, t; x', t')$, which is the conditional probability density of the stochastic variable $x(t)$, allows us to define with precision the autocorrelation (4.189). It is defined by

$$\langle x(t)x(t')\rangle = \int \int xK(x, t; x', t')x' P(x', t')dxdx'. \tag{4.190}$$

Using the result (4.129), we can write the time correlation in the following form

$$\langle x(t)x(t')\rangle = \int \int xe^{(t-t')\mathscr{W}} x P(x, t')dx. \tag{4.191}$$

Notice that the average $\langle x(t)\rangle$ of x at time t is given by

$$\langle x(t)\rangle = \int xP(x, t)dx. \tag{4.192}$$

If the waiting time is large enough, then $P(x, t')$ approaches the stationary distribution $P(x)$ so that the time correlation of x in the stationary regime depends only on the difference $t - t'$, that is, $\langle x(t)x(t')\rangle = \langle x(t - t')x(0)\rangle$ and is given by

$$\langle x(t)x(0)\rangle = \int \int xK(x, t; x', 0)x' P(x')dxdx'. \tag{4.193}$$

which can yet be written as

$$\langle x(t)x(0)\rangle = \int xe^{t\mathscr{W}} x P(x)dx. \tag{4.194}$$

Moreover, the average of the variable x,

$$\langle x(0)\rangle = \int xP(x)dx, \tag{4.195}$$

in this regime is independent of t. Notice that, in formulas (4.193) and (4.195), the origin of times was moved inside the stationary regime. In other words, the waiting time occurs inside the stationary regime and, as a consequence, also the observation time. In the stationary regime, we see thus that the average of a state function becomes independent of time and the time correlation of two functions depends only on the difference between the waiting time and the observation time.

Example 4.11 We determine the correlation $\langle x(t)x(0)\rangle$ in the stationary regime for the situation of Example 4.5, where $f(x) = -\nu x$. Using the transition probability presented in this example, we have

$$\langle x(t)x(0)\rangle = \int x\frac{1}{\sqrt{2\pi b(t)}}e^{-(x-x'a(t))^2/2b(t)} x'\frac{1}{\sqrt{2\pi b(0)}}e^{-(x')^2/2b(0)} dxdx'. \tag{4.196}$$

Integrating first in x,

$$\langle x(t)x(0)\rangle = a(t)\int (x')^2 \frac{1}{\sqrt{2\pi b(0)}} e^{-(x')^2/2b(0)} dx', \qquad (4.197)$$

and then in x',

$$\langle x(t)x(0)\rangle = a(t)b(0) = \frac{\Gamma}{2v} e^{-vt}. \qquad (4.198)$$

Exercises

1. Show that the solution (4.99) of the Fokker-Planck Eq. (4.83), related to the Brownian motion of a particle subject to a constant force c and absorbing wall in $x = 0$, can be constructed as follows. We start by observing that the probability distribution (4.12), which we denote by $P(x,t;x_0)$, solves the Fokker-Planck Eq. (4.83) for any value of x_0. A possible solution can then be constructed by the linear combination $P(x,t) = AP(x,t;x_0) + BP(x,t,x_1)$. Determine A, B and x_1 so that $P(x,t)$ vanishes in $x = 0$ and with the initial condition $P(x,0) = \delta(x - x_0)$.

2. Consider the Brownian motion of a free particle along the x-axis restricted to the region $x \geq 0$, that is, there is a wall in $x = 0$. At the initial time $t = 0$ the particle is in $x = x_0$. Show that the probability distribution is given by

$$P(x,t) = \frac{1}{\sqrt{2\pi\Gamma t}}\{e^{-(x-x_0)^2/2\Gamma t} \pm e^{-(x+x_0)^2/2\Gamma t}\},$$

   where the $+$ sign is valid for the reflecting wall and the $-$ sign for the absorbing wall. In both cases, use the method of the previous exercise attending the boundary conditions. That is, $J(0,t) = 0$ for the reflecting wall and $P(0,t) = 0$ for the absorbing wall.

3. Determine the eigenfunctions and the eigenvalues of the evolution operator $\mathcal{W}$ given by (4.34), and related to the force $f(x) = c > 0$ for $x < 0$, $f(x) = 0$ for $x = 0$, and $f(x) = -c$ for $x > 0$.

4. Determine the stationary distribution corresponding to the Fokker-Planck equations associated to the Langevin equations of the Exercise 8 of Chap. 3. Compare them with the histogram obtained in that exercise.

5. Use the Fokker-Planck Eq. (4.4) to derive Eq. (3.81), which gives the time evolution of the moments $\langle x^\ell \rangle$. To this end, multiply both sides of the Fokker-Planck equation by x^ℓ and integrate in x. Next, do an integration by parts, assuming that the probability distribution an its derivatives vanish at the boundary of x.

Chapter 5
Fokker-Planck Equation II

5.1 Equation in Several Variables

In the previous chapter we studied the Fokker-Planck equation in just one variable. In this chapter we analyze the Fokker-Planck equation in several variables. We consider a system described by N variables x_1, x_2, x_3, ..., x_N. The equation of motion for this system is given by the set of equations

$$\frac{dx_i}{dt} = f_i(x) + \zeta_i(t), \tag{5.1}$$

for $i = 1, 2, \ldots, N$, where $x = (x_1, x_2, \ldots, x_N)$, and the stochastic variables $\zeta_1(t)$, $\zeta_2(t)$, ..., $\zeta_N(t)$ have the following properties

$$\langle \zeta_i(t) \rangle = 0, \tag{5.2}$$

$$\langle \zeta_i(t)\zeta_j(t') \rangle = \Gamma_i \delta_{ij} \delta(t - t'), \tag{5.3}$$

where $\Gamma_1, \Gamma_2, \ldots, \Gamma_N$ are constants.

We want to determine the time evolution of the probability distribution $P(x, t) = P(x_1, x_2, \ldots, x_N, t)$ of the variables $x_1, x_2, \ldots, x_N$. Using a procedure analogous to that used for the case of one variable, seen in Chap. 3, we can shown that this probability distribution obeys the equation

$$\frac{\partial P}{\partial t} = -\sum_{i=1}^{N} \frac{\partial}{\partial x_i}(f_i P) + \frac{1}{2}\sum_{i=1}^{N} \Gamma_i \frac{\partial^2 P}{\partial x_i^2}, \tag{5.4}$$

which we call Fokker-Planck equation in several variables.

© Springer International Publishing Switzerland 2015
T. Tomé, M.J. de Oliveira, *Stochastic Dynamics and Irreversibility*, Graduate Texts in Physics,
DOI 10.1007/978-3-319-11770-6_5

Example 5.1 Diffusion equation. When $f_i = 0$, the Fokker-Planck equation (5.4) is called diffusion equation and is given by

$$\frac{\partial P}{\partial t} = D \sum_{i=1}^{n} \frac{\partial^2 P}{\partial x_i^2}, \tag{5.5}$$

where $\Gamma_i = 2D$ and D is the diffusion constant and n is the number of variables. The solution of this equation is obtained from the characteristic function

$$G(k,t) = \int e^{ik \cdot x} P(x,t) dx, \tag{5.6}$$

where $k = (k_1, \ldots, k_n)$. Deriving $G(k,t)$ with respect to t and using the diffusion equation, we get the equation

$$\frac{\partial G}{\partial t} = -Dk^2 G, \tag{5.7}$$

where $k^2 = \sum_i k_i^2$, whose solution is

$$G(k,t) = e^{-Dtk^2} G(k,0). \tag{5.8}$$

As initial condition, we use $P(x,0) = \prod_i \delta(x_i - a_i)$, which gives us $G(k,0) = e^{ik \cdot a}$, where $a = (a_1, \ldots a_n)$, and

$$G(k,t) = e^{ik \cdot a - Dtk^2}. \tag{5.9}$$

This condition means that the particle is in the position $x = a$ at $t = 0$. As is well known $G(k,t)$ is the Fourier transform of the multidimensional Gaussian

$$P(x,t) = \prod_i \frac{1}{\sqrt{4\pi Dt}} e^{(x_i - a_i)^2 / 4Dt}. \tag{5.10}$$

The variables x_i are independent, with variance

$$\langle x_i^2 \rangle - \langle x_i \rangle^2 = 2Dt. \tag{5.11}$$

The Fokker-Planck equation can still be written in the form of a continuity equation

$$\frac{\partial P}{\partial t} = -\sum_{i=1}^{N} \frac{\partial J_i}{\partial x_i}, \tag{5.12}$$

where J_i, the i-th component of the probability current, is given by

$$J_i = f_i P - \frac{\Gamma_i}{2} \frac{\partial P}{\partial x_i}. \tag{5.13}$$

The solutions of the Fokker-Planck equation have to be determined according to the boundary conditions given a priori. These conditions concerns the surface that delimits the region of the space defined by the variables $x_1, x_2, \ldots, x_N$. Here we consider boundary conditions such that, in the points of the surface, the normal component of the probability current vanish. In addition, we will consider natural boundary conditions, that is, such that the boundary are placed at infinity. In this case, the probability distribution and its derivatives vanish rapidly when $|x_i| \to \infty$ so that $J_i \to 0$ in this limit.

In the stationary regime, the probability density is time independent and fulfills the equation

$$-\sum_{i=1}^{N} \frac{\partial}{\partial x_i}(f_i P) + \frac{1}{2} \sum_{i=1}^{N} \Gamma_i \frac{\partial^2 P}{\partial x_i^2} = 0, \tag{5.14}$$

which can be written in the form

$$\sum_{i=1}^{N} \frac{\partial J_i}{\partial x_i} = 0, \tag{5.15}$$

For just one variable, as we have seen previously, this equation implies a position independent probability current which therefore must vanish. For more than one variable, however, this might not happen. The fact that the normal component is zero at the points of the surface is not sufficient to guarantee that the current is zero everywhere. In fact, the currents might be circular and tangent to the boundary surface.

Next, we examine the conditions that the forces must satisfy so that, in the stationary regime, the current vanishes everywhere, that is,

$$J_i(x) = 0. \tag{5.16}$$

When the current vanishes in all points, the stationary regime corresponds to the thermodynamic equilibrium and the condition (5.16) expresses the microscopic reversibility. If $J_i = 0$, then Eq. (5.13) gives us

$$f_i = \frac{\Gamma}{2} \frac{\partial}{\partial x_i} \ln P. \tag{5.17}$$

We assume that the constant Γ_i is the same for all particles, $\Gamma_i = \Gamma$. From this equation follows the condition sought

$$\frac{\partial f_i}{\partial x_j} = \frac{\partial f_j}{\partial x_i}, \tag{5.18}$$

which must be fulfilled for any pair i, j.

If the condition (5.18) is satisfied, then f_i must be the gradient of a potential $V(x)$, that is,

$$f_i = -\frac{\partial V}{\partial x_i}. \tag{5.19}$$

After V is determined, we may write

$$\ln P = -\frac{2}{\Gamma}V + \text{const}, \tag{5.20}$$

or yet

$$P(x) = A\,e^{-2V(x)/\Gamma}, \tag{5.21}$$

where A is a constant that must be determined by the normalization of P.

Summarizing, when the stationary solution of the Fokker-Planck equation is such that each component of the probability current vanishes, we say that the system holds microscopic reversibility. The condition for this to occur is that the forces are conservative, that is, they are generated by a potential. In this case, the stationary probability distribution has the form of an equilibrium distribution, that is to say, a Boltzmann-Gibbs distribution, given by (5.21), and we say that the system is found in thermodynamic equilibrium. Due to this result, we can say that the system described by the Langevin equation (5.1) or equivalently by the Fokker-Planck equation (5.4) is in contact with a heat reservoir at a temperature proportional to Γ.

5.2 Solenoidal Forces

When the forces f_i are conservatives, that is, when they obey the condition (5.18), and $\Gamma_i = \Gamma$, we have seen that the stationary solution is of the form (5.21). In this section we examine the special case in which, in spite of f_i be nonconservative, it is still possible to write the stationary solution in the form (5.21). To this end, we start by writing the force f_i as a sum of two parts

$$f_i = f_i^C + f_i^D, \tag{5.22}$$

where f_i^C is the conservative part, which fulfills the condition

$$\frac{\partial f_i^C}{\partial x_j} = \frac{\partial f_j^C}{\partial x_i}, \tag{5.23}$$

for any pair (i, j) and therefore can be generated by a potential V, that is,

$$f_i^C = -\frac{\partial V}{\partial x_i}. \tag{5.24}$$

The other part f_i^D is nonconservative, that is, $\partial f_i^D/\partial x_j \neq \partial f_j^D/\partial x_i$ for at least one pair (i, j) and has null divergence (solenoidal force)

$$\sum_{i=1}^{N} \frac{\partial f_i^D}{\partial x_i} = 0. \tag{5.25}$$

In addition, we assume that the two parts are orthogonal, that is,

$$\sum_i f_i^C f_i^D = 0. \tag{5.26}$$

If the conditions (5.23), (5.25) and (5.26) are fulfilled, then the stationary solution has the form

$$P(x) = A e^{-2V(x)/\Gamma}, \tag{5.27}$$

where V is the potential of the conservative force f_i^C. To show this result, we start from the Fokker-Planck equation which in the stationary regime is given by

$$-\sum_{i=1}^{N} \frac{\partial}{\partial x_i}(f_i^D P) - \sum_{i=1}^{N} \frac{\partial}{\partial x_i}(f_i^C P) + \frac{\Gamma}{2} \sum_{i=1}^{N} \frac{\partial^2}{\partial x_i^2} P = 0. \tag{5.28}$$

Inserting in this expression the distribution $P(x)$ given by (5.27) we find that the two last terms of (5.28) vanish remaining the equation

$$\sum_{i=1}^{N} \left(\frac{\partial}{\partial x_i} f_i^D + \frac{2}{\Gamma} f_i^D f_i^C \right) = 0, \tag{5.29}$$

which in fact is satisfied in virtue of the properties (5.25) and (5.26).

As an example of a system that satisfy the condition (5.23), (5.25) and (5.26), we examine the electrical circuit, studies in Sect. 3.9, composed by a resistor of resistance R, a capacitor of capacity C and an inductor of inductance L, in which we suppose there are fluctuations in both the voltage and the current. The fluctuations in the voltage are due to the contact with a heat reservoir whereas the fluctuations in the current arise from the contact with a charge reservoir, electrically neutral. The equations for the charge Q of the capacitor and for the current I in the circuit are given by

$$\frac{dQ}{dt} = -\gamma Q + I + \zeta_1(t), \tag{5.30}$$

$$L\frac{dI}{dt} = -\frac{1}{C}Q - RI + V_r(t), \tag{5.31}$$

where $\zeta_1(t)$ and $V_r(t)$ are the random current and random voltage, respectively, considered to be independent. These two stochastic variables have zero mean and time correlation with the following properties

$$\langle \zeta_1(t)\zeta_1(t') \rangle = \Gamma\delta(t - t'), \tag{5.32}$$

$$\langle V_r(t)V_r(t') \rangle = B\delta(t - t'). \tag{5.33}$$

Equation (5.30) is already in the form (5.1). To transform Eq. (5.31) into the form (5.1), we divide both sides of (5.31) by L to get

$$\frac{dI}{dt} = -\frac{1}{LC}Q - \frac{R}{L}I + \zeta_2(t), \tag{5.34}$$

where $\zeta_2 = V_r/L$. Compelling $\zeta_2(t)$ to have a time correlation of the form

$$\langle \zeta_2(t)\zeta_2(t') \rangle = \Gamma\delta(t - t'), \tag{5.35}$$

and to have zero mean, then B and Γ must be related by $B = L^2\Gamma$.

Making the identification $x_1 = Q$ and $x_2 = I$, we see that the forces are given by

$$f_1 = -\gamma Q + I, \qquad f_2 = -\frac{1}{LC}Q - \frac{R}{L}I. \tag{5.36}$$

The conservative and solenoidal parts are given by

$$f_1^C = -\gamma Q, \qquad f_2^C = -\frac{R}{L}I, \tag{5.37}$$

$$f_1^D = I, \qquad f_2^D = -\frac{1}{LC}Q. \tag{5.38}$$

In the stationary regime, we postulate that the probability distribution is the equilibrium distribution, given by (5.27). For this, we assume that the condition (5.26) is fulfilled, what leads us to the following relation $\gamma = R/CL^2$. Therefore, the stationary distribution is given by (5.27) with a potential $V(x)$ given by

$$V = \frac{\gamma}{2}Q^2 + \frac{R}{2L}Y^2 = \frac{R}{L^2}\left(\frac{1}{2C}Q^2 + \frac{L}{2}I^2\right), \tag{5.39}$$

that is,

$$P = A\exp\{-\frac{2R}{B}\left(\frac{1}{2C}Q^2 + \frac{L}{2}I^2\right)\}, \tag{5.40}$$

where we have taken into account the relation $B = L^2\Gamma$. Comparing with the Gibbs distribution

$$P = A\exp\{-\frac{1}{k_BT}\left(\frac{1}{2C}Q^2 + \frac{L}{2}I^2\right)\}, \tag{5.41}$$

we reach the following relation between the coefficient B and the temperature

$$B = 2Rk_BT, \tag{5.42}$$

which is a result due to Nyquist.

5.3 Linear System

We study here the general solution of the Fokker-Planck equation such that the functions f_i are linear,

$$f_i = \sum_{j=1}^{N} A_{ij}x_j. \tag{5.43}$$

The coefficients A_{ij} are understood as the entries of a $N \times N$ square matrix. If the forces f_i are conservatives, A is a symmetrical matrix. The Fokker-Planck equation reads

$$\frac{\partial}{\partial t}P = -\sum_{i,j=1}^{N} A_{ij}\frac{\partial}{\partial x_i}(x_j P) + \frac{1}{2}\sum_{i=1}^{N} \Gamma_i\frac{\partial^2}{\partial x_i^2}P, \tag{5.44}$$

whose solution is obtained by means of the characteristic function $G(k,t)$ defined by

$$G(k,t) = \int e^{ik^\dagger x}P(x,t)dx, \tag{5.45}$$

where x is a column matrix with entries x_i and $k^\dagger$ is the transpose of the column matrix k whose entries are k_i. Deriving G with respect to time and using the Fokker-Planck equation, we obtain the following equation for the characteristic function

$$\frac{\partial G}{\partial t} = \sum_{i,j=1}^{N} A_{ij} k_i \frac{\partial G}{\partial k_j} - \frac{1}{2} \sum_{i=1}^{N} \Gamma_i k_i^2 G. \tag{5.46}$$

Each term to the right is obtained by integration by parts and assuming that P and its derivative vanish in the limits of integration. The solution of this equation is reached by assuming the following form for G,

$$G(k,t) = e^{ia^\dagger k - k^\dagger C k/2}, \tag{5.47}$$

where $a^\dagger$ is the transpose of the column matrix a whose entries are a_i and C is the square matrix with entries C_{ij}. Both, a and C depend on t. The characteristic function G corresponds to the N-dimensional Gaussian distribution

$$P(x,t) = \frac{1}{\sqrt{(2\pi)^N \det C}} e^{(x^\dagger - a^\dagger) C^{-1}(x-a)}, \tag{5.48}$$

where a_i and C_{ij} are the means and the covariances,

$$a_i = \langle x_i \rangle, \qquad\qquad C_{ij} = \langle x_i x_j \rangle - \langle x_i \rangle \langle x_j \rangle. \tag{5.49}$$

Inserting the form (5.47) into (5.46), we get the following equations for a and C

$$\frac{da_i}{dt} = \sum_j A_{ij} a_j, \tag{5.50}$$

$$\frac{dC_{ij}}{dt} = \sum_\ell C_{i\ell} A_{j\ell} + \sum_\ell A_{i\ell} C_{\ell j} + \Gamma_i \delta_{ij}, \tag{5.51}$$

which can be written in the matrix form

$$\frac{da}{dt} = Aa, \tag{5.52}$$

$$\frac{dC}{dt} = CA^\dagger + AC + \Gamma, \tag{5.53}$$

where Γ is the diagonal matrix whose elements are $\Gamma_i \delta_{ij}$. This last equation coincides with Eq. (3.153) of Sect. 3.8.

5.4   Entropy and Entropy Production

Here we study the time evolution of the entropy of a system described by the Fokker-Planck equation (5.4). The entropy S of a system described by the probability distribution $P(x,t)$ is defined by

$$S(t) = -\int P(x,t)\ln P(x,t)dx, \qquad (5.54)$$

where $dx = dx_1 dx_2 \ldots dx_N$. Using the Fokker-Planck equation in the form (5.12), we determine the time variation of entropy, which is given by

$$\frac{dS}{dt} = \sum_i \int \ln P \frac{\partial J_i}{\partial x_i} dx. \qquad (5.55)$$

Performing an integration by parts,

$$\frac{dS}{dt} = -\sum_i \int J_i \frac{\partial}{\partial x_i} \ln P dx. \qquad (5.56)$$

From the definition (5.13) of the probability current, it follows that

$$\frac{\partial}{\partial x_i} \ln P = \frac{2f_i}{\Gamma_i} - \frac{2J_i}{\Gamma_i P}. \qquad (5.57)$$

Therefore,

$$\frac{dS}{dt} = -\sum_i \frac{2}{\Gamma_i} \int f_i J_i dx + \sum_i \frac{2}{\Gamma_i} \int \frac{J_i^2}{P} dx. \qquad (5.58)$$

The second term is nonnegative and we identify as the entropy production rate Π, given by

$$\Pi = \sum_i \frac{2}{\Gamma_i} \int \frac{J_i^2}{P} dx. \qquad (5.59)$$

The other term is the entropy flux Φ,

$$\Phi = \sum_i \frac{2}{\Gamma_i} \int f_i j_i dx, \qquad (5.60)$$

so that we may write the time variation of the entropy in the form advanced by Prigogine

$$\frac{dS}{dt} = \Pi - \Phi. \tag{5.61}$$

Next, we multiply the current density, defined in (5.13), by f_i and integrate to get

$$\int f_i J_i dx = \int f_i^2 P dx - \frac{\Gamma_i}{2} \int f_i \frac{\partial P}{\partial x_i} dx. \tag{5.62}$$

Integrating the last term by parts, we reach the result

$$\int f_i J_i dx = \int f_i^2 P dx + \frac{\Gamma_i}{2} \int f_{ii} P dx = \langle f_i^2 \rangle + \frac{\Gamma_i}{2} \langle f_{ii} \rangle, \tag{5.63}$$

where $f_{ii} = \partial f_i / \partial x_i$. From this result we obtain the following expression for the entropy flux

$$\Phi = \sum_i \left(\frac{2}{\Gamma_i} \langle f_i^2 \rangle + \langle f_{ii} \rangle \right). \tag{5.64}$$

It is important to remark that in the stationary state $dS/dt = 0$ so that $\Phi = \Pi$. If, in addition, there is microscopic reversibility, $J_i = 0$, then $\Phi = \Pi = 0$ as can be seen in Eq. (5.59).

5.5   Kramers Equation

In this section we analyze the Fokker-Planck in two variables, called Kramers equation. A particle performs a Brownian motion along the x-axis and is subject to an external force $F(x)$, in addition to a dissipative force and a random force. The Newton equation for this particle is

$$m \frac{dv}{dt} = -\alpha v + F(x) + F_a(t). \tag{5.65}$$

The random force $F_a(t)$ has the properties

$$\langle F_a(t) \rangle = 0, \tag{5.66}$$

$$\langle F_a(t) F_a(t') \rangle = B \delta(t - t'). \tag{5.67}$$

Equation (5.65) together with

$$\frac{dx}{dt} = v \tag{5.68}$$

constitute the equation of motion of the particle. Dividing both sides of Eq. (5.65) by m we get the equation

$$\frac{dv}{dt} = -\gamma v + f(x) + \zeta(t), \tag{5.69}$$

where $\gamma = \alpha/m$, $f(x) = F(x)/m$ and $\zeta(t) = F_a(t)/m$. The noise $\zeta(t)$ has the properties

$$\langle \zeta(t) \rangle = 0, \tag{5.70}$$

$$\langle \zeta(t)\zeta(t') \rangle = \Gamma\delta(t - t'). \tag{5.71}$$

with $\Gamma = B/m^2$.

Using a procedure analogous to that used in Chap. 3, it is possible to show, from (5.68) and (5.69), that the probability distribution $P(x, v, t)$ of x and v obeys the following equation

$$\frac{\partial P}{\partial t} = -\frac{\partial}{\partial x}(vP) - \frac{\partial}{\partial v}[(f - \gamma v)P] + \frac{\Gamma}{2}\frac{\partial^2 P}{\partial v^2}, \tag{5.72}$$

or yet

$$\frac{\partial P}{\partial t} = -v\frac{\partial P}{\partial x} - \frac{F}{m}\frac{\partial P}{\partial v} + \frac{\alpha}{m}\frac{\partial}{\partial v}(vP) + \frac{B}{2m^2}\frac{\partial^2 P}{\partial v^2}. \tag{5.73}$$

which is called Kramers equation.

The Kramers equation can be written in the form of a continuity equation

$$\frac{\partial P}{\partial t} = -\frac{\partial J^x}{\partial x} - \frac{\partial J^v}{\partial v}, \tag{5.74}$$

where J^x and J^v are the probability current in directions x and v, respectively, and given by

$$J^x = vP, \tag{5.75}$$

$$J^v = (f - \gamma v)P - \frac{\Gamma}{2}\frac{\partial P}{\partial v}. \tag{5.76}$$

We should notice that the real current $J(x)$ is obtained by integrating $J^x(x, v)$ over the velocities, that is,

$$J(x) = \int J^x(x, v)dv. \tag{5.77}$$

Next we determine the stationary solution $P(x, v)$ of Kramers equation, which satisfies the equation

$$\frac{\partial J^x}{\partial x} + \frac{\partial J^v}{\partial v} = 0. \tag{5.78}$$

In the present case we cannot assume, as we did before for the case of one variable, that the currents $J^x(x, v)$ and $J^v(x, v)$ vanish. In the stationary regime the microscopic reversibility is expressed by the conditions

$$J^x(x, -v) = -J^x(x, v), \tag{5.79}$$

$$J^v(x, -v) = J^v(x, v). \tag{5.80}$$

The first condition implies the property

$$P(x, -v) = P(x, v), \tag{5.81}$$

that is, $P(x, v)$ must be an even function in v. This property implies $J(x) = 0$, that is, the real current vanish in the stationary state.

The second condition together with this last property gives us the following equation

$$-\gamma v - \frac{\Gamma}{2} \frac{\partial P}{\partial v} = 0, \tag{5.82}$$

whose solution is

$$P(x, v) = Q(x)e^{-\gamma v^2/\Gamma}, \tag{5.83}$$

where $Q(x)$ depends only on x. Replacing this result into the stationary equation (5.78), we obtain an equation for $Q(x)$,

$$\gamma f Q - \frac{\Gamma}{2} \frac{\partial Q}{\partial x} = 0. \tag{5.84}$$

Defining $V(x)$ as the potential associated to the force $f(x)$, that is, as the function such that $f(x) = -\partial V(x)/\partial x$, then the solution is

$$Q(x) = \frac{1}{Z} e^{-2\gamma V(x)/\Gamma}, \tag{5.85}$$

so that

$$P(x,v) = \frac{1}{Z} \exp\{-\frac{2\gamma}{\Gamma}[\frac{1}{2}v^2 + V(x)]\}, \tag{5.86}$$

where Z is a normalization constant, which is equivalent to

$$P(x,v) = \frac{1}{Z} \exp\{-\frac{1}{k_B T}[\frac{1}{2}mv^2 + U(x)]\}, \tag{5.87}$$

where $U(x) = mV(x)$ is the potential associated to the force $F(x)$, that is, $F(x) = -\partial U(x)/\partial x$, and Γ is related to the temperature by

$$\Gamma = \frac{2\gamma k_B T}{m}, \tag{5.88}$$

and therefore $B = 2\alpha k_B T$. This last result and the result (5.87) allow us to interpret (5.65) as the equation that describes a particle subject to a force $F(x)$ and in contact with a heat reservoir at temperature T. Both forces, $-\alpha v$ and F, describe the coupling to the heat reservoir. This interpretation is valid as long as B is proportional not only to temperature but also to α, or γ, which is the quantity that describes the intensity of the coupling of the particle to the reservoir. It is worth to note that the equilibrium probability distribution (5.87) does not depend on α. In other terms, the thermodynamic equilibrium occurs no matter what the intensity of the coupling with the reservoir. The quantity α is related only to the relaxation to equilibrium.

5.6 Kramers Equation for Several Particles

In this section we enlarge the idea of microscopic reversibility to include the cases in which the system of particles presents as stochastic variables not only the position of the particles but also their velocities. We denote by x the set of positions, that is, $x = (x_1, x_2, \ldots, x_N)$ and by v the set of velocities, that is, $v = (v_1, v_2, \ldots, v_N)$. We assume the following equations of motion

$$m\frac{dv_i}{dt} = F_i(x) - \alpha v_i + F_i^a(t), \tag{5.89}$$

$$\frac{dx_i}{dt} = v_i, \tag{5.90}$$

where $F_i^a(t)$ are random forces with the properties

$$\langle F_i^a(t) \rangle = 0, \tag{5.91}$$

$$\langle F_i^a(t) F_j^a(t') \rangle = B_i \delta_{ij} \delta(t - t'). \tag{5.92}$$

and $F_i(x)$ depends only on x.

Dividing both sides by the mass m of the particles, we get the following Langevin equation

$$\frac{dv_i}{dt} = f_i(x) - \gamma v_i + \zeta_i(t), \tag{5.93}$$

where $f_i(x) = F_i(x)/m$, $\gamma = \alpha/m$ and $\zeta_i(t) = F_i^a(t)$. The noise ζ_i has the properties

$$\langle \zeta_i(t) \rangle = 0, \tag{5.94}$$

$$\langle \zeta_i(t) \zeta_j(t') \rangle = \Gamma_i \delta_{ij} \delta(t - t'), \tag{5.95}$$

where $\Gamma_i = B_i/m^2$.

The Fokker-Planck equation associated to the Langevin equations (5.93) and (5.90), which gives the time evolution of the probability distribution $P(x, v, t)$, is given by

$$\frac{\partial P}{\partial t} = -\sum_i \frac{\partial J_i^x}{\partial x_i} - \sum_i \frac{\partial J_i^v}{\partial v_i}, \tag{5.96}$$

where

$$J_i^x = v_i P, \tag{5.97}$$

$$J_i^v = (f_i - \gamma v_i) P - \frac{\Gamma_i}{2} \frac{\partial P}{\partial v_i}, \tag{5.98}$$

are the components of the probability current related to the positions and velocities, respectively.

Next, we determine the conditions that we should impose on the forces f_i so that in the stationary regime the system is found in thermodynamic equilibrium. We assume that the system of particles is in contact with the same heat reservoir so that $\Gamma_i = \Gamma$, independent of i. We admit that in equilibrium the components of the probability current have the properties

$$J_i^x(x, -v) = -J_i^x(x, v), \tag{5.99}$$

$$J_i^v(x, -v) = J_i^v(x, v), \tag{5.100}$$

which constitute, in the present case, the expression of the reversibility in the place of those given by (5.16). The property (5.99) combined with (5.97) implies

$P(x, -v) = P(x, v)$, that is, the stationary probability distribution must be an even function of the velocity. The property (5.100) combined with (5.98) entail

$$\gamma v_i P + \frac{\Gamma}{2} \frac{\partial P}{\partial v_i} = 0, \tag{5.101}$$

where we have taken into account that $P(x, v)$ is an even function in v. Equation (5.101) can be solved leading to the result

$$P(x, v) = Q(x) \exp\{-\sum_i \frac{\gamma}{\Gamma} v_i^2\}. \tag{5.102}$$

To determine $Q(x)$, it suffices to insert this form into

$$\frac{\partial J_i^x}{\partial x_i} + \frac{\partial J_i^v}{\partial v_i} = 0. \tag{5.103}$$

First, however, if we use the result (5.101) we see that J_i^v given by (5.98) is reduced to $J_i^v = f_i P$, so that (5.103) is equivalent to

$$v_i \frac{\partial P}{\partial x_i} + \frac{\partial}{\partial v_i} (f_i P) = 0. \tag{5.104}$$

From this we get the following equation for $Q(x)$

$$\frac{\partial Q}{\partial x_i} - \frac{2\gamma}{\Gamma} f_i Q = 0. \tag{5.105}$$

or

$$\frac{\partial}{\partial x_i} \ln Q = \frac{2\gamma}{\Gamma} f_i. \tag{5.106}$$

Again this equation entails the property

$$\frac{\partial f_j}{\partial x_i} = \frac{\partial f_i}{\partial x_j}, \tag{5.107}$$

which must be valid for any pair (i, j). Therefore, f_i must be generated by a potential, $V(x)$, that is,

$$f_i(x) = -\frac{\partial}{\partial x_i} V(x), \tag{5.108}$$

so that

$$Q(x) = \frac{1}{Z} e^{-2\gamma V(x)/\Gamma}.$$ (5.109)

Finally

$$P(x,v) = \frac{1}{Z} \exp\{-\frac{2\gamma}{\Gamma}[V(x) + \frac{1}{2}\sum_i v_i^2]\},$$ (5.110)

which is equivalent to

$$P(x,v) = \frac{1}{Z} \exp\{-\frac{1}{k_B T}[U(x) + \frac{m}{2}\sum_i v_i^2]\},$$ (5.111)

where we use the relation $\Gamma = 2\gamma k_B T/m$ between Γ and the temperature T, or yet, $B = 2\alpha k_B T$.

5.7 Kramers Linear System

We study now the set of Kramers coupled equations such that the forces f_i are linear,

$$f_i = \sum_j A_{ij} x_j.$$ (5.112)

The Fokker-Planck equation that gives the time evolution of the probability distribution $P(x,v,t)$ is the one given by

$$\frac{\partial P}{\partial t} = -\sum_i \frac{\partial}{\partial x_i}(v_i P) - \sum_{ij} A_{ij} \frac{\partial}{\partial v_i}(x_j P) + \sum_i \gamma_i \frac{\partial}{\partial v_i}(v_i P) + \sum_i \frac{\Gamma_i}{2} \frac{\partial^2 P}{\partial v_i^2}.$$ (5.113)

We assume that the dissipation coefficients γ_i are distinct and that the intensity of the random forces Γ_i are also distinct. The Fokker-Planck equation can be understood as the equation that describes a set of harmonic oscillators, each one in contact with a heat reservoir at a certain temperature, provided Γ_i is proportional to the product $\gamma_i T_i$, where T_i is the temperature of the heat reservoir at which the i-th particle is coupled and γ_i is understood as the intensity of the coupling with the reservoir.

To solve the Fokker-Planck equation (5.113), we start by defining the characteristic function $G(k,q,t)$,

$$G(k,q,t) = \int e^{i(k^\dagger x + q^\dagger v)} P(x,v,t) dx dv,$$ (5.114)

where $q^\dagger$ is the transpose of the column matrix q whose entries are q_i. Deriving G with respect to t and using the Fokker-Planck equation, we get the following equation for G

$$\frac{\partial G}{\partial t} = \sum_i k_i \frac{\partial G}{\partial q_i} + \sum_{ij} A_{ij} q_i \frac{\partial G}{\partial k_j} - \gamma \sum_i q_i \frac{\partial G}{\partial q_i} - \frac{1}{2} \sum_i \Gamma_i q_i^2 G. \qquad (5.115)$$

Each term to the right is obtained by an integration by parts and assuming that P and its derivatives vanish in the limits of integration.

Next, we assume the following form for the characteristic function

$$G(k,t) = \exp\{-\frac{1}{2}(k^\dagger X k + 2k^\dagger Z q + q^\dagger Y q)\}, \qquad (5.116)$$

where X, Y, and Z are square matrices that depend of t, X and Y being symmetric. This form is the characteristic function of a multidimensional Gaussian distribution in the variables x_i and v_i such that $\langle x_i \rangle = 0$ and $\langle v_i \rangle = 0$ and

$$X_{ij} = \langle x_i x_j \rangle, \qquad Y_{ij} = \langle v_i v_j \rangle, \qquad Z_{ij} = \langle x_i v_j \rangle. \qquad (5.117)$$

Replacing the form (5.116) into (5.115), we get the following equations for X, Y and Z,

$$\frac{dx}{dt} = Z + Z^\dagger, \qquad (5.118)$$

$$\frac{dY}{dt} = AZ^\dagger + ZA^\dagger - BY - YB + \Gamma, \qquad (5.119)$$

$$\frac{dZ}{dt} = Y + XA^\dagger - ZB, \qquad (5.120)$$

where B is the diagonal matrix defined by $B_{ij} = \gamma_i \delta_{ij}$ and Γ is the diagonal matrix whose entries are $\Gamma_i \delta_{ij}$.

Defining the $2N \times 2N$ matrices C, M and Ω by

$$C = \begin{pmatrix} X & Z \\ Z^\dagger & Y \end{pmatrix} \qquad M = \begin{pmatrix} 0 & I \\ A & -B \end{pmatrix} \qquad \Omega = \begin{pmatrix} 0 & 0 \\ 0 & \Gamma \end{pmatrix}, \qquad (5.121)$$

then Eqs. (5.118)–(5.120) can be written in compact form as

$$\frac{dC}{dt} = MC + CM^\dagger + \Omega. \qquad (5.122)$$

5.8   Entropy Production

Now we determine the production of entropy for a system described by the Fokker-Planck seen in the previous section. We assume that the probability distribution and its derivatives vanish at the boundaries of a region R of the space (x, v). We start by the expression for the entropy of this system, given by

$$S(t) = -\int P(x, v, t) \ln P(x, v, t) dx dv. \tag{5.123}$$

Deriving with respect to time, then

$$\frac{dS}{dt} = -\int \ln P \frac{\partial P}{\partial t} dx dv, \tag{5.124}$$

and using the Fokker-Planck equation in the form (5.96), we get

$$\frac{dS}{dt} = \sum_i A_i + \sum_i B_i, \tag{5.125}$$

where

$$A_i = \int \ln P \frac{\partial J_i^x}{\partial x_i} dx dv \tag{5.126}$$

and

$$B_i = \int \ln P \frac{\partial J_i^v}{\partial v_i} dx dv. \tag{5.127}$$

Performing the integral by parts, the quantity A_i becomes

$$A_i = -\int J_i^x \frac{\partial}{\partial x_i} \ln P \, dx dv = -\int v_i \frac{\partial P}{\partial x_i} dx dv = 0, \tag{5.128}$$

where we use the definition $J_i^x = v_i P$ and the fact that P vanishes at the boundary of R. Integrating by parts, the quantity B_i becomes

$$B_i = -\int J_i^v \frac{\partial}{\partial v_i} \ln P \, dx dv, \tag{5.129}$$

where we use the fact that P vanishes at the boundary of R. Defining the current $J_i(x, v)$ by

$$J_i = -\gamma v_i P - \frac{\Gamma_i}{2} \frac{\partial P}{\partial v_i}, \tag{5.130}$$

we see that $J_i^v = f_i P + J_i$, which inserted in the integral gives

$$B_i = -\frac{2}{\Gamma_i} \int f_i \frac{\partial P}{\partial v_i} \, dx dv - \frac{2}{\Gamma_i} \int \frac{J_i}{P} \frac{\partial P}{\partial v_i} \, dx dv. \tag{5.131}$$

The first integral vanishes since f_i depends only on x. Using the definition of J_i,

$$B_i = \frac{2}{\Gamma_i} \int \frac{J_i^2}{P} dx dv + \frac{2\gamma}{\Gamma_i} \int v_i J_i dx dv, \tag{5.132}$$

Using again the definition of J_i, the second integral becomes

$$\int v_i J_i dx dv = -\gamma \int v_i^2 P \, dx dv + \frac{\Gamma_i}{2} \int P \, dx dv = -\gamma \langle v_i^2 \rangle + \frac{\Gamma_i}{2}, \tag{5.133}$$

where the second integral on the right-hand side was obtained by an integration by parts. Therefore,

$$B_i = -\frac{2\gamma^2}{\Gamma_i} \langle v_i^2 \rangle + \gamma + \frac{2}{\Gamma_i} \int \frac{J_i^2}{P} dx dv. \tag{5.134}$$

Thus, we conclude that the entropy variation can be written in the form introduced by Prigogine

$$\frac{dS}{dt} = \Pi - \Phi, \tag{5.135}$$

where Π is the entropy production rate, given by

$$\Pi = \sum_i \frac{2}{\Gamma_i} \int \frac{J_i^2}{P} dx dv, \tag{5.136}$$

a nonnegative quantity, and Φ is the flux of entropy, given by

$$\Phi = \sum_i \left(\frac{2\gamma^2}{\Gamma_i} \langle v_i^2 \rangle - \gamma \right). \tag{5.137}$$

It is worth to note that if f_i is conservative, the current J_i vanishes in the stationary regime, as seen when we introduced the equilibrium distribution (5.110) in the definition of J_i, given by (5.130). Therefore the entropy production rate Π vanishes and, since $dS/dt = 0$, the entropy flux Φ also vanishes in equilibrium.

5.9   Dissipated Power

Here we analyze the relation between the power dissipated by the nonconservative forces and the production of entropy. We assume that the system is in contact with only one heat reservoir so that $\Gamma_i = \Gamma$. The variation of the average kinetic energy of the particles,

$$E = \sum_i \frac{1}{2} m \langle v_i^2 \rangle = \frac{1}{2} m \sum_i \int v_i^2 P \, dx dv, \qquad (5.138)$$

is given by

$$\frac{dE}{dt} = \frac{1}{2} m \sum_i \int v_i^2 \frac{\partial P}{\partial t} dx dv. \qquad (5.139)$$

Using the Fokker-Planck equation in the form (5.96) and integrating by parts, we obtain the result

$$\frac{dE}{dt} = m \sum_i \int v_i J_i^v dx dv. \qquad (5.140)$$

Using now the definition of current J_i^v

$$\frac{dE}{dt} = m \sum_i \int v_i \left(f_i P - \gamma v_i P - \frac{\Gamma}{2} \frac{\partial P}{\partial v_i} \right) dx dv. \qquad (5.141)$$

Integrating the last term by parts, we get

$$\frac{dE}{dt} = m \sum_i \left(\langle v_i f_i \rangle - \gamma \langle v_i^2 \rangle + \frac{\Gamma}{2} \right). \qquad (5.142)$$

The power dissipated by the force $F_i(x)$ is given by

$$\mathscr{P} = \sum_i \langle F_i v_i \rangle = m \sum_i \langle f_i v_i \rangle. \qquad (5.143)$$

Therefore,

$$\frac{dE}{dt} = \mathscr{P} + m \sum_i \left(-\gamma \langle v_i^2 \rangle + \frac{\Gamma}{2} \right). \qquad (5.144)$$

Using this result we can write the entropy flux in the form

$$\Phi = \frac{1}{k_B T} \left(\mathscr{P} - \frac{dE}{dt} \right), \tag{5.145}$$

where we use the following relation between Γ, γ and the temperature T, previously obtained, $\Gamma m / 2\gamma = k_B T$.

In the stationary regime, $dS/dt = 0$ and $dE/dt = 0$ so that

$$\Pi = \Phi = \frac{\mathscr{P}}{k_B T} = \frac{2}{\Gamma} \sum_i \int \frac{(J_i)^2}{P} dx dv. \tag{5.146}$$

If the forces are conservative, then in the stationary regime the system exhibits microscopic reversibility $J_i = 0$ and is described by the equilibrium distribution. In this case $\Pi = \Phi = 0$ since $J_i = 0$. We can show yet explicitly that the power $\mathscr{P}$ vanishes in equilibrium. To this end, we use the result that the equilibrium is established if the forces are conservative, that is, if they are generated by a potential $V(x)$. Therefore

$$\frac{d}{dt} V(x) = -\sum_i f_i(x) v_i, \tag{5.147}$$

and we see that the power is

$$\mathscr{P} = -m \frac{d}{dt} \langle V(x) \rangle. \tag{5.148}$$

But in the stationary state the right-hand side vanishes so that $\mathscr{P} = 0$.

Exercises

1. Show that the stationary solution of the Kramers equation (5.73) is of the type $P = A \exp\{-\beta[V(x) + mv^2/2]\}$, where $V(x)$ is such that $F = -dV/dx$. Determine the constant β. For the elastic force $F(x) = -Kx$, normalize P and find the constant A.
2. Show that the time-dependent solution of the Kramers equation (5.73) for $F(x) = -Kx$ is of the type $P = A \exp\{-ax^2/2 - bv^2/2 - cxv\}$, where a, b, c and A depend on time. Show that for this probability density,

$$\langle x^2 \rangle = \frac{b}{ab - c^2}, \qquad \langle v^2 \rangle = \frac{a}{ab - c^2}, \qquad \langle xv \rangle = \frac{-c}{ab - c^2}.$$

To determine a, b and c as functions of time, invert these equations to find a, b and c in terms of $\langle x^2 \rangle$, $\langle v^2 \rangle$ and $\langle xv \rangle$. Next, use the time dependent expressions of these last quantities obtained in the Exercise 5 of Chap. 3.

3. Show explicitly that in the stationary regime the right-hand side of Eq. (5.64) vanishes for conservative forces, that is, when the stationary state is a state of equilibrium.

Chapter 6
Markov Chains

6.1 Stochastic Processes

A random variable that depends on a parameter t is called a random function or, if t stands for time, a stochastic variable. We consider here stochastic processes at discrete time and such that the stochastic variable is also discrete. Suppose that a stochastic variable x_t takes integer values and that t takes the values $0, 1, 2, 3, \ldots$. A stochastic process becomes entirely defined up to time ℓ by the joint probability distribution

$$\mathscr{P}_\ell(n_0, n_1, n_2, \ldots, n_\ell) \tag{6.1}$$

of x_t taking the value n_0 at time $t = 0$, the value n_1 at time $t = 1$, the value n_2 at time $t = 2, \ldots$, and the value n_ℓ at time $t = \ell$.

Next, consider the conditional probability

$$\mathscr{P}_{\ell+1}(n_{\ell+1} | n_0, n_1, n_2, \ldots, n_\ell) \tag{6.2}$$

of x_t taking the value $n_{\ell+1}$ at time $t = \ell + 1$, given that it has taken the value n_0 at time $t = 0$, the value n_1 at time $t = 1$, the value n_2 at time $t = 2, \ldots$, and the value n_ℓ at time $t = \ell$. If it is equal to the conditional probability

$$\mathscr{P}_{\ell+1}(n_{\ell+1} | n_\ell) \tag{6.3}$$

of x_t taking the value $n_{\ell+1}$ at time $t = \ell + 1$, given that it has taken the value n_ℓ at time $t = \ell$, then the stochastic process is a Markovian process. In other terms, a Markovian process is the one for which the conditional probability of x_t taking a certain value, in a given time, depends only on the value that it has taken in the previously instant of time.

© Springer International Publishing Switzerland 2015
T. Tomé, M.J. de Oliveira, *Stochastic Dynamics and Irreversibility*, Graduate Texts in Physics, DOI 10.1007/978-3-319-11770-6_6

From the definition of conditional probability, we obtain the following formula

$$\mathscr{P}_\ell(n_0, n_1, n_2, \ldots, n_\ell) = \mathscr{P}_\ell(n_\ell|n_{\ell-1}) \ldots \mathscr{P}_2(n_2|n_1)\mathscr{P}_1(n_1|n_0)\mathscr{P}_0(n_0). \quad (6.4)$$

We see thus that the Markovian process becomes entirely defined by the conditional probabilities given by (6.3) and by the initial probability $\mathscr{P}_0(n_0)$.

We define now the probability $P_\ell(n_\ell)$ that the variable x_t takes the value n_ℓ at time $t = \ell$ independently of the values it has taken in the previous instants of time. It is given by

$$P_\ell(n_\ell) = \sum \mathscr{P}_\ell(n_0, n_1, n_2, \ldots, n_\ell), \quad (6.5)$$

where the sum is over $n_0, n_1, \ldots, n_{\ell-1}$ but not over n_ℓ. If we use Eq. (6.4), we get the following recurrence equation

$$P_\ell(n_\ell) = \sum_{n_{\ell-1}} \mathscr{P}_\ell(n_\ell|n_{\ell-1}) P_{\ell-1}(n_{\ell-1}). \quad (6.6)$$

Therefore, given $P_0(n_0)$, we can obtain $P_\ell(n_\ell)$ at any time.

The conditional probability $\mathscr{P}_\ell(n_\ell|n_{\ell-1})$ is interpreted as the probability of transition from state $n_{\ell-1}$ to state n_ℓ. In principle it can depend on time. That is, given two states, the transition probability between them could be different for each instant of time. However, we will consider only Markovian processes whose transition probability do not vary in time. In this case we write

$$\mathscr{P}_\ell(n_\ell|n_{\ell-1}) = T(n_\ell, n_{\ell-1}), \quad (6.7)$$

so that Eq. (6.6) becomes

$$P_\ell(n_\ell) = \sum_{n_{\ell-1}} T(n_\ell, n_{\ell-1}) P_{\ell-1}(n_{\ell-1}). \quad (6.8)$$

6.2 Stochastic Matrix

We have seen that a Markovian process becomes completely defined by the transition probability and by the initial probability. Writing the previous equation in the simplified form

$$P_\ell(n) = \sum_m T(n, m) P_{\ell-1}(m) \quad (6.9)$$

we interpret $T(n, m)$ as the entries of a matrix T. It has the following properties

$$T(n, m) \geq 0, \tag{6.10}$$

since $T(n, m)$ is a (conditional) probability, and

$$\sum_n T(n, m) = 1 \tag{6.11}$$

due to normalization. That is, the entries of the matrix T must be nonnegative and the sum of the entries of any column must be equal to unity. Note that the summation is done over the first variable, which denote the row index of a matrix. Any square matrix that possesses these two properties is called a stochastic matrix.

The simplest example of a Markov chain consists in a sequence of two states, denoted by the letters A and B. A possible sequence is ABBABABAABB. In the table below, we show the transition probabilities. Each entry of the table corresponds to the probability of occurrence of one letter, shown in the first column, given the occurrence of a letter in the preceding position, shown in the first row:

	A	B
A	p_1	p_2
B	q_1	q_2

where the sum of the entries of a column equals one, $p_i + q_i = 1$. From this table we determine the stochastic matrix T given by

$$T = \begin{pmatrix} p_1 & p_2 \\ q_1 & q_2 \end{pmatrix}. \tag{6.12}$$

Note that the sum of the entries of the same column is equal to unity.

If we define the matrix P_ℓ as the column matrix whose entries are $P_\ell(n)$, then Eq. (6.9) can be written in the form of a product of matrices, that is,

$$P_\ell = T P_{\ell-1}. \tag{6.13}$$

Thus, given the initial column matrix P_0, we get P_ℓ through

$$P_\ell = T^\ell P_0 \tag{6.14}$$

and the problem of determining $P_\ell(n)$ is reduced to the calculation of the ℓ-th power of the stochastic matrix T. This equation can be written in the form

$$P_\ell(n) = \sum_m T^\ell(n, m) P_0(m), \tag{6.15}$$

where the matrix entry $T^\ell(n,m)$ is interpreted as the transition probability from the state m to state n in ℓ time steps, that is, it is the probability of the variable x_t to take the value n at time t given that it has taken the value m at a previous time $t - \ell$.

6.3 Chain of Higher Order

In the previous sections we considered Markov chains in which the conditional probability of state n_ℓ,

$$\mathscr{P}_\ell(n_\ell|n_0, n_1, n_2, \ldots, n_{\ell-1}), \tag{6.16}$$

depends only on $n_{\ell-1}$ and is therefore equal to

$$\mathscr{P}_\ell(n_\ell|n_{\ell-1}). \tag{6.17}$$

We may imagine Markov chain in which the conditional probability depends not only on $n_{\ell-1}$ but also on $n_{\ell-2}$. In this case the conditional probability becomes equal to

$$\mathscr{P}_\ell(n_\ell|n_{\ell-1}, n_{\ell-2}). \tag{6.18}$$

and we say that the Markov chain has range two. It is possible to consider Markov chains of range three, four, etc. Here we focus on those of range two.

For range two, the probability $\mathscr{P}_\ell(n_{\ell+1}, n_\ell)$ of occurrence of state $n_{\ell+1}$ at time $\ell + 1$ and state n_ℓ at time ℓ can be obtained recursively from the equation

$$\mathscr{P}_\ell(n_\ell, n_{\ell-1}) = \sum_{n_{\ell-2}} \mathscr{P}_\ell(n_\ell|n_{\ell-1}, n_{\ell-2})\mathscr{P}_{\ell-1}(n_{\ell-1}, n_{\ell-2}). \tag{6.19}$$

If the conditional probability does not depend on time we can write this equation in the form

$$\mathscr{P}_\ell(n, n') = \sum_{n''} \mathscr{P}(n|n', n'')\mathscr{P}_{\ell-1}(n', n''), \tag{6.20}$$

which gives the probability of two states at consecutive instants of time in a recursive manner. Being a conditional probability $\mathscr{P}(n|n', n'')$ has the properties

$$\sum_n \mathscr{P}(n|n', n'') = 1, \qquad \mathscr{P}(n|n', n'') \geq 0. \tag{6.21}$$

The probability of state n at time ℓ is obtained from

$$\mathscr{P}_\ell(n) = \sum_{n'} \mathscr{P}_\ell(n, n'). \tag{6.22}$$

Next we introduce the quantity $T(n, n'; m, m')$, defined by

$$T(n, n'; m, m') = \delta(m, n') \mathscr{P}(n|n', m'), \tag{6.23}$$

so that the recurrence Eq. (6.20) can be written in the form

$$\mathscr{P}_\ell(n, n') = \sum_{m, m'} T(n, n'; m, m') \mathscr{P}_{\ell-1}(m, m'). \tag{6.24}$$

Interpreting $T(n, n'; m, m')$ as the entries of a square matrix T and $\mathscr{P}_\ell(n, n')$ as the entries of a column matrix $\mathscr{P}_\ell$, the above equation can be written in the form of a product of matrices

$$\mathscr{P}_\ell = T \mathscr{P}_{\ell-1}. \tag{6.25}$$

The entries of the matrix T have the properties

$$\sum_{n, n'} T(n, n'; m, m') = 1, \qquad T(n, n'; m, m') \geq 0. \tag{6.26}$$

which follow directly from (6.21).

In the table below, we present an example of a Markov chain of range two consisting of a sequence of two states, denoted by the letters A and B. Each entry of the table corresponds to the probability of the occurrence of the letter, show in the firs column, given the occurrence of two letters in the two preceding positions, shown in the first row:

	AA	AB	BA	BB
A	p_1	p_2	p_3	p_4
B	q_1	q_2	q_3	q_4

where the sum of the two entries of a column is equal to one, $p_i + q_i = 1$. From this table we determine the stochastic matrix T given by

$$T = \begin{pmatrix} p_1 & 0 & p_3 & 0 \\ q_1 & 0 & q_3 & 0 \\ 0 & p_2 & 0 & p_4 \\ 0 & q_2 & 0 & q_4 \end{pmatrix}. \tag{6.27}$$

Note that the sum of the entries of the same column is equal to unity.

6.4 Irreducible Matrices

Given two states n and m of a Markov chain, we may ask whether n can be reached from m or, in other terms, whether the probability of reaching n from m is nonzero. In one step, this probability is simply the entry $T(n,m)$ of the stochastic matrix. In ℓ steps this probability is $T^\ell(n,m)$. Therefore, n can be reached from m if there exists ℓ such that $T^\ell(n,m) \neq 0$, that is, if the entry labeled (n,m) of any power of the stochastic matrix is nonzero.

A Markov chain can be represented by a transition diagram, as shown in Fig. 6.1, where a transition probability from m to n is represented by an arrow from m to n. The two transition diagrams of Fig. 6.1 correspond to the two Markov chains comprising four states, denoted by 1, 2, 3, and 4, and described by the following matrices

$$C = \begin{pmatrix} 0 & 0 & 0 & 1 \\ 0 & 0 & 1 & 0 \\ 1 & 0 & 0 & 0 \\ 0 & 1 & 0 & 0 \end{pmatrix}, \qquad R = \begin{pmatrix} 0 & p & 0 & a \\ 1 & 0 & 1 & 0 \\ 0 & 0 & 0 & b \\ 0 & q & 0 & 0 \end{pmatrix}, \qquad (6.28)$$

where p, q, a and b are positive, $p + q = 1$ and $a + b = 1$. The existence of a path formed by consecutive arrows between two states, indicates that a state can be reached from another. The number of arrows along the path equals the number of steps. For example, in the C process the state 4 is reached from state 1 in three steps through the path $1 \to 3 \to 2 \to 4$, and in the R process, in two steps through $1 \to 2 \to 4$.

In the two examples shown above, we can verify by inspection that any state can be reached from any other state. Equivalently, we can say that for each pair (m,n) there exists a power ℓ such that $T^\ell(m,n) > 0$. The exponent ℓ need not be necessarily the same for all pairs. However, this property is not general. It is valid

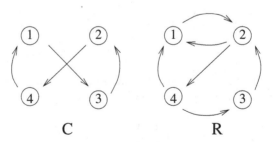

Fig. 6.1 Transition diagrams related to the irreducible matrices C and R, defined in (6.28), which are examples of cyclic and regular matrices, respectively. *Each arrow* represents a transition between two states and corresponds to a nonzero entry of the stochastic matrix

only for the Markov chain described by stochastic irreducible matrices. A reducible matrix has the form

$$T = \begin{pmatrix} X & Z \\ O & Y \end{pmatrix}, \tag{6.29}$$

or can be put in this form by permutation of the indices. Here we denote by X and Y square matrices and by Z and O rectangular matrices, the matrix O being the null matrix, that is, with all entries equal to zero.

Any power of the matrix T has the same form of the original matrix T, that is,

$$T^\ell = \begin{pmatrix} X^\ell & Z' \\ O & Y^\ell \end{pmatrix}, \tag{6.30}$$

From this result, we conclude that there always exists a pair (n, m) such that $T^\ell(n, m)$ is zero for any ℓ. In other terms, the state n cannot be reached from m, in any number of steps. Therefore, if a matrix is such that any state can be reached from any other state, then the matrix is irreducible. On the other hand, if a matrix is irreducible it is possible to show that any state can be obtained from any state.

Among the irreducible matrices we distinguish those called regular. A stochastic matrix is regular if all elements of some power of T are strictly positive. That is, if there exists ℓ such that $T^\ell(n, m) > 0$ for any n and m. Notice that all the entries must be strictly positive for the same ℓ. An example of a regular matrix is the matrix R of (6.28) whose transition diagram is shown in Fig. 6.1. We can check that the sixth power of this matrix has all the entries strictly positive. It is worth to note that an irreducible matrix that has at least one nonzero diagonal entry is regular.

The irreducible matrices that are not regular constitute a class of matrices called cyclic. The matrix C of (6.28) is an example of a cyclic stochastic matrix with period four. A regular matrix is equivalent to an irreducible acyclic matrix.

Next, we analyze the reducible stochastic matrices. Examples of reducible matrices are

$$A = \begin{pmatrix} 1 & 0 & 0 & a \\ 0 & 0 & 1 & 0 \\ 0 & p & 0 & b \\ 0 & q & 0 & 0 \end{pmatrix}, \quad B = \begin{pmatrix} 1 & 0 & 0 & a \\ 0 & 1 & p & 0 \\ 0 & 0 & 0 & b \\ 0 & 0 & q & 0 \end{pmatrix}, \quad C = \begin{pmatrix} 0 & p & 0 & a \\ 1 & q & 0 & 0 \\ 0 & 0 & 0 & b \\ 0 & 0 & 1 & 0 \end{pmatrix}, \tag{6.31}$$

where p, q, a and b are positive and $p + q = 1$ and $a + b = 1$. The transition diagrams are shown in Fig. 6.2. As we see, the three matrices are of the type (6.29).

The matrices A and B exhibit states called absorbing states. The absorbing states are such that once we reach them it is not possible to escape from them. If n is an absorbing state, then $T(n, n) = 1$. The matrix A has only one absorbing state whereas the matrix B has two absorbing states. The others are transient. Starting

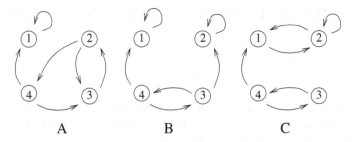

Fig. 6.2 Transition diagram related to the reducible matrices A, B and C, defined in (6.31). In A, the state 1 is absorbing and the others are transient. In B and C, only the states 3 and 4 are transient. In B, the state 1 and the state 2 are absorbing

from a transient state, the absorbing state is always reached, or in other terms, the probability of reaching the absorbing state in ℓ steps approaches 1 when $\ell \to \infty$. It is worth to note that this result is valid for Markov chains with a finite number of states.

6.5 Perron-Frobenius Theorem

The fundamental problem of Markovian processes is to determine the properties of the stochastic matrix T such that

$$\lim_{\ell \to \infty} P_\ell = P, \tag{6.32}$$

where P is the stationary solution, that is, the one that obeys the equation

$$TP = P. \tag{6.33}$$

With this purpose, we present some general properties of the stochastic matrices.

1. The stochastic matrix has an eigenvalue equal to unity.
2. Any eigenvalue λ of T fulfills the condition $|\lambda| \leq 1$, that is, in the complex plane, the eigenvalues are located in the disk of radius equal to one.
3. To the eigenvalue $\lambda = 1$ corresponds an eigenvector with nonnegative entries. Notice that in general the eigenvalues can be complex and that the eigenvalue $\lambda = 1$ may be associated to more than one eigenvector, that is, the eigenvalue $\lambda = 1$ can be degenerate.
4. Perron-Frobenius theorem. The eigenvalue $\lambda = 1$ of an irreducible matrix is nondegenerate. Moreover, the corresponding eigenvector has all entries strictly positive. This theorem states that the stationary solution of (6.33) is unique and that $P(n) > 0$. It is worth mentioning that it is possible to have eigenvalues such

that $|\lambda| = 1$ in addition to the eigenvalue $\lambda = 1$. This in fact happens with the cyclic matrices.
5. With the exception of the eigenvalue $\lambda = 1$, the absolute values of all eigenvalues of a regular matrix are strictly less than unity, that is, $|\lambda| < 1$.
6. When $\ell \rightarrow \infty$, T^ℓ converges to a matrix whose columns are all equal to P. From this, we conclude that $P_\ell = T^\ell P_0$ converges to P for any P_0.

For the limit P_ℓ to exist when $\ell \rightarrow \infty$, it is necessary that there is no complex eigenvalue on the circumference of unit radius, except $\lambda = 1$. However, for the limit to be independent of the initial probability, the stationary probability must be unique, that is, the eigenvalue $\lambda = 1$ must be nondegenerate. This occurs with the regular matrices. However, there may be matrices that are not regular but fulfill this property. An example of a matrix of this type is the one corresponding to a stochastic process with an absorbing state.

Example 6.1 The stochastic matrix

$$T = \begin{pmatrix} 0 & 0 & p \\ 1 & 0 & q \\ 0 & 1 & 0 \end{pmatrix}, \tag{6.34}$$

where $p + q = 1$, has the following eigenvalues

$$\lambda_0 = 1, \qquad \lambda_1 = \frac{1}{2}(-1 + \sqrt{1 - 4p}), \qquad \lambda_2 = \frac{1}{2}(-1 - \sqrt{1 - 4p}). \tag{6.35}$$

The eigenvalues are found on the disk of unit radius as seen in Fig. 6.3. If $p \neq 0$ and $q \neq 0$, the matrix is regular and the eigenvalues are found inside the disk except the eigenvalue $\lambda_0 = 1$. If $p = 1$, the matrix is cyclic of period three and the three eigenvalues are found at the circumference of unit radius.

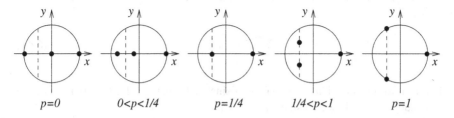

Fig. 6.3 Eigenvalues, represented by *full circles*, of the stochastic matrix (6.34) in the complex plane $x = \text{Re}\lambda$, $y = \text{Im}\lambda$. The circumference has unit radius and the *dashed line* represents the *straight line* $x = -1/2$

6.6 Microscopic Reversibility

For a Markov chain whose transition probabilities are time independent, the probability of the trajectory $n_0 \to n_1 \to \ldots \to n_\ell$ is given by

$$P(n_0, n_1, \ldots, n_\ell) = T(n_\ell, n_{\ell-1}) \ldots T(n_1, n_0) P(n_0). \tag{6.36}$$

Here we assume that T is a regular matrix and we consider a trajectory occurring in the stationary regime so that the probability at time t_0 is the stationary probability $P(n)$, that obeys the equation

$$P(n) = \sum_m T(n, m) P(m). \tag{6.37}$$

Next, we consider the reverse trajectory $n_\ell \to \ldots \to n_1 \to n_0$. The probability of this trajectory is

$$P(n_\ell, n_{\ell-1}, \ldots, n_1, n_0) = T(n_0, n_1) \ldots T(n_{\ell-1}, n_\ell) P(n_\ell), \tag{6.38}$$

which can be written in the following form

$$P(n_\ell, n_{\ell-1}, \ldots, n_1, n_0) = \tilde{T}(n_\ell, n_{\ell-1}) \ldots \tilde{T}(n_1, n_0) P(n_0), \tag{6.39}$$

where

$$\tilde{T}(n, m) = T(m, n) \frac{P(n)}{P(m)}. \tag{6.40}$$

Equation (6.39) tell us that the reverse trajectories of a Markovian process are also Markovian processes with transition probabilities given by $\hat{T}(n, m)$. However, this statement is valid only if $\hat{T}$ is a stochastic matrix. Using the condition (6.37), we see that

$$\sum_n \tilde{T}(n, m) = 1. \tag{6.41}$$

Taking into account that $\tilde{T}(n, m) \geq 0$ since $T(n, m) \geq 0$ and $P(n) > 0$, we conclude that $\tilde{T}$ is indeed a stochastic matrix.

A Markovian process has microscopic reversibility when

$$\tilde{T}(n, m) = T(n, m), \tag{6.42}$$

that is, when

$$T(n, m) P(m) = T(m, n) P(n), \tag{6.43}$$

for any pair m, n of states. The left hand side is interpreted as the transition probability from m to n, while the right hand side is the transition probability from n to m.

The stationary probability $P(n)$ satisfies Eq. (6.37), which can be written in the form

$$\sum_m \{T(n, m) P(m) - T(m, n) P(n)\} = 0, \tag{6.44}$$

since $\sum_m T(m, n) = 1$, which we call global balance. If the Markovian process is reversible, than the condition (6.43) implies that each term in Eq. (6.44) vanishes. For this reason the microscopic reversibility (6.43) is called detailed balance condition. In this case, we say that $P(n)$ is the probability of thermodynamic equilibrium. However, the microscopic reversibility is a property of the stochastic matrix and therefore of the stochastic process we are considering. Some stochastic processes have microscopic reversibility at the stationary state, others do not.

In principle, it is possible to know if reversibility takes place without determining the stationary probability $P(n)$. Consider any three states n, n', and n''. In the stationary regime, the probability of occurrence of the trajectory $n \to n' \to n'' \to n$ is given by

$$T(n, n'')T(n'', n')T(n', n)P(n), \tag{6.45}$$

while the probability of occurrence of the reverse trajectory $n \to n'' \to n' \to n$ is given by

$$T(n, n')T(n', n'')T(n'', n)P(n). \tag{6.46}$$

If microscopic reversibility takes place, these two expressions must be equal, implying

$$T(n, n')T(n', n'')T(n'', n) = T(n, n'')T(n'', n')T(n', n). \tag{6.47}$$

Therefore, a process with microscopic reversibility must satisfy these equation for any triplets of states. Analogous expressions can be written also for four or more states.

An important property of the stochastic matrices that possess microscopic reversibility is that their eigenvalues are all real. Define the matrix $\hat{T}$ by

$$\hat{T}(m, n) = \frac{1}{\chi(m)} T(m, n)\chi(n), \tag{6.48}$$

where $\chi(n) = \sqrt{P(n)}$. If we divide both sides of Eq. (6.43) by $\chi(n)\chi(m)$, we see that the matrix $\hat{T}$ is symmetric. Being symmetric (Hermitian), it has real eigenvalues. It suffices to shown now that $\hat{T}$ has the same eigenvalues of T. Indeed, let ψ_k be an eigenvector of T and λ_k the corresponding eigenvalue, that is,

$$\sum_n T(m,n)\psi_k(n) = \lambda_k \psi_k(m). \tag{6.49}$$

Dividing both sides by $\chi(m)$, we get

$$\sum_n \hat{T}(m,n)\frac{1}{\chi(n)}\psi_k(n) = \lambda_k \frac{1}{\chi(m)}\psi_k(m). \tag{6.50}$$

Therefore, $\psi_k(n)/\chi(n)$ is an eigenvector of $\hat{T}$ with the same eigenvalue λ_k of T.

6.7 Monte Carlo Method

According to statistical mechanics, the properties of a system in thermodynamic equilibrium are obtained from the Boltzmann-Gibbs probability distribution $P(n)$, which gives the probability of occurrence of the microscopic states n of the system. It is given by

$$P(n) = \frac{1}{Z} e^{-\beta E(n)}, \tag{6.51}$$

where β is proportional to the inverse of the absolute temperature and Z is the partition function,

$$Z = \sum_n e^{-\beta E(n)}. \tag{6.52}$$

The Monte Carlo method provides an estimate of the average

$$\langle f \rangle = \sum_n f(n) P(n). \tag{6.53}$$

of any state function $f(n)$ and consists in the following. Suppose that a certain number M of states are generated according to the probability $P(n)$. Then, we can say that the arithmetic mean,

$$\frac{1}{M} \sum_{i=1}^{M} f(n_i), \tag{6.54}$$

is an estimate of $\langle f \rangle$, where $n_1, n_2, \ldots, n_M$ are the states generated. The estimate will be better the larger the number M of states generated. Next, we should solve the problem of generate states with the desired probability $P(n)$. The solution is found in the construction of a Markovian process whose stationary probability is $P(n)$, that is, in the construction of a stochastic matrix such that

$$\sum_{n'} T(n, n') P(n') = P(n). \tag{6.55}$$

This problem is the opposite of that which is usual in Markovian processes. In general T is known and we wish to determine P. Here P is known and we want to know T. In general there are more than one solution for this problem, which is convenient from the computational point of view.

The approach commonly used in the construction of the stochastic matrix is by the use of the detailed balance condition,

$$T(n, n') P(n') = T(n', n) P(n). \tag{6.56}$$

Once this condition is fulfilled, Eq. (6.55) becomes satisfied provided

$$\sum_{n'} T(n', n) = 1. \tag{6.57}$$

One of the algorithms used to build the stochastic matrix is the Metropolis algorithm. For each state n we define a set of neighboring states such that $T(n', n) = 0$ when n' does not belong to the neighborhood of n. This means to say that the transition from n to a state out of the neighborhood is forbidden. It is important to define the neighborhoods so that if one state n' does not belong to the neighborhood of n, then n does not belong to the neighborhood of n'. All neighborhoods are chosen with the same number of states, which we denote by N.

The stochastic matrix is defined by

$$T(n', n) = \frac{1}{N} e^{-\beta[E(n') - E(n)]}, \qquad \text{if} \qquad E(n') > E(n), \tag{6.58}$$

$$T(n', n) = \frac{1}{N}, \qquad \text{if} \qquad E(n') \leq E(n), \tag{6.59}$$

for a state n' that belongs to the neighborhood of n, provided n' is distinct from n. The diagonal entry is given by

$$T(n, n) = 1 - \sum_{n'(\neq n)} T(n', n). \tag{6.60}$$

To show that the detailed balance condition (6.56) is fulfilled, consider two states n_1 and n_2 such that $E(n_1) > E(n_2)$. In this case, according to Eqs. (6.58) and (6.59), we have

$$T(n_1, n_2) = \frac{1}{N} e^{-\beta[E(n_1)-E(n_2)]}, \qquad T(n_2, n_1) = \frac{1}{N}. \qquad (6.61)$$

On the other hand,

$$P(n_1) = \frac{1}{Z} e^{-\beta E(n_1)}, \qquad P(n_2) = \frac{1}{Z} e^{-\beta E(n_2)}. \qquad (6.62)$$

Replacing these results in (6.56), we see that the detailed balance condition (6.56) is satisfied. We remark that in the construction of $T(n', n)$ we do not need to know Z but only the difference between $E(n')$ and $E(n)$.

If the neighborhoods are chosen so that any state can be reached from any other state, we guarantee that the stochastic matrix is regular and therefore for ℓ large enough, the state n_ℓ will be chosen with the equilibrium probability $P(n_\ell)$.

Computationally, we start from any state n_0. From this state, we generate a sequence of states $n_1, n_2, n_3, \ldots$ as follows. Suppose that at the ℓ-th time step the state is n_ℓ. In the next step we choose randomly a state in the neighborhood of n_ℓ, say the state n'_ℓ. We then calculate the difference $\Delta E = E(n'_\ell) - E(n_\ell)$.

(a) If $\Delta E \leq 0$, then the new state will be n'_ℓ, that is, $n_{\ell+1} = n'_\ell$.
(b) If $\Delta E > 0$, we calculate $p = e^{-\beta \Delta E}$ and generate a random number ξ uniformly distributed in the interval $[0, 1]$. If $\xi \leq p$, then $n_{\ell+1} = n'_\ell$, otherwise $n_{\ell+1} = n_\ell$, that is, the state remains the same.

After discarding the first D states, we use the following M states to estimate the average $\langle f \rangle$ of any state function by means of

$$\frac{1}{M} \sum_{\ell=1}^{M} f(n_{\ell+D}). \qquad (6.63)$$

As an example of the application of the Monte Carlo method, we examine here the thermal properties of a one-dimensional quantum oscillator, whose equilibrium probability distribution is given by (6.51) with

$$E(n) = \alpha n, \qquad n = 0, 1, 2, \ldots, \qquad (6.64)$$

where α is a constant. The neighborhood of state n is composed by the states $n + 1$ and $n - 1$, with the exception of state $n = 0$ which will be treated shortly. The Metropolis algorithm is constructed as follows. Suppose that at a certain time step the state is n. In the next step:

(a) We choose one of the states $n + 1$ or $n - 1$ with equal probability, in this case, $1/2$.

(b) If the state is $n + 1$, then $\Delta E = \alpha(n + 1) - \alpha n = \alpha > 0$. This state will be the next state with probability $e^{-\beta\alpha}$.
(c) If the state is $n - 1$, then $\Delta E = \alpha(n - 1) - \alpha n = -\alpha < 0$ and the new state will be the state $n - 1$.

Thus, the corresponding stochastic matrix is given by

$$T(n + 1, n) = \frac{1}{2}q, \tag{6.65}$$

$$T(n - 1, n) = \frac{1}{2}, \tag{6.66}$$

where $q = e^{-\beta\alpha}$. The diagonal entry is

$$T(n, n) = 1 - T(n + 1, n) - T(n - 1, n) = \frac{1}{2}p, \tag{6.67}$$

where $p = 1 - q$. These equations are valid for $n = 1, 2, 3, \ldots$.
When $n = 0$, the neighborhood is simply the state $n = 1$. To satisfy the detailed balance condition, we should have

$$T(1, 0)P(0) = T(0, 1)P(1), \tag{6.68}$$

from which we get

$$T(1, 0) = \frac{1}{2}q. \tag{6.69}$$

Therefore, when $n = 0$, the new state will be $n = 1$ with probability $q/2$, that is,

$$T(0, 0) = 1 - \frac{1}{2}q. \tag{6.70}$$

Equations (6.65)–(6.67), (6.69) and (6.70) define the stochastic matrix $T(m, n)$ which has as the equilibrium probability distribution the expression (6.51) with $E(n) = \alpha n$.

6.8 Expansion in Eigenfunctions

Given the transition matrix T and the initial probability P_0, we can obtain P_ℓ by $P_\ell = T^\ell P_0$. To this end we should determine T^ℓ, what we will do by the use of a method that consists in the determination of the eigenvectors and eigenvalues of T.

We examine the situation in which the eigenvalues of T are nondegenerate. Since T is in general not symmetric, we should bear in mind that the components of right and left eigenvectors of the same eigenvalue might be distinct. Let $\{\psi_k\}$ and $\{\phi_k\}$ the right and left eigenvectors, respectively, and $\{\lambda_k\}$ the corresponding eigenvalues, that is,

$$T\psi_k = \lambda_k \psi_k, \tag{6.71}$$

$$\phi_k T = \lambda_k \phi_k. \tag{6.72}$$

Notice that ψ_k is a column matrix and that ϕ_k is a row matrix. The eigenvectors form a complete and orthonormalized set, that is,

$$\phi_j \psi_k = \delta_{jk}, \tag{6.73}$$

$$\sum_k \psi_k \phi_k = I, \tag{6.74}$$

where I is the identity matrix.

It is worth to note that P is a right eigenvector with eigenvalue equal to one. We write then $\psi_0 = P$. The corresponding left eigenvector ϕ_0 is the row matrix with all entries equal to unity, that is, $\phi_0(n) = 1$ for any n. Indeed, the equation

$$\sum_n T(n, m) = 1, \tag{6.75}$$

written in the form

$$\sum_n \phi_0(n) T(n, m) = \phi_0(m), \tag{6.76}$$

implies $\phi_0 T = \phi_0$. The normalization $P = \psi_0$ gives us $\phi_0 \psi_0 = 1$.

Consider now the power T^ℓ of the matrix T. Multiplying both sides of Eq. (6.74) by T^ℓ and taking into account that $T^\ell \psi_k = \lambda_k^\ell \psi_k$, we get the following expression

$$T^\ell = \sum_k \lambda_k^\ell \psi_k \phi_k. \tag{6.77}$$

Using (6.14), $P_\ell = T^\ell P_0$, we get

$$P_\ell = \sum_k \lambda_k^\ell \psi_k \phi_k P_0 = P + \sum_{k \neq 0} \lambda_k^\ell \psi_k \phi_k P_0 \tag{6.78}$$

since $\lambda_0 = 1$, $\psi_0 = P$, the stationary probability, and $\phi_0 P_0 = 1$ since P_0 is normalized. As the eigenvalues fulfill the inequality $|\lambda_k| < 1$ for $k \neq 0$, then $|\lambda_k|^\ell \to 0$ when $\ell \to \infty$ so that all terms of the summation vanish. Then $P_\ell \to P$ when $\ell \to \infty$, no matter what is the initial probability distribution P_0.

Example 6.2 Consider the 2×2 stochastic matrix

$$T = \begin{pmatrix} 1-b & q \\ b & 1-q \end{pmatrix}. \tag{6.79}$$

The eigenvectors and the eigenvalues are

$$\phi_0 = \begin{pmatrix} 1 & 1 \end{pmatrix}, \qquad \psi_0 = \frac{1}{q+b} \begin{pmatrix} q \\ b \end{pmatrix}, \qquad \lambda_0 = 1, \tag{6.80}$$

$$\phi_1 = \frac{1}{q+b} \begin{pmatrix} b & -q \end{pmatrix}, \qquad \psi_1 = \begin{pmatrix} 1 \\ -1 \end{pmatrix}, \qquad \lambda_1 = 1-b-q. \tag{6.81}$$

Thus we get

$$T^\ell = \lambda_0^\ell \psi_0 \phi_0 + \lambda_1^\ell \psi_1 \phi_1, \tag{6.82}$$

or,

$$T^\ell = \frac{1}{q+b} \begin{pmatrix} q & q \\ b & b \end{pmatrix} + \frac{(1-q-b)^\ell}{q+b} \begin{pmatrix} b & -q \\ -b & q \end{pmatrix}. \tag{6.83}$$

For the initial probability

$$P_0 = \begin{pmatrix} p_1 \\ p_2 \end{pmatrix}, \tag{6.84}$$

we get

$$P_\ell = T^\ell P_0 = \frac{1}{q+b} \begin{pmatrix} q \\ b \end{pmatrix} + \frac{(1-q-b)^\ell}{q+b} \begin{pmatrix} bp_1 - qp_2 \\ -bp_1 + qp_2 \end{pmatrix}. \tag{6.85}$$

If $q + b \neq 0$, then, in the limit $\ell \to \infty$, P_ℓ approaches the stationary probability ψ_0 no matter what the initial condition.

6.9 Recurrence

We have seen that the entry $T^\ell(n, m)$ of the ℓ-th power of the stochastic matrix gives us the probability of occurrence of state n after ℓ steps, from the state m. This probability is determined by taking into account the several trajectories that start at m and reach n in ℓ steps. The intermediate states can be any state including the final state n, that is, the trajectories may pass through n before they reach n at time ℓ.

Now we are interested in determining the probability of occurrence of the state n after ℓ steps, starting at m, taking into account only trajectories that have not passed through the state n. Or, in other terms, the probability of occurrence of the state n, after ℓ steps, starting from m, without the occurrence of n at intermediate time steps. This probability of first passage we denote by $R_\ell(n, m)$. According to Kac, there is a relation between the stochastic matrix T and the probability of first passage, which is given by

$$T_\ell(n, m) = \sum_{j=1}^{\ell} T_{\ell-j}(n, n) R_j(n, m), \tag{6.86}$$

valid for $\ell \geq 1$, where $T_j(n, m) = T^j(n, m)$ for $j \neq 0$ and $T_0(n, m) = \delta(n, m)$. This relation can be understood as follows. Starting from the state m, the state n can be reached, in ℓ time steps, through ℓ mutually excluding ways. Each way, which corresponds to one term in the summation (6.86), is labeled by the index j that indicates the number of steps in which n is reached for the first time. Thus, the state n will occur after ℓ steps if the state n occurs for the first time in j steps and going back to state n after $\ell - j$ steps.

Before solving Eq. (6.86) we present some properties of $R_j(n, m)$. Assigning the value $\ell = 1$ in (6.86), taking into account that $T_0(n, n) = 1$ and that $T_1(n, m) = T(n, m)$, then

$$R_1(n, m) = T(n, m), \tag{6.87}$$

which is a expected result. From (6.86) we obtain the following relation

$$R_{\ell+1}(n, n) = \sum_{m(\neq n)} R_\ell(n, m) T(m, n), \tag{6.88}$$

valid for $\ell \geq 1$.

To solve Eq. (6.86), we begin by defining the generating functions

$$G(n, m, z) = \sum_{\ell=1}^{\infty} T_\ell(n, m) z^\ell + \delta(n, m), \tag{6.89}$$

$$H(n, m, z) = \sum_{\ell=1}^{\infty} R_\ell(n, m) z^\ell. \tag{6.90}$$

Multiplying Eq. (6.86) by z^ℓ, summing over ℓ and using the definitions above, we get

$$G(n, m, z) = G(n, n, z) H(n, m, z) + \delta(n, m), \tag{6.91}$$

so that, for $n \neq m$,

$$H(n,m,z) = \frac{G(n,m,z)}{G(n,n,z)}, \tag{6.92}$$

and, for $m = n$,

$$H(n,n,z) = 1 - \frac{1}{G(n,n,z)}. \tag{6.93}$$

It is worth to note that this last relation can also be obtained from relation (6.88) and using (6.92).

From $H(n,m,z)$ we obtain $R_\ell(n,m)$. However, what interest particularly us here is to determine the recurrence probability of state n, which we denote by $\mathscr{R}(n)$, and given by

$$\mathscr{R}(n) = \sum_{\ell=1}^{\infty} R_\ell(n,n) = H(n,n,1), \tag{6.94}$$

that is,

$$\mathscr{R}(n) = 1 - \frac{1}{G(n,n,1)}. \tag{6.95}$$

If $G(n,n,1)$ diverges, then $\mathscr{R}(n) = 1$ and the state is recurrent. In this case, we can define the mean time of recurrence $\langle \ell \rangle$ by

$$\langle \ell \rangle = \sum_{\ell=1}^{\infty} \ell R_\ell(n,n) = \lim_{z \to 1} \frac{d}{dz} H(n,n,z). \tag{6.96}$$

If $G(n,n,1)$ is finite then the probability $\mathscr{R}(n)$ is less than one and the state n may never be reached.

Using relation (6.77) in the form

$$T_\ell(n,m) = \sum_k \lambda_k^\ell \psi_k(n)\phi_k(m), \tag{6.97}$$

and the definition (6.89), we get

$$G(n,m,z) = \sum_k \frac{\psi_k(n)\phi_k(m)}{1 - z\lambda_k}. \tag{6.98}$$

To know if n is recurrent we should look at $G(n, n, z)$, which we write as

$$G(n, n, z) = \frac{P(n)}{1 - z} + \sum_{k \neq 0} \frac{\psi_k(n)\phi_k(n)}{1 - z\lambda_k} \tag{6.99}$$

since $\lambda_0 = 1$, $\psi_0(n) = P(n)$, which is the stationary probability, and $\phi_0(n) = 1$.

As long as $P(n) > 0$, which occurs when the number of states is finite (recall the Perron-Frobenius theorem), then $G(n, n, z) \to \infty$ when $z \to 1$. We conclude that $\mathscr{R} = 1$ and the state n (or any other) is always recurrent. For values of z around one, $G(n, n, z)$ is dominated by the first term and we write $G(n, n, z) = P(n)(1 - z)^{-1}$, so that $H(n, n, z) = 1 - (1 - z)/P(n)$. Therefore $\langle \ell \rangle = 1/P(n)$.

6.10   Ehrenfest Model

Consider two urns A and B and a certain number N of balls numbered from 1 to N. Initially, the balls are placed on urn A. Next, one of the balls is chosen at random and transferred to the other urn. This procedure is then repeated at each time step. We wish to determined the probability $P_\ell(n)$ of having n balls in urn A at time ℓ.

Suppose that at a certain time step urn A has n balls. The probability of the decreasing the number to $n-1$ is equal to the probability of choosing one of the balls of urn A (which is taken off from A and transferred to B), that is, n/N. Similarly, the probability of increasing the number of balls in A to $n + 1$ is equal to the probability of choosing one of the balls of urn B (which is transferred to A), that is, $(N - n)/N$. Therefore, the transition probability $T(m, n)$ from n to m is

$$T(n - 1, n) = \frac{n}{N}, \tag{6.100}$$

$$T(n + 1, n) = \frac{N - n}{N}, \tag{6.101}$$

$n = 0, 1, 2, \ldots, N$. In other case, $T(m, n) = 0$. The transition diagram is shown in Fig. 6.4.

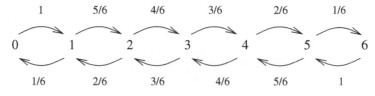

Fig. 6.4 Transition diagram of the Ehrenfest urn model for 6 balls. The states numbered from 0 to 6 represent the number of balls in one of the urns. The fractions are the transition probabilities

Here however we study the more general case in which we permit that the number of balls in urn A may not change. The probability that the number remains the same is p. Therefore,

$$T(n-1,n) = q\frac{n}{N}, \tag{6.102}$$

$$T(n,n) = p, \tag{6.103}$$

$$T(n+1,n) = q\frac{N-n}{N}, \tag{6.104}$$

where $q = 1 - p$ and $n = 0, 1, 2, \ldots, N$. The original Ehrenfest model is recovered when $p = 0$.

Inserting $T(m,n)$ into the time evolution equation

$$P_{\ell+1}(m) = \sum_{n=0}^{N} T(m,n) P_\ell(n), \tag{6.105}$$

we get

$$P_{\ell+1}(n) = q(1 - \frac{n-1}{N}) P_\ell(n-1) + p P_\ell(n) + q(\frac{n+1}{N}) P_\ell(n+1), \tag{6.106}$$

equation valid for $n = 0, 1, 2, \ldots, N$, provided we set $P_\ell(N+1) = 0$ and $P_\ell(-1) = 0$. This equation will be solved for the initial condition $P_0(n) = \delta_{n,N}$ which corresponds to having all balls in urn A.

The stationary probability $P(n)$ fulfills the equation

$$P(n) = (1 - \frac{n-1}{N}) P(n-1) + (\frac{n+1}{N}) P(n+1), \tag{6.107}$$

whose solution is

$$P(n) = 2^{-N} \binom{N}{n}. \tag{6.108}$$

Before we determine $P_\ell(n)$, we set up the equation for the time evolution of the average number of balls in urn N, given by

$$X_\ell = \sum_{n=0}^{N} n P_\ell(n). \tag{6.109}$$

From Eq. (6.106), we get

$$X_{\ell+1} = X_\ell + q(1 - \frac{2}{N}X_\ell), \qquad (6.110)$$

which must be solved with the initial condition $X_0 = N$. The solution is

$$X_\ell = \frac{N}{2} + \frac{N}{2}(1 - \frac{2}{N}q)^\ell. \qquad (6.111)$$

Therefore, X_ℓ approaches exponentially the stationary solution $N/2$.

Similarly, we can determine the second moment, defined by

$$Y_\ell = \sum_{n=0}^{N} n^2 P_\ell(n). \qquad (6.112)$$

The time evolution of the second moment is given by

$$Y_{\ell+1} = (1 - \frac{4}{N}q)Y_\ell + 2qX_\ell + q \qquad (6.113)$$

which must be solved with the initial condition $Y_0 = N^2$. Using the solution X_ℓ, we get

$$Y_\ell = \frac{N(N+1)}{4} + \frac{N^2}{2}(1 - \frac{2}{N}q)^\ell + \frac{N(N-1)}{4}(1 - \frac{4}{N}q)^\ell. \qquad (6.114)$$

From the previous results we get the variance

$$Y_\ell - X_\ell^2 = \frac{N}{4} - \frac{N^2}{4}(1 - \frac{2}{N}q)^{2\ell} + \frac{N(N-1)}{4}(1 - \frac{4}{N}q)^\ell, \qquad (6.115)$$

which also approaches its stationary value $N/4$, exponentially.

Next, we obtain the detailed solution of the Ehrenfest model defined by the matrix given by Eqs. (6.102)–(6.104). Consider the right eigenvalue equation

$$\sum_{n=0}^{N} T(m,n)\psi(n) = \lambda\psi(m), \qquad (6.116)$$

or yet

$$q(1 - \frac{n-1}{N})\psi(n-1) + p\psi(n) + q(\frac{n+1}{N})\psi(n+1) = \lambda\psi(n), \qquad (6.117)$$

valid for $n = 0, 1, 2, \ldots, N$, provided we set $\psi(-1) = 0$ and $\psi(N + 1) = 0$. Next, define the function $f(x)$ by

$$f(x) = \sum_{n=0}^{N} \psi(n)x^n, \tag{6.118}$$

which is a polynomial of degree N. Multiplying both sides of Eq. (6.117) by x^n and summing over n, we conclude that $f(x)$ must obey the equation

$$q(1 - x^2)f'(x) = N(\lambda - p - qx)f(x), \tag{6.119}$$

whose solution is

$$f(x) = A(1 + x)^{N-k}(1 - x)^k, \tag{6.120}$$

where A is a constant and $k = -N(\lambda - 1)/2q$.

As the function $f(x)$ must be a polynomial of degree N, it follows that k has to be an integer greater or equal to zero and less or equal to N. Therefore, the eigenvalues are given by

$$\lambda_k = 1 - \frac{2q}{N}k, \qquad k = 0, 1, 2, \ldots, N. \tag{6.121}$$

To each eigenvalue λ_k corresponds an eigenvector ψ_k whose entries are the coefficients of $f_k(x)$, given by

$$f_k(x) = \sum_{n=0}^{N} \psi_k(n)x^n. \tag{6.122}$$

Since the eigenvector ψ_0 must be identified as the stationary probability vector P, which is normalized, then we should have $f_0(1) = 1$, from which we obtain $A = 2^{-N}$. Thus

$$f_k(x) = 2^{-N}(1 + x)^{N-k}(1 - x)^k, \tag{6.123}$$

which is the generating function of the eigenvectors of T. In particular $f_0(x) = 2^{-N}(1 + x)^N$ whose expansion gives the stationary distribution $P(n)$ shown by Eq. (6.108).

Using relations (6.73) and (6.74) between the right eigenvectors ψ_k and left eigenvectors ϕ_k, we obtain the following relation,

$$\phi_k(n) = 2^N \psi_n(k). \tag{6.124}$$

Notice the change between k and n.

Suppose that initially all balls are in urn A. Then $P_0(n) = \delta_{nN}$ so that

$$P_\ell(n) = \sum_{m=0}^{N} T^\ell(n,m) P_0(m) = T^\ell(n,N), \qquad (6.125)$$

or, using the spectral theorem (6.77),

$$P_\ell(n) = \sum_{k=0}^{N} \lambda_k^\ell \psi_k(n) \phi_k(N). \qquad (6.126)$$

If we wish to determine $\langle x^n \rangle$, we multiply this equation by x^n, sum over n and use the generating function. The result is

$$\langle x^n \rangle = \sum_{n=0}^{N} x^n P_\ell(n) = \sum_{k=0}^{N} \lambda_k^\ell f_k(x) \phi_k(N). \qquad (6.127)$$

Using the result $\phi_k(N) = (-1)^k \binom{N}{k}$ and Eq. (6.123), we get

$$\langle x^n \rangle = 2^{-N}(1+x)^N + \sum_{k=1}^{N} \lambda_k^\ell (-1)^k \binom{N}{k} 2^{-N}(1+x)^{N-k}(1-x)^k. \qquad (6.128)$$

Deriving this equation with respect to x and setting $x = 1$, we get

$$\langle n \rangle = \frac{N}{2} + \frac{N}{2}\lambda_1^\ell = \frac{N}{2} + \frac{N}{2}\left(1 - \frac{2}{N}q\right)^\ell, \qquad (6.129)$$

which is the result obtained previously. Similarly, we obtain the other moments.

Suppose that we are in the stationary regime and that at time ℓ there are n balls in urn A. The probability of having $n - 1$ balls in urn A at time $\ell - 1$ and the same number of balls at time $\ell + 1$ is given by

$$P_{--} = \frac{T(n-1,n)T(n,n-1)P(n-1)}{P(n)} = \frac{n^2}{N^2}. \qquad (6.130)$$

In same manner, the probability of having $n - 1$ balls in urn A at time $\ell - 1$ and $n + 1$ at the previous time $\ell + 1$ is given by

$$P_{-+} = \frac{T(n+1,n)T(n,n-1)P(n-1)}{P(n)} = \frac{n}{N}\left(1 - \frac{n}{N}\right). \qquad (6.131)$$

The probability of having $n+1$ balls in urn A at time $\ell-1$ and $n-1$ at time $\ell+1$ is

$$P_{+-} = \frac{T(n-1,n)T(n,n+1)P(n+1)}{P(n)} = \frac{n}{N}(1-\frac{n}{N}). \tag{6.132}$$

Finally, the probability of having $n+1$ balls in urn A at time $\ell-1$ and the same number of balls at time $\ell+1$ is given by

$$P_{++} = \frac{T(n+1,n)T(n,n+1)P(n+1)}{P(n)} = (1-\frac{n}{N})^2. \tag{6.133}$$

Therefore, if n is around N, the larger of these four quantities is P_{--}. This has the following meaning. Do a series of simulations of the Ehrenfest model and plot the number of balls of urn A as a function of time. Of all curves that pass through n at a certain time ℓ, the ones with greatest frequencies are those that at previous time $\ell-1$ and that at later time $\ell+1$ pass through $n-1$, that is, those in the form of a "Λ" at point (ℓ, n).

6.11 Random Walk

Let us return to the problem of a random walk introduced in Chap. 2 and recast it in terms of a Markov process. At each time step, a particle, which moves along a straight line, jumps a unit distance to the left or to the right with equal probability. The possible positions of the particle are $n = 0, 1, 2, \ldots, N-1$. To simplify the problem we use periodic boundary conditions meaning that, when the particle is in position $n = N-1$, it may jump to $n = 0$ and vice-versa. The probability of the transition $T(m, n)$ from n to m is $T(n-1, n) = T(n+1, n) = 1/2$ and $T(m, n) = 0$ otherwise.

Here however we permit that the particle may remain in the same place with a certain probability p. In this case, the stochastic matrix is

$$T(n-1,n) = T(n+1,n) = \frac{1}{2}q, \tag{6.134}$$

$$T(n,n) = p, \tag{6.135}$$

where $q = 1 - p$. In other cases, $T(m, n) = 0$. Replacing this in the equation

$$P_{\ell+1}(n) = \sum_m T(n,m)P_\ell(m), \tag{6.136}$$

which gives the evolution of the probability $P_\ell(n)$ of finding the particle at position n at time ℓ, then

$$P_{\ell+1}(n) = \frac{1}{2}qP_\ell(n+1) + pP_\ell(n) + \frac{1}{2}qP_\ell(n-1), \qquad (6.137)$$

where $n = 0, 1, 2, \ldots, N-1$ and we are using periodic boundary conditions so that $P_\ell(N+n) = P_\ell(n)$. Notice that the stationary probability is $P(n) = 1/N$.

The stochastic matrix above is an example of a Toeplitz matrix, whose entries of a certain diagonal are all equal. That is, a Toeplitz matrix T has the property

$$T(n,m) = f(n-m), \qquad (6.138)$$

where $f(n)$ is periodic, $f(n+N) = f(n)$. Notice that $f(n)$ has the properties $f(n) \geq 0$ and

$$\sum_n f(n) = 1. \qquad (6.139)$$

For the case under consideration, $f(1) = f(-1) = q/2$, $f(0) = p$ and zero otherwise. The evolution equation for $P_\ell(n)$ is thus

$$P_{\ell+1}(n) = \sum_m f(n-m)P_\ell(m). \qquad (6.140)$$

Any Toeplitz matrix has the following eigenvectors

$$\psi_k(n) = \frac{1}{N}e^{ikn}, \qquad k = \frac{2\pi}{N}j, \qquad j = 0, 1, 2, \ldots, N-1. \qquad (6.141)$$

To see this, it suffices to replace these eigenvectors into the eigenvalue equation

$$\sum_m T(n,m)\psi_k(m) = \lambda_k \psi_k(n). \qquad (6.142)$$

Taking into account the properties of $f(n)$, we get

$$\lambda_k = \sum_n e^{ikn} f(n) = p + q\cos k. \qquad (6.143)$$

The left eigenvectors are given by $\phi_k(n) = e^{-ikn}$.

Using relation (6.77) in the form

$$T_\ell(n,m) = \sum_k \lambda_k^\ell \psi_k(n)\phi_k(m), \qquad (6.144)$$

we get

$$T^\ell(n,m) = \frac{1}{N} \sum_k \lambda_k^\ell e^{ik(n-m)}. \tag{6.145}$$

Assuming that at time zero the particle that performs the random motion is at the origin, then $P_0(n) = \delta_{n0}$ and

$$P_\ell(n) = \sum_m T^\ell(n,m) P_0(m) = T^\ell(n,0). \tag{6.146}$$

Therefore the probability distribution is given by

$$P_\ell(n) = \frac{1}{N} \sum_k \lambda_k^\ell e^{ikn} = \frac{1}{N} \sum_k (p + q \cos k)^\ell e^{ikn}. \tag{6.147}$$

In the limit $N \to \infty$, the summation becomes an integral,

$$P_\ell(n) = \frac{1}{2\pi} \int_{-\pi}^{\pi} (p + q \cos k)^\ell e^{ikn} dk. \tag{6.148}$$

For ℓ large enough, we obtain the result

$$P_\ell(n) = \frac{1}{2\pi} \int_{-\infty}^{\infty} e^{-q\ell k^2/2} e^{ikn} dk. \tag{6.149}$$

which gives us

$$P_\ell(n) = \frac{1}{\sqrt{2\pi q\ell}} e^{-n^2/2q\ell}. \tag{6.150}$$

6.12 Recurrence in the Random Walk

Next, we study the recurrence related to the random walk. According to the analysis made in Sect. 6.9, as long as N is finite any state is recurrent and the mean time of recurrence is $\langle \ell \rangle = 1/P(n) = N$. When $N \to \infty$, the first term in (6.99) vanishes and the integral

$$G(n,n,z) = \frac{1}{2\pi} \int_{-\pi}^{\pi} \frac{dk}{1 - z\lambda_k} = \frac{1}{2\pi} \int_{-\pi}^{\pi} \frac{dk}{1 - z + zq(1 - \cos k)}. \tag{6.151}$$

Performing the integral,

$$G(n,n,z) = \frac{1}{\sqrt{(1-z)(1-z+2qz)}}. \tag{6.152}$$

We recall that the probability of recurrence of state n is

$$\mathscr{R}(n) = 1 - \frac{1}{G(n,n,1)}. \tag{6.153}$$

When $z \to 1$, the integral diverges according to $G(n,n,z) \sim (1-z)^{-1/2}$. Thus, $\mathscr{R}(n) = 1$ and any state is recurrent. However, the mean time of recurrence is infinite. To see this, it suffices to use the formula (6.96).

Next, we study the recurrence of the random walk in two dimensions. A particle moves in a plane and at each time step it can jump a unit distance east, west, north or south with equal probability. We denote thus the position of the particle by the vector $n = (n_1, n_2)$. The nonzero transition probabilities are

$$T(n,m) = \frac{1}{4}q \quad \text{if} \quad n_1 = m_1 \pm 1 \quad \text{or} \quad n_2 = m_2 \pm 1, \tag{6.154}$$

$$T(n,n) = p, \tag{6.155}$$

where we used periodic boundary conditions in both directions. The function $f(n) = f(n_1, n_2)$ is thus $f(n) = \frac{1}{4}q$, if $|n| = 1$ and $f(0) = p$, and zero otherwise so that the eigenvalues are given by

$$\lambda_k = \sum_n e^{ik \cdot n} f(n) = p + \frac{1}{2}q(\cos k_1 + \cos k_2), \tag{6.156}$$

where $k = (k_1, k_2)$. The function $G(n,n,z)$ for a finite system is now given by

$$G(n,n,z) = \frac{1}{(2\pi)^2} \int_{-\pi}^{\pi} \int_{-\pi}^{\pi} \frac{dk_1 dk_2}{1 - z\lambda_k}. \tag{6.157}$$

or

$$G(n,n,z) = \frac{1}{(2\pi)^2} \int_{-\pi}^{\pi} \int_{-\pi}^{\pi} \frac{dk_1 dk_2}{1 - z + qz(2 - \cos k_1 + \cos k_2)/2}. \tag{6.158}$$

When $z \to 1$, this integral also diverges. To see this, we analyze the integral in a neighborhood of the origin. We choose a neighborhood defined by $|k| \le \epsilon$, where $\epsilon \ll 1$. The integral above, determined in this region, is proportional to

$$\int \frac{dk_1 dk_2}{a + k_1^2 + k_2^2} = \int_0^{\epsilon} \frac{2\pi k dk}{a + k^2} = \pi \ln \frac{a + \epsilon^2}{a}, \tag{6.159}$$

where $a = (1 - z)/qz$. We see thus that this integral diverges when $z \to 1$ and the integral (6.158) also diverges according to $G(n, n, z) = \pi |\ln(1 - z)|$. Therefore, $\mathscr{R}(n) = 1$ and, in the plane, any state is recurrent. Here, also, the mean time of recurrence is infinite. To see this it suffices to use formula (6.96).

In three or more dimensions we should analyze the following integral

$$G(n, n, z) = \frac{1}{(2\pi)^d} \int_{-\pi}^{\pi} \cdots \int_{-\pi}^{\pi} \frac{dk_1 dk_2 \ldots dk_d}{1 - z\lambda_k}, \tag{6.160}$$

where d is the space dimension in which the random walk takes place and

$$\lambda_k = p + \frac{1}{d} q (\cos k_1 + \cos k_2 + \ldots + \cos k_d), \tag{6.161}$$

When $z \to 1$, this integral becomes

$$G(n, n, 1) = \frac{d}{q(2\pi)^d} \int_{-\pi}^{\pi} \int_{-\pi}^{\pi} \cdots \int_{-\pi}^{\pi} \frac{dk_1 dk_2 \ldots dk_d}{d - (\cos k_1 + \cos k_2 + \ldots + \cos k_d)} \tag{6.162}$$

and again we should look at the behavior of the integral in a certain region $|k| \leq \epsilon \ll 1$ around the origin. The integral in this region is

$$\int \frac{dk_1 dk_2 \ldots dk_d}{k_1^2 + k_2^2 + \ldots + k_d^2} = \int_0^\epsilon \frac{ck^{d-1}}{k^2} dk = c \int_0^\epsilon k^{d-3} dk, \tag{6.163}$$

which is finite for $d \geq 3$, where c is a constant. Thus, $\mathscr{R} < 1$, for $d \geq 3$.

The above results allow us to make the following statements due to Pólya: in one and two dimensions any state is recurrent ($\mathscr{R} = 1$), that is, the particle comes back to the starting point with probability one; in three or more dimensions however this does not occur; in these cases, the recurrence probability is smaller than one ($\mathscr{R} < 1$) and the particle may never return to the starting point.

In three dimensions

$$G(n, n, 1) = \frac{1}{q(2\pi)^3} \int_{-\pi}^{\pi} \int_{-\pi}^{\pi} \int_{-\pi}^{\pi} \frac{3}{3 - (\cos k_1 + \cos k_2 + \cos k_3)} dk_1 dk_2 dk_3. \tag{6.164}$$

Performing the integral,

$$G(n, n, 1) = \frac{1}{q} 1.516386 \ldots \tag{6.165}$$

which gives $\mathscr{R} = 1 - 0.659462 \ldots q < 1$, while $q > 0$. For $q = 1, \mathscr{R} = 0.340537 \ldots$.

Exercises

1. A very long chain is constructed by a succession of three types of atoms, A, B, and C. In this chain, two atoms of the same type never appear together. An atom of type B succeeds one of type A with probability $1/3$ and an atom of type A succeeds one of type B with probability $1/3$. After an atom of type C always comes an atom of type A. Determine the concentration of each type of atom and the fraction of the possible pairs of neighboring atoms.

2. A Markov chain is defined by the stochastic matrix T and by the initial probability vector P_0, given by

$$T = \begin{pmatrix} 0 & 1/3 & 1 \\ 1/3 & 0 & 0 \\ 2/3 & 2/3 & 0 \end{pmatrix}, \qquad P_0 = \begin{pmatrix} 1 \\ 0 \\ 0 \end{pmatrix}.$$

   Determine the eigenvectors and the eigenvalues of T and from them find the probability vector P_ℓ at any time ℓ.

3. The nonzero entries of a stochastic matrix T, corresponding to a Markov chain among three states $n = 1, 2, 3$, are given by $T(2, 1) = 1$, $T(3, 2) = 1$, $T(1, 3) = p$ and $T(2, 3) = q = 1 - p$. Determine the probability $P_\ell(n)$ at any time ℓ for any initial condition.

4. A very long chain is constituted by a succession of atoms of two types, A and B. In this chain, a pair of equal atoms are always succeeded by a distinct atom. A pair AB is succeeded by an atom A with probability $1/2$ and a pair BA is succeeded by an atom B with probability $1/3$. Determine the probabilities of the pairs of neighboring atoms AA, AB, BA, and BB as well as the probabilities of A and B.

5. A written text can be understood as a succession of signs, which we consider to be the letters of the alphabet and the space between them. The text can be approximated by a Markov chain of several ranges by determining the number $N(x)$ of each sign x, the number of pairs $N(x, y)$ of neighboring signs xy, the number of triplets $N(x, y, z)$ of signs in a row xyz, etc. Show that the stochastic matrices of various ranges can be obtained from these numbers. Use a text of your choice, for example, Pushkin's *Eugene Onegin*, to determine the stochastic matrix of the Markov chain of range one, two, etc. From the stochastic matrix generate a text by numerical simulation.

6. Simulate the Ehrenfest model with N balls considering as initial condition that all balls are in urn A. Do several runs and determine the average number of balls $\langle n \rangle$ in urn A as a function of time ℓ. Make a plot of $\langle n \rangle / N$ versus ℓ / N. From a single run, build the histogram $h(n)$ of the number of balls in urn A, in the stationary state. Make a normalized plot of the histogram versus n/N.

7. Use the Monte Carlo method to determine the average energy and the heat capacity of a quantum harmonic oscillator. Do the same for a quantum rotor.

Chapter 7
Master Equation I

7.1 Introduction

Consider the stochastic matrix $T(n,m)$ of a Markov chain. Suppose that the transitions occur at each time interval τ and that the stochastic matrix is given by

$$T(n,m) = \tau W(n,m), \qquad n \neq m, \tag{7.1}$$

$$T(n,n) = 1 - \tau \Omega(n). \tag{7.2}$$

Suppose moreover that τ is small so that the probability of permanence in the same state is very close to unity. The property

$$\sum_m T(m,n) = 1 \tag{7.3}$$

implies

$$\Omega(n) = \sum_{m(\neq n)} W(m,n). \tag{7.4}$$

Next, we examine the evolution of the probability $P_\ell(n)$ of the system being at the state n at the ℓ-th time interval, which we write as

$$P_{\ell+1}(n) = \sum_{m(\neq n)} T(n,m) P_\ell(m) + T(n,n) P_\ell(n), \tag{7.5}$$

© Springer International Publishing Switzerland 2015
T. Tomé, M.J. de Oliveira, *Stochastic Dynamics and Irreversibility*, Graduate Texts in Physics,
DOI 10.1007/978-3-319-11770-6_7

or yet, using Eqs. (7.1) and (7.2),

$$P_{\ell+1}(n) = \tau \sum_{m(\neq n)} W(n,m) P_\ell(m) + P_\ell(n) - \tau \Omega(n) P_\ell(n). \tag{7.6}$$

Defining the probability of state n at time $t = \ell\tau$ by $P(n,t) = P_\ell(n)$, then

$$\frac{P(n,t+\tau) - P(n,t)}{\tau} = \sum_{m(\neq n)} W(n,m) P(m,t) - \Omega(n) P(n,t). \tag{7.7}$$

In the limit $\tau \to 0$, the left-hand side becomes the time derivative of $P(n,t)$ so that

$$\frac{d}{dt} P(n,t) = \sum_{m(\neq n)} W(n,m) P(m,t) - \Omega(n) P(n,t). \tag{7.8}$$

Using Eq. (7.4), we can still write

$$\frac{d}{dt} P(n,t) = \sum_{m(\neq n)} \{W(n,m) P(m,t) - W(m,n) P(n,t)\}, \tag{7.9}$$

which is the master equation. The quantity $W(n,m)$ is the transition probability from m to n per unit time, or yet, the transition rate from m to n.

The results above show us that a Markovian stochastic process in continuous time in a discretized space becomes defined completely by the transition rates $W(n,m)$, in addition to the initial probability distribution $P(n,0)$. Given the transition rates, our aim is to determine the probability distribution $P(n,t)$, solution of the master equation (7.9), and in particular determine the stationary solution $P_e(n)$, which fulfills the equation

$$\sum_{m(\neq n)} \{W(n,m) P_e(m) - W(m,n) P_e(n)\} = 0. \tag{7.10}$$

7.2   Evolution Matrix

Examining the right-hand side of the master equation (7.9), we notice that the summation extends only over the states m distinct from n. Since the element $W(n,n)$ does not take part of this equation we can use it according to our convenience. We define $W(n,n)$ so that

$$\sum_{m} W(m,n) = 0, \tag{7.11}$$

that is.

$$W(n,n) = - \sum_{m(\neq n)} W(m,n) = -\Omega(n). \qquad (7.12)$$

Taking into account that, now, all the elements of $W(n,m)$ are determined, we define the square matrix W, called evolution matrix, as the one whose entries are $W(n,m)$. It has the following properties:

(a) Any non-diagonal entry is great or equal to zero,

$$W(n,m) \geq 0, \qquad n \neq m, \qquad (7.13)$$

(b) The sum of the entries of a column vanishes, Eq. (7.11).

It is clear that the diagonal entries must be negative or zero.

Let ψ be a column matrix whose entries are $\psi(n)$. Then, using Eq. (7.12), we see that

$$\sum_m W(n,m)\psi(m) = \sum_{m(\neq n)} \{W(n,m)\psi(m) - W(m,n)\psi(n)\}. \qquad (7.14)$$

Thus, the master equation can be written in the form

$$\frac{d}{dt}P(n,t) = \sum_m W(n,m)P(m,t), \qquad (7.15)$$

or yet,

$$\frac{d}{dt}P(t) = WP(t), \qquad (7.16)$$

where $P(t)$ is the column matrix whose entries are $P(n,t)$. The solution of Eq. (7.16) with the initial condition $P(n,0)$ is given by

$$P(t) = e^{tW}P(0), \qquad (7.17)$$

where $P(0)$ is the column matrix whose entries are $P(n,0)$. These results follow directly from the definition of the matrix e^{tW}, given by

$$e^{tW} = I + tW + \frac{t^2}{2!}W^2 + \frac{t^3}{3!}W^3 + \frac{t^4}{4!}W^4 + \dots \qquad (7.18)$$

where I is the identity matrix.

The vector P_e, corresponding to the stationary probability distribution $P_e(n)$, is the solution of

$$WP_e = 0. \qquad (7.19)$$

Given the transition rates, that is, the evolution matrix W, we wish to determine the conditions that we should impose on W so that the solution P_e is unique and

$$\lim_{t \to \infty} P(t) = P_e, \qquad (7.20)$$

that is, so that $P(t)$ approaches the stationary solution for large times. To this end, we discretize the time $t = \ell \Delta t$ in intervals equal to Δt and define the matrix

$$T = I + \Delta t W, \qquad (7.21)$$

where I is the identity matrix. For Δt small enough, T is a stochastic matrix an therefore it defines a Markov chain so that $P(t)$ is given by

$$P(t) = T^\ell P(0) = (I + \Delta t W)^\ell P(0), \qquad (7.22)$$

which reduces to Eq. (7.17) when $\Delta t \to 0$.

If the matrix T fulfills the requirements of the Perron-Frobenius theorem, the stationary solution is unique and it will be reached in the limit $t \to \infty$ for any initial condition. Applying the results of Sect. 6.5 to the matrix T, given by (7.21), we can make the following statements of general character:

1. The evolution matrix W has zero eigenvalue.
2. The real part of any eigenvalue of W is negative or zero.
3. To the zero eigenvalue corresponds an eigenvector with components non-negative.

    If any state can be reached from any other state, then T is irreducible. Being T irreducible, it is also regular since it always have nonzero diagonal entries, a result that follows from (7.21) because ΔT is small enough. Applying the results of Sect. 6.5 concerning the regular matrices T, we can make still the following statements about the matrix W:

4. The zero eigenvalue is non-degenerate and the corresponding eigenvector has all components strictly positive. In other words, the stationary state is unique and $P_e(n) > 0$.
5. All the other eigenvalues have real part strictly negative.
6. When $t \to \infty$ the probability $P(n,t)$ converges to $P_e(n)$ no matter what is the initial condition $P(n,0)$. Notice that this result is valid for finite matrices.

Define the matrix $K(t, t')$ by

$$K(t, t') = e^{(t-t')W}. \qquad (7.23)$$

The probability distribution at time t is related to the probability distribution at an earlier time $t' < t$ by

$$P(t) = K(t, t')P(t') \tag{7.24}$$

or explicitly

$$P(n, t) = \sum_m K(n, t; n', t')P(n', t'). \tag{7.25}$$

The element $K(n, t; n', t')$ of the matrix $K(t, t')$ is interpreted as the conditional probability of occurrence of state n at time t, given the occurrence of state n' at time t', and which we call transition probability. Notice that $K(n, t; n', t') \rightarrow \delta(n, n')$ when $t \rightarrow t'$. It is worth to note that $K(n, t; n', t')$ fulfills the master equation related to the variables n and t. Indeed, deriving (7.23) with respect to t,

$$\frac{\partial}{\partial t}K(t, t') = WK(t, t'), \tag{7.26}$$

or in explicit form

$$\frac{\partial}{\partial t}K(n, t; n', t') = \sum_{n''} W(n, n'')K(n'', t; n', t'), \tag{7.27}$$

which is the master equation. Therefore, a way of obtaining the transition probability $K(n, t; n', t')$ consists in solving the master equation with a generic initial condition, that is, such that $n = n'$ at $t = t'$.

Taking into account that $WK(t, t') = K(t, t')W$, then

$$\frac{\partial}{\partial t}K(t, t') = K(t, t')W, \tag{7.28}$$

or explicitly

$$\frac{\partial}{\partial t}K(n, t; n', t') = \sum_{n''} K(n, t; n'', t')W(n'', n'). \tag{7.29}$$

From (7.23) we see that

$$K(t, t') = K(t, t'')K(t'', t'), \tag{7.30}$$

or in explicit form,

$$K(n, t; n't') = \sum_{n''} K(n, t; n'', t'')K(n'', t''; n', t'), \tag{7.31}$$

which is the Chapman-Kolmogorov equations for Markovian processes with discrete states.

7.3   Expansion in Eigenvectors

The solution of the master equation can be obtained from the eigenvectors and eigenvalues of the evolution matrix W. Denote by $\{\psi_k\}$ and $\{\phi_k\}$ the right and left eigenvectors, respectively, and by $\{\lambda_k\}$ the corresponding eigenvalues, that is,

$$W\psi_k = \lambda_k \psi_k, \qquad\qquad \phi_k W = \lambda_k \phi_k. \qquad (7.32)$$

We assume that the eigenvectors form a complete and orthogonalized set, that is, they have the following properties

$$\phi_j \psi_k = \delta_{jk}, \qquad\qquad \sum_k \psi_k \phi_k = I, \qquad (7.33)$$

where I is the identity matrix.

We point out some fundamental properties of W. The stationary probability P_e is an eigenvector with zero eigenvalue. Denoting by ψ_0 the eigenvector corresponding to the zero eigenvalue $\lambda_0 = 0$, we see that they coincide, $\psi_0 = P_e$. Moreover, the corresponding left eigenvector ϕ_0 is a row matrix with all entries equal to unity. Indeed, Eq. (7.11) can be written as

$$\sum_n \phi_0(n)W(n,m) = 0, \qquad (7.34)$$

or yet $\phi_0 W = 0$. The normalization of P_e gives us $\phi_0 P_e = 1$ that is $\phi_0 \psi_0 = 1$.

Consider now the following expansion

$$e^{tW} = e^{tW} I = e^{tW} \sum_k \psi_k \phi_k = \sum_k e^{t\lambda_k} \psi_k \phi_k, \qquad (7.35)$$

from which we obtain

$$P(t) = \sum_k e^{t\lambda_k} \psi_k \phi_k P(0). \qquad (7.36)$$

Since one of the eigenvalues is zero, we can still write

$$P(t) = P_e + \sum_{k(\neq 0)} e^{t\lambda_k} \psi_k \phi_k P(0), \qquad (7.37)$$

where we used the fact that $\phi_0 P(0) = 1$ since $P(0)$ is normalized. We have seen above that the eigenvalues of the matrix W are negative or zero. We have also seen that one class of matrices W has non-degenerate zero eigenvalue. From these properties, it follows that $\lim P(t) = P_e$ when $t \to \infty$.

In explicit form, Eq. (7.37) is written as

$$P(n,t) = P_e(n) + \sum_{k(\neq 0)} e^{t\lambda_k} \psi_k(n) \sum_{n'} \phi_k(n') P(n',0). \tag{7.38}$$

If at the initial time $P(n,0) = \delta_{nn_0}$, then

$$P(n,t) = P_e(n) + \sum_{k(\neq 0)} e^{t\lambda_k} \psi_k(n)\phi_k(n_0). \tag{7.39}$$

Using the expansion in eigenvectors, given by (7.35), we obtain an explicit form for the transition probability $K(n,t;n',t')$. From (7.35), and the definition (7.23), we get

$$K(t,t') = \sum_k e^{(t-t')\lambda_k} \psi_k \phi_k, \tag{7.40}$$

Taking into account that ψ_k is a column matrix and ϕ_k is a row matrix, then we may conclude that the elements of $K(n,t;n',t')$ are given explicitly by

$$K(n,t;n',t') = \sum_k e^{(t-t')\lambda_k} \psi_k(n)\phi_k(n'). \tag{7.41}$$

Example 7.1 Consider the evolution matrix W given by

$$W = \begin{pmatrix} -a & b \\ a & -b \end{pmatrix}, \tag{7.42}$$

where a and b are transition rates. The eigenvectors and eigenvalues are given by

$$\phi_0 = \begin{pmatrix} 1 & 1 \end{pmatrix}, \qquad \psi_0 = \frac{1}{a+b}\begin{pmatrix} b \\ a \end{pmatrix}, \qquad \lambda_0 = 0, \tag{7.43}$$

$$\phi_1 = \frac{1}{a+b}\begin{pmatrix} a & -b \end{pmatrix}, \qquad \psi_1 = \begin{pmatrix} 1 \\ -1 \end{pmatrix}, \qquad \lambda_1 = -(a+b). \tag{7.44}$$

The eigenvector ψ_0 represents the stationary probability. From

$$e^{tW} = e^{t\lambda_0}\psi_0\phi_0 + e^{t\lambda_1}\psi_1\phi_1, \tag{7.45}$$

we get

$$e^{tW} = \frac{1}{a+b}\begin{pmatrix} b & b \\ a & a \end{pmatrix} + \frac{e^{-t(a+b)}}{a+b}\begin{pmatrix} a & -b \\ -a & b \end{pmatrix}. \qquad (7.46)$$

For the initial probability

$$P(0) = \begin{pmatrix} p_1 \\ p_2 \end{pmatrix}, \qquad (7.47)$$

we reach the result

$$P(t) = e^{tW}P(0) = \frac{1}{a+b}\begin{pmatrix} b \\ a \end{pmatrix} + \frac{e^{-t(a+b)}}{a+b}(ap_1 - bp_2)\begin{pmatrix} 1 \\ -1 \end{pmatrix}, \qquad (7.48)$$

which gives the probability of the two states at any time. When $t \to \infty$, we obtain the stationary probability.

7.4 Recurrence

Here we analyze the problem of first passage and recurrence. That is, we are interested in determining the probability that a state is reached for the first time starting from a certain state. More precisely, starting from the state m at $t = 0$, we want to determine the probability of the occurrence of state n between t and $t + \Delta t$, without it having been reached before time t. The ratio between this probability and Δt we denote by $R(n, t; m, 0)$. Extending formula (6.86) of Sect. 6.9 to continuous time, we get the following equality, valid for $n \neq m$,

$$K(n, t; m, 0) = \int_0^t K(n, t; n, t') R(n, t'; m, 0) dt', \qquad (7.49)$$

where $K(n, t; m, t')$ is the conditional probability of occurrence of state n at time t given the occurrence of state m at time t'. We have seen in Sect. 7.2 that it is the solution of the master equation with the initial condition $n = m$ at time $t = t'$.

Defining the Laplace transforms

$$\hat{K}(n, m; s) = \int_0^\infty K(n, t; m, 0) e^{-st} dt, \qquad (7.50)$$

$$\hat{R}(n, m; s) = \int_0^\infty R(n, t; m, 0) e^{-st} dt, \qquad (7.51)$$

then Eq. (7.49) becomes

$$\hat{K}(n, m; s) = \hat{K}(n, n; s)\hat{R}(n, m; s), \tag{7.52}$$

from which we get

$$\hat{R}(n, m; s) = \frac{\hat{K}(n, m; s)}{\hat{K}(n, n; s)}, \tag{7.53}$$

a result valid for $n \neq m$.

To determine $R(n, t'; m, 0)$ when $n = m$, we extend formula (6.88) of Sect. 6.9 to continuous time, which becomes

$$\alpha(n) R(n, n; t) = \sum_{m(\neq n)} R(n, m; t) W(m, n), \tag{7.54}$$

where $\alpha(n) = \sum_{m(\neq n)} W(m, n)$. Taking the Laplace transform

$$\alpha(n)\hat{R}(n, n; s) = \sum_{m(\neq n)} W(m, n)\hat{R}(n, m; s). \tag{7.55}$$

Using (7.53) and the result (7.29), we reach the following formula

$$\hat{R}(n, n; s) = 1 - \frac{1}{\alpha(n)} \left(\frac{1}{\hat{K}(n, n; s)} - s \right). \tag{7.56}$$

The probability of recurrence $\mathscr{R}(n)$ of state n is

$$\mathscr{R}(n) = \int_0^\infty R(n, n, t) dt = \hat{R}(n, n, 0), \tag{7.57}$$

and therefore

$$\mathscr{R}(n) = 1 - \frac{1}{\alpha(n)\hat{K}(n, n; 0)}. \tag{7.58}$$

If $\hat{K}(n, n; 0) \to \infty$, then $\mathscr{R}(n) = 1$ and state n is recurrent.

7.5 Absorbing State

Consider a stochastic process described by the evolution matrix $W(n, m)$ such that, if $W(n, m) \neq 0$, then $W(m, n) \neq 0$. In other terms, if the transition $m \to n$ is allowed, so is the reverse $n \to m$. The conditional probability $K(n, m, t)$ of occurrence of state n at time t, given the occurrence of sate m at time zero, is the solution of the master equation

$$\frac{d}{dt}K(n,m,t) = \sum_{n'} W(n,n')K(n',m,t), \tag{7.59}$$

where

$$W(n,n) = -\sum_{n'(\neq n)} W(n',n), \tag{7.60}$$

and recall that $K(n,m,0) = \delta(n,m)$.

Consider now another process, that differ from the previous only through the transition that involves a single state, which we choose to be the state $n = 0$. The rate of transition from $n = 0$ to any other is zero, which means that the transitions from $n = 0$ to any other state are forbidden and therefore the state $n = 0$ is an absorbing state. The master equation, that governs the time evolution of the probability $P^*(n,t)$ of occurrence of state n at time t, is given by

$$\frac{d}{dt}P^*(n,t) = \sum_{n'(\neq 0)} W(n,n')P^*(n',t). \tag{7.61}$$

Notice that the probability $P^*(0,t)$ do not appear in any equation except in equation for $n = 0$, which we rewrite as

$$\frac{d}{dt}P^*(0,t) = \sum_{n'(\neq 0)} W(0,n')P^*(n',t). \tag{7.62}$$

Therefore the probabilities $P^*(n,t)$, for $n \neq 0$, can be obtained using only Eqs. (7.61), for $n \neq 0$. With this purpose, we introduce a set of auxiliary variables $P(n,t)$, that coincide with the probabilities $P^*(n,t)$, except $P(0,t)$, which we choose as being identically zero. The equation for $P(n,t)$ is given by

$$\frac{d}{dt}P(n,t) = \sum_{n'} W(n,n')P(n',t) - \delta(n,0)\Phi(t), \tag{7.63}$$

where $\Phi(t)$ is a parametric function that should be chosen so that $P(0,t) = 0$. With this restriction, we see that Eqs. (7.63) and (7.61) are indeed equivalent for $n \neq 0$ and hence

$$P^*(n,t) = P(n,t) \qquad n \neq 0. \tag{7.64}$$

Moreover, the application of restriction $P(0, t) = 0$ into Eq. (7.63), for $n = 0$, shows us that $\Phi(t)$ becomes identified with the right-hand side of (7.62) entailing the result

$$\frac{d}{dt} P^*(0, t) = \Phi(t), \tag{7.65}$$

which tell us that $\Phi(t)$ must be interpreted as the flux of probability to the absorbing state, that is, the rate of increase of the absorbing state probability.

The next step is to solve Eqs. (7.63). The initial condition we choose is $P(n, 0) = P^*(n, 0) = \delta(n, m)$, where $m \neq 0$. To this end, we start by taking the Laplace transforms of Eqs. (7.59) and (7.63),

$$s\hat{K}(n, m, s) - \delta(n, m) = \sum_{n'} W(n, n')\hat{K}(n', m, s), \tag{7.66}$$

$$s\hat{P}(n, s) - \delta(n, m) = \sum_{n'} W(n, n')\hat{P}(n', s) - \delta(n, 0)\hat{\Phi}(s), \tag{7.67}$$

With the help of (7.66), we see that the solution of (7.67) is given by

$$\hat{P}(n, s) = \hat{K}(n, m, s) - \hat{\Phi}(s)\hat{K}(n, 0) \tag{7.68}$$

what can be verified by substitution. It remains to determine $\hat{\Phi}(s)$. Recalling that $\hat{P}(0, s) = 0$, since we should impose the restriction $P(0, t) = 0$, we find

$$\hat{\Phi}(s) = \frac{\hat{K}(0, m, s)}{\hat{K}(0, 0, s)}. \tag{7.69}$$

The probabilities $P^*(n, t)$, for $n \neq 0$, are determined since, in this case, $P^*(n, s) = P(n, s)$, which is given by (7.68). To determine $P^*(0, t)$ we take the Laplace transform of (7.65), to get

$$\hat{P}^*(0, s) = \frac{\hat{\Phi}(s)}{s}, \tag{7.70}$$

recalling that $P^*(0, 0) = 0$ since the initial state is distinct from the state $n = 0$.

When we compare formulas (7.69) and (7.53) we see that the problem of first passage through a certain state of a stochastic process is equivalent to the same stochastic process in which this state is converted into an absorbing state.

7.6 Time Series Expansion

We have seen that the solution of the master equation in the matrix form

$$\frac{d}{dt}P(t) = WP(t),$$
(7.71)

with the initial condition $P(0)$, is given by

$$P(t) = e^{tW}P(0),$$
(7.72)

or in explicit form

$$P(t) = \{I + tW + \frac{t^2}{2!}W^2 + \frac{t^3}{3!}W^3 + \ldots\}P(0).$$
(7.73)

If W^ℓ is determined for any ℓ, then (7.73) gives the solution of the master equation in the form of a time series.

Example 7.2 Consider the simplest case of a system with two states $n = 1, 2$ with the evolution matrix W given by

$$W = \begin{pmatrix} -\gamma/2 & \gamma/2 \\ \gamma/2 & -\gamma/2 \end{pmatrix}.$$
(7.74)

It is easy to see that

$$W^\ell = (-\gamma)^{\ell-1}W, \qquad \ell = 1, 2, 3, \ldots,$$
(7.75)

so that

$$e^{tW} = I + \sum_{\ell=1}^{\infty}\frac{t^\ell}{\ell!}W^\ell = I + W\sum_{\ell=1}^{\infty}\frac{t^\ell}{\ell!}(-\gamma)^{\ell-1} = I + \frac{1}{\gamma}(1 - e^{-t\gamma})W,$$
(7.76)

where I is the 2×2 identity matrix. Hence

$$P(t) = P(0) + \frac{1}{\gamma}(1 - e^{-t\gamma})WP(0),$$
(7.77)

or, explicitly,

$$P(1, t) = \frac{1}{2}(1 - e^{-t\gamma}) + e^{-t\gamma}P(1, 0)$$
(7.78)

and

$$P(2,t) = \frac{1}{2}(1 - e^{-t\gamma}) + e^{-t\gamma} P(2,0).$$ (7.79)

In the limit $t \to \infty$, we get $P(1,t) \to 1/2$ and $P(2,t) \to 1/2$ no matter the initial condition.

In the example above all terms of the time series could be calculated explicitly. However, in many cases, it is impossible from the practical point of view to calculate all terms. If a certain number of them can be calculated, then the truncated series may be extrapolated by other means. If this is possible, then the time series becomes a useful tool to obtain the time dependent properties. This approach requires that the relevant properties may be developed in time series.

To determine the time expansion of the average

$$\langle F \rangle = \sum_n F(n)P(n,t),$$ (7.80)

of a state function $F(n)$, we proceed as follows. We start by introducing the row matrix Ω, which we call reference vector, whose components are all equal to one, $\Omega(n) = 1$. Then, the matrix product $\Omega P(t)$ is given by

$$\Omega P(t) = \sum_n \Omega(n)P(n,t) = \sum_n P(n,t) = 1.$$ (7.81)

An important property of the reference vector Ω is

$$\Omega W = 0,$$ (7.82)

which is obtained from the property (7.11).

Next, we define a square matrix F whose nondiagonal entries vanish and the diagonal entries are $F(n)$. With this definition, the average $\langle F \rangle$ can be calculated by the formula

$$\langle F \rangle = \Omega F P(t)$$ (7.83)

since

$$\Omega F P(t) = \sum_n \Omega(n)F(n)P(n,t) = \sum_n F(n)P(n,t).$$ (7.84)

Using the time expansion (7.73) of $P(t)$ in (7.83), we get

$$\langle F \rangle = \Omega F \{I + tW + \frac{t^2}{2!}W^2 + \frac{t^3}{3!}W^3 + \ldots\}P(0).$$ (7.85)

Therefore

$$\langle F \rangle = f_0 + \sum_{\ell=1}^{\infty} t^\ell f_\ell, \tag{7.86}$$

where the coefficients are given by

$$f_0 = \Omega F P(0), \tag{7.87}$$

which is the average of F at time $t = 0$, and

$$f_\ell = \frac{1}{\ell!} \Omega F W^\ell P(0), \qquad \ell \geq 1. \tag{7.88}$$

Now we examine the Laplace transform of $P(t)$ given by

$$\hat{P}(z) = \int_0^\infty P(t) e^{-zt} dt. \tag{7.89}$$

Using the time expansion (7.73) and taking into account the identity

$$\frac{1}{\ell!} \int_0^\infty t^\ell e^{-zt} dt = \frac{1}{z^{\ell+1}}, \tag{7.90}$$

then

$$\hat{P}(z) = \{\frac{1}{z} I + \frac{1}{z^2} W + \frac{1}{z^3} W^2 + \frac{1}{z^4} W^3 + \ldots\} P(0). \tag{7.91}$$

But the sum between brackets is identified with the inverse of the matrix $(zI - W)$, which we denote by $(zI - W)^{-1}$, that is,

$$(zI - W)^{-1} = \frac{1}{z} I + \frac{1}{z^2} W + \frac{1}{z^3} W^2 + \frac{1}{z^4} W^3 + \ldots \tag{7.92}$$

so that

$$\hat{P}(z) = (zI - W)^{-1} P(0). \tag{7.93}$$

Similarly, we obtain the Laplace transform of the average $\langle F \rangle$ given by

$$\int_0^\infty \langle F \rangle e^{-zt} dt = \Omega F \hat{P}(z) = \Omega F (zI - W)^{-1} P(0), \tag{7.94}$$

which is obtained by using Eqs. (7.83) and (7.92). The stationary probability $P(\infty)$ is determined by means of the formula $\lim_{t\to\infty} P(t) = \lim_{z\to 0} z \hat{P}(z)$.

7.7 Perturbation Expansion

Suppose that we wish to calculate the stationary vector P corresponding to the evolution matrix W, that is, we want to know the solution of

$$WP = 0. \tag{7.95}$$

To do a series expansion, we imagine that W can be written in the form

$$W = W_0 + \lambda V, \tag{7.96}$$

where W_0 is the nonperturbed evolution matrix and λV the perturbation. We intend to obtain the properties of the system described by the operator W as power series in λ. We suppose that the sets of the right eigenvectors $\{\psi_n\}$, the left eigenvectors $\{\phi_n\}$ and the eigenvalues $\{\Lambda_n\}$ of the evolution matrix W_0 are known. They obey the equations

$$W_0 \psi_n = \Lambda_n \psi_n, \qquad \phi_n W_0 = \Lambda_n \phi_n. \tag{7.97}$$

We denote by $\Lambda_0 = 0$ the zero eigenvalue of W_0. The corresponding right eigenvector ψ_0 is identified as the stationary vector P_0 of W_0, and the left eigenvector ϕ_0 as the reference vector Ω, that is,

$$\psi_0 = P_0, \qquad \phi_0 = \Omega. \tag{7.98}$$

In addition, we have the following properties

$$\phi_n \psi_{n'} = \delta_{nn'}, \qquad \sum_n \psi_n \phi_n = I, \tag{7.99}$$

where I is the identity matrix. From these results, it is easy to see that W_0 has the expansion

$$W_0 = \sum_{n(\neq 0)} \psi_n \Lambda_n \phi_n, \tag{7.100}$$

where the term $n = 0$ has been excluded since $\Lambda_0 = 0$. To see this, it suffices to use the identity $W_0 = W_0 I$ and the properties contained in the expressions (7.97) and (7.99).

Next, we assume that P can be developed in a power series in λ, that is,

$$P = P_0 + \lambda P_1 + \lambda^2 P_2 + \lambda^3 P_3 + \dots. \tag{7.101}$$

Inserting this in (7.95) and taking into account (7.96), we obtain the following equation

$$(W_0 + \lambda V)(P_0 + \lambda P_1 + \lambda^2 P_2 + \lambda^3 P_3 + \ldots) = 0. \tag{7.102}$$

As the coefficients of the various powers of λ must vanish, we conclude that

$$W_0 P_\ell = -V P_{\ell-1}, \qquad\qquad \ell \geq 1. \tag{7.103}$$

Multiplying both sides of Eq. (7.101) by Ω and taking into account that $\Omega P = 1$ and that $\Omega P_0 = \phi_0 \psi_0 = 1$, we obtain one more property

$$\Omega P_\ell = 0, \qquad\qquad \ell \neq 0. \tag{7.104}$$

Next, we define the matrix R by

$$R = \sum_{n(\neq 0)} \psi_n \frac{1}{\Lambda_n} \phi_n, \tag{7.105}$$

which has the property

$$RW_0 = W_0 R = I - \psi_0 \phi_0 = I - P_0 \Omega. \tag{7.106}$$

That is, the matrix R is the inverse of the matrix W_0 inside the subspace whose vectors are orthogonal to the vector ψ_0. To check this property, is suffices to multiply the expression of the definition of R by the expansion of W_0, given by (7.100), and use the orthogonality (7.99) among the eigenvectors of W_0.

Multiplying both sides of Eq. (7.103) by R and using the property (7.106), we get

$$(I - P_0 \Omega) P_\ell = -RV P_{\ell-1}, \qquad\qquad \ell \geq 1. \tag{7.107}$$

But, from (7.104), $\Omega P_\ell = 0$ for $\ell \geq 1$ so that

$$P_\ell = -RV P_{\ell-1}, \qquad\qquad \ell \geq 1. \tag{7.108}$$

which is the equation that gives P_ℓ recursively. From this equation, we get

$$P_\ell = (-RV)(-RV) \ldots (-RV) P_0 = (-RV)^\ell P_0, \tag{7.109}$$

which gives P_ℓ from P_0. Replacing in the expansion (7.101), we finally obtain

$$P = P_0 + \sum_{\ell=1}^{\infty} (-\lambda RV)^\ell P_0. \tag{7.110}$$

In view of

$$(I + \lambda RV)^{-1} = I + \sum_{\ell=1}^{\infty}(-\lambda RV)^{\ell} \qquad (7.111)$$

we may write

$$P = (I + \lambda RV)^{-1} P_0. \qquad (7.112)$$

The average $\langle F \rangle$, given by

$$\langle F \rangle = \Omega FP, \qquad (7.113)$$

can be calculated by using expression (7.110) for P, that is,

$$\langle F \rangle = \Omega FP_0 + \sum_{\ell=1}^{\infty} \Omega F(-\lambda RV)^{\ell} P_0, \qquad (7.114)$$

or

$$\langle F \rangle = f_0 + \sum_{\ell=1}^{\infty} \lambda^{\ell} f_{\ell}, \qquad (7.115)$$

where

$$f_0 = \Omega FP_0, \qquad f_{\ell} = \Omega F(-RV)^{\ell} P_0. \qquad (7.116)$$

Thus, if we are able to determine the coefficients f_{ℓ}, we have the development of the average $\langle F \rangle$ in powers of λ.

7.8 Boltzmann H Theorem

Define the function $H(t)$ by

$$H(t) = \sum_{n} P_n^e f(x_n), \qquad x_n(t) = \frac{P_n(t)}{P_n^e}, \qquad (7.117)$$

where $f(x)$ is a differentiable and convex function such that $f(x) \geq 0$, what implies $H(t) \geq 0$. We use the notation $P_n(t)$ and P_n^e in the place of $P(n,t)$ and $P_e(n)$, respectively.

Deriving the function $H(t)$, we get

$$\frac{dH}{dt} = \sum_n f'(x_n) \frac{dP_n}{dt} \tag{7.118}$$

and therefore

$$\frac{dH}{dt} = \sum_n f'(x_n) \sum_{m(\neq n)} \{W_{nm} x_m P_m^e - W_{mn} x_n P_n^e\}. \tag{7.119}$$

Multiplying both sides of Eq. (7.10) by an arbitrary function A_n and summing in n we obtain

$$\sum_n A_n \sum_{m(\neq n)} \{W_{nm} P_m^e - W_{mn} P_n^e\} = 0, \tag{7.120}$$

which added to the previous equation results in

$$\frac{dH}{dt} = \sum_{n,m(n \neq m)} \{(x_m f'(x_n) + A_n) W_{nm} P_m^e - (x_n f'(x_n) + A_n) W_{mn} P_n^e\}, \tag{7.121}$$

or

$$\frac{dH}{dt} = \sum_{n,m(n \neq m)} \{(x_m f'(x_n) + A_n) - (x_m f'(x_m) + A_m)\} W_{nm} P_m^e. \tag{7.122}$$

If we choose $A_n = f(x_n) - x_n f'(x_n)$ we obtain

$$\frac{dH}{dt} = \sum_{n,m(n \neq m)} \{(x_m - x_n) f'(x_n) + f(x_n) - f(x_m)\} W_{nm} P_m^e. \tag{7.123}$$

For a convex function, the expression between brackets is always strictly negative, if $x_m \neq x_n$. If $x_m \neq x_n$ for some pair m, n such that $W(n, m) \neq 0$, then $dH/dt < 0$ and the function $H(t)$ must decrease. Since H is bounded from below, $H(t) \geq 0$, it must approach its minimum value, that is, it must vanish when $t \to \infty$, what occur if, for any pair n, m such that $W_{nm} \neq 0$, we have $x_n(\infty) = x_m(\infty)$. If the evolution matrix W is such that any state can be reached from any other state, then $x_n(\infty)$ must have the same value for all possible values of n, from which we conclude that $P_n(\infty)$ is proportional to P_n^e. Since $P_n(\infty)$ is normalized, then $P_n(\infty) = P_n^e$.

The choice $f(x) = x \ln x - x + 1$ results in the function H originally introduced by Boltzmann,

$$H(t) = \sum_n P_n(t) \ln \frac{P_n(t)}{P_n^e}. \tag{7.124}$$

7.9 Microscopic Reversibility

When the transition rates $W(n, m)$ are such that the stationary probability $P_e(n)$ fulfills the equation

$$W(n, m) P_e(m) - W(m, n) P_e(n) = 0, \qquad (7.125)$$

for any pair of states m and n, we say that they obey detailed balance. In this case, we say that $P_e(n)$, in addition to be the stationary probability, is also the thermodynamic equilibrium probability. However, we should remark that the fulfillment of the detailed balance is a property of the evolution equation W. Some matrices, that is, some processes obey detailed balance, others do not.

The detailed balance condition is equivalent to microscopic reversibility. The transition probability $m \rightarrow n$ in a small time interval Δt, in the stationary state, is equal to $(\Delta t) W(n, m) P_e(m)$, while the probability of the reverse transition $n \rightarrow m$ is equal to $(\Delta t) W(m, n) P_e(n)$. If Eq. (7.125) is satisfied, then these two probabilities are equal for any pair of states m and n.

We can in principle establish whether a certain process obey the microscopic reversibility without appealing to Eq. (7.125), that is, without knowing a priori, the stationary probability. Consider any three states n, n', and n'' but distinct. In the stationary state, the probability of occurrence of the closed trajectory $n \rightarrow n' \rightarrow n'' \rightarrow n$ is

$$\Delta t \, W(n, n'') \Delta t \, W(n'', n') \Delta t \, W(n', n) P_e(n), \qquad (7.126)$$

while the occurrence of the reverse trajectory is

$$\Delta t \, W(n, n') \Delta t \, W(n', n'') \Delta t \, W(n'', n) P_e(n). \qquad (7.127)$$

If reversibility takes place, these two probabilities are equal, so that

$$W(n, n'') W(n'', n') W(n', n) = W(n, n') W(n', n'') W(n'', n), \qquad (7.128)$$

which is the sought condition.

An important property of the evolution matrices W that fulfills the detailed balance is that their eigenvalues are all real. Defining the matrix $\hat{W}$ by

$$\hat{W}(m, n) = \frac{1}{\chi(m)} W(m, n) \chi(n), \qquad (7.129)$$

where $\chi(n) = \sqrt{P_e(n)}$, we divide both sides of Eq. (7.125) by $\chi(m)\chi(n)$ and conclude that $\hat{W}$ is symmetric. Now, dividing both sides of the eigenvalue equation

$$\sum_n W(m, n) \psi_k(n) = \lambda_k \psi_k(m) \qquad (7.130)$$

by $\chi(m)$ and using Eq. (7.129), we get

$$\sum_n \hat{W}(m,n) \frac{\psi_k(n)}{\chi(n)} = \lambda_k \frac{\psi_k(m)}{\chi(m)}, \qquad (7.131)$$

that is, $\hat{W}$ has the same eigenvalues of W and its eigenvectors have components $\psi_k(n)/\chi(n)$. Since $\hat{W}$ is symmetric and real, its eigenvalues are real. Therefore, the eigenvalues of W are real.

7.10   Entropy and Entropy Production

One way of characterizing the systems that follow a stochastic dynamics can be done by means of the averages of the various state functions such as, for example, the energy of the system. Denoting by E_n a state function, for example the energy, its average $U = \langle E_n \rangle$ is

$$U(t) = \sum_n E_n P_n(t). \qquad (7.132)$$

Another quantity, that may also be used to characterize the system, but cannot be considered as an average of a state function, is the entropy S, defined according to Boltzmann and Gibbs by

$$S(t) = -k \sum_n P_n(t) \ln P_n(t), \qquad (7.133)$$

where k is the Boltzmann constant.

We determine now the time average of the quantities above. First, we have

$$\frac{dU}{dt} = \sum_n E_n \frac{d}{dt} P_n. \qquad (7.134)$$

In the stationary state we see that $dU/dt = 0$ so that the average of a state function remains constant in the stationary state. Using the master equation,

$$\frac{dU}{dt} = \sum_{nm} E_n \{W_{nm} P_m - W_{mn} P_n\}, \qquad (7.135)$$

or

$$\frac{dU}{dt} = \sum_{nm} (E_n - E_m) W_{nm} P_m. \qquad (7.136)$$

Next, we determine the time variation of entropy,

$$\frac{dS}{dt} = -k \sum_n (\ln P_n) \frac{d}{dt} P_n. \tag{7.137}$$

There is a second term $\sum_n dP_n/dt$ which vanishes identically since P_n is normalized. In the stationary state we see that $dS/dt = 0$ and hence the entropy remains constant in the stationary sate. Using the master equation,

$$\frac{dS}{dt} = -k \sum_{nm} \ln P_n \{ W_{nm} P_m - W_{mn} P_n \}, \tag{7.138}$$

which can be written as

$$\frac{dS}{dt} = -k \sum_{nm} W_{nm} P_m \ln \frac{P_n}{P_m}. \tag{7.139}$$

According to the laws of thermodynamics, the energy is a conserved quantity. The variation in energy of a system must be, thus, equal to the flux of energy from the environment to the system. Therefore, we write Eq. (7.136) in the form

$$\frac{dU}{dt} = -\Phi_e, \tag{7.140}$$

where

$$\Phi_e = \sum_{nm} (E_n - E_m) W_{mn} P_n. \tag{7.141}$$

and we interpret Φ_e as the flux of energy per unit time from the system to the environment.

On the other hand, still in accordance with the laws of thermodynamics, the entropy is not a conserved quantity. It remains constant or increase. Therefore, the variation of entropy must be equal to the flux of entropy from the environment plus the internal production of entropy. Therefore, we write Eq. (7.139) in the form

$$\frac{dS}{dt} = \Pi - \Phi, \tag{7.142}$$

where Φ is the flux of entropy from the system to the environment and Π is the entropy production rate.

The entropy production rate Π is a part of the expression on the right-hand side of (7.139). According to the laws of thermodynamics, the entropy production rate Π must have the fundamental properties: (1) it must be positive or zero; and (2) it must vanish in the thermodynamic equilibrium, which we assume to be the regime for which microscopic reversibility holds. Thus, we assume the following expression for the entropy production rate

$$\Pi = k \sum_{nm} W_{mn} P_n \ln \frac{W_{mn} P_n}{W_{nm} P_m}. \tag{7.143}$$

Comparing the right-hand side of (7.139), which must be equal to $\Pi - \Phi$, we get the following expression for the flux of entropy from the system to the environment

$$\Phi = k \sum_{nm} W_{mn} P_n \ln \frac{W_{mn}}{W_{nm}}. \tag{7.144}$$

Expression (7.143) has the two properties mentioned above. To see this, we rewrite (7.143) in the equivalent form

$$\Pi = \frac{k}{2} \sum_{nm} \{W_{mn} P_n - W_{nm} P_m\} \ln \frac{W_{mn} P_n}{W_{nm} P_m}. \tag{7.145}$$

Each term of the summation is of the type $(x - y) \ln(x/y)$, which is always greater or equal to zero so that $\Pi \geq 0$. In thermodynamic equilibrium, that is, when microscopic reversibility takes place, $W_{mn} P_n = W_{nm} P_m$ for any pair of states and therefore $\Pi = 0$. As a consequence, $\Phi = 0$. The flux of entropy Φ can also be written in a form similar to the entropy production rate given by (7.145),

$$\Phi = \frac{k}{2} \sum_{nm} \{W_{mn} P_n - W_{nm} P_m\} \ln \frac{W_{mn}}{W_{nm}}. \tag{7.146}$$

In the stationary state, that is, when

$$\sum_{m} \{W_{mn} P_n - W_{nm} P_m\} = 0, \tag{7.147}$$

then $\Pi = \Phi$ since $dS/dt = 0$. If the system is in a stationary state which is not equilibrium, then the entropy production rate is strictly positive, $\Pi = \Phi > 0$.

A system in contact with a heat reservoir at a certain temperature T is described by the following transition rate

$$W_{nm} = A_{nm} e^{-(E_n - E_m)/2kT}, \tag{7.148}$$

where $A_{mn} = A_{nm}$ so that

$$\frac{W_{nm}}{W_{mn}} = e^{-(E_n - E_m)/kT}. \tag{7.149}$$

In this case, the entropy flux is

$$\Phi = \frac{1}{T} \sum_{nm} W_{mn} P_n (E_n - E_m) = \frac{1}{T} \Phi_e. \tag{7.150}$$

Using the relations (7.140) and (7.142), we reach the result

$$\frac{dU}{dt} - T \frac{dS}{dt} = -T\Pi. \tag{7.151}$$

Defining the free energy $F = U - TS$ we see that

$$\frac{dF}{dt} = -T\Pi, \tag{7.152}$$

from which we conclude that $dF/dt \leq 0$ so that $F(t)$ is a monotonic decreasing function of time.

For systems that in the stationary state are found in thermodynamic equilibrium, the equilibrium probability is given by the Gibbs distribution

$$P_n^e = \frac{1}{Z} e^{-E_n/kT}. \tag{7.153}$$

In this case, the Boltzmann H function is directly related with the free energy $F = U - TS$. Indeed, from (7.124), we see that

$$H = \sum_n P_n(t) \ln P_n + \frac{1}{kT} \sum_n P_n E_n + \ln Z, \tag{7.154}$$

and therefore

$$H = -\frac{S}{k} + \frac{U}{kT} + \ln Z = \frac{F}{kT} - \frac{F_0}{kT}, \tag{7.155}$$

where $F_0 = -kT \ln Z$ is the equilibrium free energy. Therefore $H = (F - F_0)/kT$ and the property $dH/dt \leq 0$ follows directly from $dF/dt \leq 0$, as shown above.

7.11   Transport Across a Membrane

A model for the molecular transport across a cellular membrane is sketched in Fig. 7.1. There are two types of molecules that cross the membrane: molecules of type A, represented by a large circle, and molecules of type B, represented by a small circle. The transport is intermediated by complex molecules, which we call simply complexes. These complexes are found in the membrane and can capture or

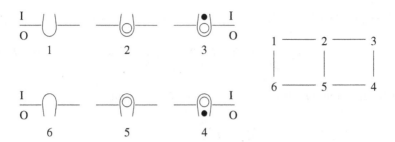

Fig. 7.1 Mechanism of transport of molecules across a cellular membrane according to Hill (1977). The *large* and *small circles* represent two types of molecules. In conformations 1, 2, and 3, the molecules communicate with the inside (I) and in conformations 4, 5, and 6, with the outside (O) of the cell. The possible transitions between the six states are indicated at the right

release molecules of both types that are inside or outside the cell. A complex can be found in two states, as shown in Fig. 7.1. In one of them, which we call state I, the complex faces the inside of the cell and therefore the complex captures and releases molecules found inside the cell. In the other, which we call O, the complex faces the outside of the cell and therefore the complex captures and releases molecules found outside the cell. We assume that the rate of I→O is equal to the rate of O→I. According to the Fig. 7.1, a complex cannot hold a molecule of type B that is alone. This property means that a molecule of type B can be captured only if the complex holds already a molecule of type A and that a molecule of type A can only be released if the complex does not hold a molecule of type B.

It is convenient to identify the membrane, constituted by the complexes and by the captured molecules, as being the system object of our analysis. We also assume that this system is open, that is, the system exchanges molecules with the environment, composed by the inside and outside of the cell. The inside and outside of the cell act like two reservoirs of molecules, which we call reservoir I and O, respectively, with chemical potentials distinct for each type of molecule. Assuming that the outside and the inside are diluted solutions with respect to both components A and B, then the concentrations of these components can be considered as proportional to $e^{\mu/kT}$, where μ is the chemical potential of the component considered, k is the Boltzmann constant and T the absolute temperature.

According to the rules above, the model of the transport of molecules across the membrane is described by the six states shown in Fig. 7.1 and by the rate of transitions among these states. The rate of transition $i \rightarrow j$ is denoted by a_{ji}. The rates a_{21}, a_{12}, a_{32} and a_{23} are assumed as being those describing the contact with the reservoir I so that they fulfill the relations

$$\frac{a_{21}}{a_{12}} = e^{\mu_A^I/kT}, \qquad \frac{a_{32}}{a_{23}} = e^{\mu_B^I/kT}, \qquad (7.156)$$

where μ_A^I and μ_B^I are the chemical potentials of molecules A and B with respect to reservoir I. Similarly, a_{56}, a_{65}, a_{54} and a_{45} describe the contact with the reservoir O and fulfill the relations

$$\frac{a_{56}}{a_{65}} = e^{\mu_A^O/kT}, \qquad \frac{a_{45}}{a_{54}} = e^{\mu_B^O/kT}, \qquad (7.157)$$

where μ_A^O and μ_B^O are the chemical potentials of molecules A and B with respect to reservoir O. As to the other rates, which describe the movement of the complex, we assume that the inversion of the complex occur with the same rate. Thus, we adopt the following relation between these rates

$$\frac{a_{16}}{a_{61}} = 1, \qquad \frac{a_{25}}{a_{52}} = 1, \qquad \frac{a_{34}}{a_{43}} = 1. \qquad (7.158)$$

The master equation that describes the evolution of the probability $P_i(t)$ of finding the system in state i is given by

$$\frac{dP_i}{dt} = \sum_j \{a_{ij} P_j - a_{ji} P_i\}. \qquad (7.159)$$

Defining J_{ij} by

$$J_{ij} = a_{ij} P_j - a_{ji} P_i, \qquad (7.160)$$

the master equation can be written as

$$\frac{dP_i}{dt} = \sum_j J_{ij}. \qquad (7.161)$$

The number of complexes in conformation i is proportional to the probability P_i.

The flux of molecules across the membrane can be written in terms of the quantities J_{ij}. The flux of molecules of type A from the system to I is denoted by J_A^I and from the system to O, by J_A^O. These fluxes are given by

$$J_A^I = J_{12}, \qquad J_A^O = J_{65}. \qquad (7.162)$$

In the stationary state, $J_A^I + J_A^O = 0$. Similarly, the flux of molecules of type B, from the system to I is denoted by J_B^I and from the system to the outside of the cell, by J_B^O. These fluxes are given by

$$J_B^I = J_{23}, \qquad J_B^O = J_{54}. \qquad (7.163)$$

In the stationary state, $J_B^I + J_B^O = 0$.

As long as the chemical potentials are distinct, the system will be found out of thermodynamic equilibrium and there will be entropy production even in the stationary state. The difference between the entropy production rate Π and the entropy flux Φ equals the variation in entropy of the system, $dS/dt = \Pi - \Phi$. According to (7.146), the entropy flux from the system to the environment is given by

$$\Phi = \frac{k}{2} \sum_{ij} J_{ij} \ln \frac{a_{ij}}{a_{ji}}. \tag{7.164}$$

Using the relations that exist between the transition rates, we reach the result

$$\Phi = -\frac{1}{T}(J_A^I \mu_A^I + J_B^I \mu_B^I + J_A^O \mu_A^O + J_B^O \mu_B^O), \tag{7.165}$$

In the stationary state $-J_A^O = J_A^I = J_A$ and $-J_B^O = J_B^I = J_B$. Moreover, $\Pi = \Phi$ and therefore we obtain the following formula for the entropy production rate in the stationary state

$$\Pi = J_A X_A + J_B X_B, \tag{7.166}$$

where X_A and X_B, given by

$$X_A = \frac{\mu_A^O - \mu_A^I}{T}, \qquad X_B = \frac{\mu_B^O - \mu_B^I}{T}, \tag{7.167}$$

are called generalized forces. Notice that J_A and J_B are the flux of molecules A and B, respectively, from the outside to the inside of the cell. Notice yet that $X_A > 0$ and $X_B > 0$ favor the flux of molecules from the outside to the inside of the cell.

The model that we are analyzing is able to describe the flux of molecules of type B against its concentration gradient. To see this, we consider that the concentration of B is larger inside than the outside the cell implying $\mu_B^I > \mu_B^O$ or $X_B < 0$. We wish to have a flux of B from the outside to the inside, $J_B > 0$. This situation can occur provided the product $J_A X_A$ is larger than $J_B|X_B|$ since $\Pi > 0$. Thus, it suffices that the concentration gradient of A is large enough. This situation is shown in Fig. 7.2, where the fluxes, obtained numerically, are shown as functions of X_B for a given value of $X_A > 0$. The values of the rates are fixed at certain values. There is an interval where, although $X_B < 0$, the flux $J_B > 0$. In this figure, we show also the entropy production rate Π versus X_B. As expected, $\Pi > 0$.

In thermodynamic equilibrium, both the forces X_A and X_B and the fluxes J_A and J_B vanish. Around equilibrium we assume that the fluxes can be expanded in terms of the forces. Up to linear terms,

$$J_A = L_{AA} X_A + L_{AB} X_B, \tag{7.168}$$

$$J_B = L_{BA} X_A + L_{BB} X_B, \tag{7.169}$$

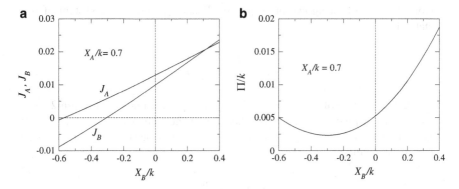

Fig. 7.2 Fluxes and entropy production related to the model represented in Fig. 7.1. (a) Fluxes of molecules J_A and J_B as functions of $X_B = \Delta\mu_B/T$ for a fixed value of $X_A = \Delta\mu_A/T$. Notice that there is an interval in which $\Delta\mu_B < 0$ and $J_B > 0$. In this situation, the flux of molecules of type B occurs in opposition to the concentration gradient. (b) Entropy production rate Π as a function of X_B. The minimum entropy production occurs for X_B corresponding to $J_B = 0$

where the coefficients L_{AA}, L_{AB}, L_{BA} and L_{BB} are the Onsager coefficients. These relation show that the flux of molecules of a certain type can be induced not only by the force related to this type but also by the force related to the other type, as long as the cross coefficients L_{BA} and L_{AB} are nonzero. However, if one coefficient vanishes so will the other because, according to Onsager, the cross coefficients are equal

$$L_{BA} = L_{AB} \tag{7.170}$$

which is called Onsager reciprocity relation.

Exercises

1. One way of solving numerically a Fokker-Planck equation is to discretize it. The spacial discretization of a Fokker-Planck equation results in a master equation. Show this result for the Smoluchowski equation using the following approximations for the first and second derivatives of the spacial function $F(x)$:

$$\frac{F(x + \Delta x) - F(x)}{\Delta x},$$

$$\frac{F(x + \Delta x) - 2F(x) + F(x - \Delta x)}{(\Delta x)^2}.$$

Find the transition rates.

2. Consider the following evolution matrix W corresponding to a system of two states 1 and 2:

$$W = \begin{pmatrix} -b & q \\ b & -q \end{pmatrix},$$

where b is the rate of the transition $1 \to 2$ and q is the rate of transition $2 \to 1$. Determine the powers W^ℓ and from them obtain the matrices e^{tW} and $(zI - W)^{-1}$. Determine then the probability vector $P(t)$ and its Laplace transform $\hat{P}(z)$, using the initial condition

$$P(0) = \begin{pmatrix} p_1 \\ p_2 \end{pmatrix}.$$

Determine $P(\infty)$ through $P(\infty) = \lim_{z \to 0} z \hat{P}(z)$.
3. Consider a system with two states described by the evolution matrix of the previous exercise. Choosing

$$W_0 = b \begin{pmatrix} -1 & 1 \\ 1 & -1 \end{pmatrix}, \qquad V = \begin{pmatrix} 0 & 1 \\ 0 & -1 \end{pmatrix}, \qquad \lambda = q - b,$$

determine the eigenvectors of W_0 and from them calculate R. Obtain then the expansion of the stationary probability vector P of W in powers of λ. Next, perform the summation.
4. Use the property $RW_0 = I - P_0\Omega$ and the equation $W = W_0 + \lambda V$ to show that $P_0 = (I + \lambda RV)P$ and thus arrive at Eq. (7.112).

Chapter 8
Master Equation II

8.1 Creation and Annihilation Process

We study here the processes that can be understood as a random walk along a straight line, but with transition rate that depend on the position where the walker is. The possible positions of the walker along the straight line are the non-negative integers. At each small time interval Δt, the walker at position n jumps to the right, to the position $n + 1$, with probability $(\Delta t)a_n$, or to the left, to position $n - 1$, with probability $(\Delta t)b_n$. Therefore, the transition rates W_{nm} are given by

$$W_{n+1,n} = a_n, \qquad W_{n-1,n} = b_n. \tag{8.1}$$

The other transition rates are zero. The master equation becomes

$$\frac{d}{dt}P_n(t) = b_{n+1}P_{n+1}(t) + a_{n-1}P_{n-1}(t) - (a_n + b_n)P_n(t). \tag{8.2}$$

When a_n and b_n are independent of n and equal, we recover the ordinary random walk. The stochastic process described by this equation is called creation and annihilation process because, if we imagine a system of particles and interpret the variable n as the number of particles, then the increase in n by one unit corresponds to the creation of a particle and the decrease of n by one unit as the annihilation of one particle. Sometimes, the process is also called birth and death process.

The time evolution of the average $\langle f(n) \rangle$ of a function of the stochastic variable n is obtained by multiplying both sides of Eq. (8.2) by $f(n)$ and summing over n. After rearranging the terms, we get

$$\frac{d}{dt}\langle f(n) \rangle = \langle [f(n+1) - f(n)]a_n \rangle + \langle [f(n-1) - f(n)]b_n \rangle. \tag{8.3}$$

© Springer International Publishing Switzerland 2015
T. Tomé, M.J. de Oliveira, *Stochastic Dynamics
and Irreversibility*, Graduate Texts in Physics,
DOI 10.1007/978-3-319-11770-6_8

From this formula we obtain the time evolution of the first moment,

$$\frac{d}{dt}\langle n \rangle = \langle a_n - b_n \rangle, \tag{8.4}$$

and of the second moment,

$$\frac{d}{dt}\langle n^2 \rangle = 2\langle n(a_n - b_n) \rangle + \langle (a_n + b_n) \rangle. \tag{8.5}$$

If the transition rates a_n or b_n are not linear in n, the equation for a certain moment will depend on the moments of higher order. From these two equations we get the evolution of the variance

$$\frac{d}{dt}\{\langle n^2 \rangle - \langle n \rangle^2\} = 2\langle n(a_n - \langle a_n \rangle - b_n + \langle b_n \rangle) \rangle + \langle (a_n + b_n) \rangle. \tag{8.6}$$

Example 8.1 Consider a process such that

$$a_n = a(N - n), \tag{8.7}$$

$$b_n = bn, \tag{8.8}$$

where a and b are positive constants and $n = 0, 1, 2, \ldots, N$. When $a = b$ this process is the continuous time version of the Ehrenfest urn model seen in Sect. 6.10. If $a \neq b$, the process can also be understood as a continuous time version of the Ehrenfest model such that one of the urns has preference over the other. The equation for the average $X = \langle n \rangle$ is given by

$$\frac{dX}{dt} = aN - cX, \tag{8.9}$$

where $c = a + b$. The solution of these equation for the initial condition $n = N$ at $t = 0$ is

$$X = N(p + qe^{-ct}), \tag{8.10}$$

where $p = a/c$ and $q = b/c = 1 - p$. The time evolution of $Y = \langle n^2 \rangle$ is given by

$$\frac{dY}{dt} = aN + (2aN - a + b)X - 2cY. \tag{8.11}$$

From this equation for Y and the equation for the average X, we get the following equation for the variance $Z = \langle n^2 \rangle - \langle n \rangle^2 = Y - X^2$

$$\frac{dZ}{dt} = -2cZ + aN - (a - b)X, \tag{8.12}$$

whose solution is

$$Z = Nq(1 - e^{-ct})(p + qe^{-ct}).$$ (8.13)

When $p = q = 1/2$,

$$Z = \frac{N}{4}(1 - e^{-2ct}).$$ (8.14)

8.2 Generating Function

In some cases, the solution of the master equation can be obtained by the use of the generating function. Consider the following master equation

$$\frac{d}{dt}P_n(t) = b(n + 1)P_{n+1}(t) - bnP_n(t),$$ (8.15)

valid for $n = 0, 1, 2 \ldots$, where b is a parameter. It describes a process in which there is only annihilation of particles. As a consequence, if initially $n = N$, then the possible values of n are $n = 0, 1, 2 \ldots, N$, only. The generating function $G(z, t)$ is defined by

$$G(z, t) = \langle z^n \rangle = \sum_{n=0}^{N} P_n(t)z^n.$$ (8.16)

The approach we use consists in setting up a differential equation for $G(z, t)$ from the master equation. Solving this equation, the expansion of the generating function in powers of z gives the probability distribution $P_n(t)$.

Deriving both sides of Eq. (8.16) with respect to time and using (8.15), we get

$$\frac{\partial G}{\partial t} = b(1 - z)\frac{\partial G}{\partial z}.$$ (8.17)

This equation is solved by the method of characteristics, which consists in comparing (8.17) with

$$\frac{\partial G}{\partial t}dt + \frac{\partial G}{\partial z}dz = 0.$$ (8.18)

The equations becomes equivalent if

$$dz + b(1 - z)dt = 0,$$ (8.19)

whose solution is $(1 - z)e^{-bt} = $ const, which is the characteristic curve. Along the characteristic curve G is constant because (8.18) tell us that $dG = 0$. Therefore, defining

$$\xi = (1 - z)e^{-bt}, \qquad (8.20)$$

the solution of Eq. (8.17) is of the form

$$G(z, t) = \Phi(\xi), \qquad (8.21)$$

where the function $\Phi(\xi)$ must be determined by the initial conditions. Since, at time $t = 0$, $n = N$, then $P(n, 0) = \delta_{n,N}$ and the generating function becomes $G(z, 0) = z^N$. But Eqs. (8.20) and (8.21) tell us that $G(z, 0) = \Phi(1 - z)$, from which we get $\Phi(1 - z) = z^N$ and therefore

$$\Phi(\xi) = (1 - \xi)^N, \qquad (8.22)$$

from which we conclude that

$$G(z, t) = [1 - (1 - z)e^{-bt}]^N. \qquad (8.23)$$

The probability distribution is obtained from the expansion of the right-hand side in powers of z. Performing the expansion and comparing with the right-hand side of (8.16), we reach the result

$$P_n = \binom{N}{n}(1 - e^{-bt})^{N-n}e^{-nbt}. \qquad (8.24)$$

From this result, we obtain the average number of particles

$$\langle n \rangle = Ne^{-bt} \qquad (8.25)$$

and the variance

$$\langle n^2 \rangle - \langle n \rangle^2 = Ne^{-bt}(1 - e^{-bt}). \qquad (8.26)$$

Next, we consider the following master equation

$$\frac{dP_n}{dt} = b(n + 1)P_{n+1} + a(N - n + 1)P_{n-1} - (bn + a(N - n))P_n, \qquad (8.27)$$

valid for $n = 0, 1, 2, \ldots, N$, where a and b are parameters, with the boundary conditions: $P_{-1} = 0$ and $P_{N+1} = 0$. When $a = 0$, we recover the problem seen previously. When $a = b$, the master equation (8.27) can be also be understood as the continuous time version of the Ehrenfest urn model, studied in Sect. 6.10.

To solve Eq. (8.27), we start by defining the generating function

$$G(z, t) = \sum_{n=0}^{N} P_n z^n. \tag{8.28}$$

Deriving G with respect to time and using Eq. (8.27),

$$\frac{\partial G}{\partial t} = (1 - z)(b + az)\frac{\partial G}{\partial z} - a(1 - z)NG. \tag{8.29}$$

Assuming as solution of the form

$$G(z, t) = (b + az)^N H(z, t), \tag{8.30}$$

we see that $H(z, t)$ fulfills the following equation

$$\frac{\partial H}{\partial t} = (1 - z)(b + az)\frac{\partial H}{\partial z}. \tag{8.31}$$

This equation can be solved by the method of characteristics with the solution

$$H(z, t) = \Phi(\xi), \qquad \xi = \frac{1 - z}{b + az}e^{-ct}, \tag{8.32}$$

where $c = a + b$ and $\Phi(\xi)$ is a function to be determined.

To find $\Phi(\xi)$ we use as initial condition $n = N$. Thus, $G(z, 0) = z^N$ so that $H(z, 0) = [z/(b + az)]^N$. Therefore

$$\Phi(\xi) = \left(\frac{z}{b + az}\right)^N, \qquad \xi = \frac{1 - z}{b + az}, \tag{8.33}$$

and the sought function is

$$\Phi(\xi) = \left(\frac{1 - b\xi}{c}\right)^N. \tag{8.34}$$

From this result we get the solution

$$G(z, t) = \left(q + pz - q(1 - z)e^{-ct}\right)^N, \tag{8.35}$$

where $p = a/c$ and $q = b/c = 1 - p$.

The expansion of $G(z, t)$ in powers of z gives the probability distribution

$$P_n(t) = \binom{N}{n}(q - qe^{-ct})^{N-n}(p + qe^{-ct})^n. \tag{8.36}$$

From the probability distribution we get the average and the variance

$$\langle n \rangle = N(p + q e^{-ct}),$$ (8.37)

$$\langle n^2 \rangle - \langle n \rangle^2 = Nq(1 - e^{-ct})(p + q e^{-ct}).$$ (8.38)

In the limit $t \to \infty$

$$P_n = \binom{N}{n} q^{N-n} p^n,$$ (8.39)

and the detailed balance $b(n+1)P_{n+1} = a(N-n)P_n$ is fulfilled, that is, the systems is in thermodynamic equilibrium. The average and variance are

$$\langle n \rangle = Np,$$ (8.40)

$$\langle n^2 \rangle - \langle n \rangle^2 = Npq.$$ (8.41)

8.3   Poisson Process

Suppose that an event occurs with a certain rate a, that is, with probability $a \Delta t$ in a small time interval Δt. We wish to determine the probability $P(n,t)$ of occurrence of n such events up to time t. To this end, it suffices to observe that $P(n,t)$ obeys the master equation

$$\frac{d}{dt} P(0,t) = -aP(0,t),$$ (8.42)

$$\frac{d}{dt} P(n,t) = aP(n-1,t) - aP(n,t),$$ (8.43)

for $n = 1, 2, \ldots$. These equations must be solved with the initial condition $P(0,0) = 1$ and $P(n,0) = 0$ for $n \neq 0$ since no events has occurred before $t = 0$. The process so defined is called Poisson process. It can also be understood as random walk completely asymmetric with transition rate to the right equal to a and transition rate to the left equal to zero.

To solve these equations, we defined the generating function

$$g(z,t) = \sum_{n=0}^{\infty} P(n,t) z^n,$$ (8.44)

and next we set up an equation for g. The equation must be solved with the initial condition $g(z, 0) = 1$ which corresponds to $P(n, 0) = \delta_{n0}$. Once the equation is solved for g, we get $P(n, t)$ from the expansion of $g(z, t)$ in powers of z. Deriving the generating function with respect to time and using Eqs. (8.42) and (8.43), we get the equation for $g(z, t)$,

$$\frac{d}{dt} g(z, t) = -a(1 - z) g(z, t),$$
(8.45)

whose solution for the initial condition $g(z, 0) = 1$ is

$$g(z, t) = e^{-a(1-z)t}.$$
(8.46)

The expansion of $g(z, t)$ gives

$$g(z, t) = e^{-at} e^{azt} = e^{-at} \sum_{n=0}^{\infty} \frac{(azt)^n}{n!},$$
(8.47)

from which we obtain

$$P(n, t) = e^{-at} \frac{(at)^n}{n!},$$
(8.48)

which is the Poisson distribution. From this distribution we get the average number of events occurring up to time t,

$$\langle n \rangle = at,$$
(8.49)

which grows linearly with time and the variance,

$$\langle n^2 \rangle - \langle n \rangle^2 = at,$$
(8.50)

which also grows linearly with time.

8.4 Asymmetric Random Walk

Here we analyze a random walk along the x-axis. The possible positions of the walker are discretized and given by $x = n$. From the present position n, the walker may jump to the right, to position $n + 1$, with transition rate a, or to the left, to position $n - 1$, with transition rate b. Therefore, when $a = b$, the random walk is symmetric and when $a \neq b$, it is asymmetric. We use periodic boundary conditions, which amounts to say that positions n and $N + n$ are the same. The elements of the evolution matrix W are $W(n+1,n) = a$, $W(n-1,n) = b$, and the time evolution of the probability $P(n,t)$ of finding the walker at position n at time t is given by

$$\frac{d}{dt}P(n,t) = aP(n-1,t) + bP(n+1,t) - (a+b)P(n,t). \tag{8.51}$$

The characteristic function,

$$G(k,t) = \langle e^{ikn} \rangle = \sum_n e^{ikn} P(n,t), \tag{8.52}$$

obeys the equation

$$\frac{d}{dt}G(k,t) = (ae^{ik} + be^{-ik} - a - b)G(k,t). \tag{8.53}$$

Which is obtained deriving (8.52) with respect to time and using (8.51). The solution is

$$G(k,t) = G(k,0)e^{t(\Gamma \cos k - ic \sin k - a - b)}, \tag{8.54}$$

where $\Gamma = a + b$ and $c = a - b$. Assuming that the walker is in position $n = 0$ at $t = 0$, then $P(n,0) = \delta_{n0}$ so that $G(k,0) = 1$ which gives

$$G(k,t) = e^{t(\Gamma \cos k - ic \sin k - a - b)}. \tag{8.55}$$

The probability distribution is obtained by taking the inverse Fourier transform,

$$P(n,t) = \frac{e^{-t\Gamma}}{2\pi} \int_{-\pi}^{\pi} e^{t\Gamma \cos k - itc \sin k + ikn} dk. \tag{8.56}$$

For large times, the integral can be approximated by

$$P(n,t) = \frac{1}{2\pi} \int_{-\infty}^{\infty} e^{-t\Gamma k^2/2 - itck + ikn} dk, \tag{8.57}$$

whose integral gives

$$P(n,t) = \frac{1}{\sqrt{2\pi\Gamma t}} e^{-(n-ct)^2/2\Gamma t}, \tag{8.58}$$

from which we obtain the average and the variance,

$$\langle n \rangle = ct, \qquad \langle n^2 \rangle - \langle n \rangle^2 = \Gamma t. \tag{8.59}$$

The probability distribution for large times is a Gaussian with a width that grows according to $\sqrt{\Gamma t}$ and with a mean that displaces with a constant velocity equal to c. When $c = 0$, the mean remains stationary.

8.5   Random Walk with an Absorbing Wall

We analyze here the random walk along the x-axis but with the restriction $x \geq 0$. A particle jumps to the right with rate a and to left with rate b,

$$W(n+1,n) = a, \qquad W(n-1,n) = b, \qquad n = 1,2,3,\ldots \quad (8.60)$$

except when the particle is at $x = 0$. Since the movement is restrict to the positive semi-axis, $W(-1,0) = 0$. As to the rate $W(1,0)$, it must vanish because once the particle is at the point $x = 0$, it cannot escape from this point. In other terms, the state $x = 0$ is absorbing. We may say that there is an absorbing wall at $x = 0$. Figure 8.1 shows the corresponding transition diagram.

The probability $P(n,t)$ of finding the particle in position $x = n$ at time t fulfills the master equation

$$\frac{d}{dt}P(0,t) = bP(1,t), \qquad (8.61)$$

$$\frac{d}{dt}P(1,t) = bP(2,t) - (a+b)P(1,t), \qquad (8.62)$$

$$\frac{d}{dt}P(n,t) = aP(n-1,t) + bP(n+1,t) - (a+b)P(n,t), \qquad (8.63)$$

the last one valid for $n \geq 2$. The stationary state is $P_e(0) = 1$ and $P_e(n) = 0$ for $n \neq 0$, which describes the localization of the particle at $x = 0$. The initial state is assumed to be the one in which the particle is at the position $x = n_0$ distinct from

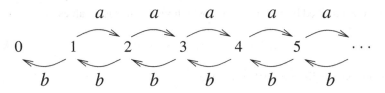

Fig. 8.1 Transition diagram for the random walk with an absorbing state. The jumping rate to the right is a and to the left is b, for any state, except $n = 0$, which is an absorbing state. When $a \leq b$, the particle will reach the absorbing state if the waiting time is large enough. When $a > b$, there is a nonzero probability that the particle never reaches the absorbing state

zero. As we will see shortly, when $a > b$ there is a probability that the particle never reaches the point $x = 0$, that is, that the particle will never be absorbed.

It is worth to note that the equations corresponding to the probabilities $P(n,t)$, for $n = 1,2,\ldots$ do not include $P(0,t)$. Thus, the procedure that we use consists in solving equations $n = 1,2,\ldots$ to determine these probabilities. The probability $P(0,t)$ can be obtained afterwards by the normalization or by the time integration of Eqs. (8.61),

$$P(0,t) = b \int_0^t P(1,t')dt' \tag{8.64}$$

According to this scheme, we start by writing the eigenvalue equations for $n = 1,2,\ldots$

$$-(a+b)\psi(1) + b\psi(2) = \lambda\psi(1), \tag{8.65}$$

$$a\psi(n-1) - (a+b)\psi(n) + b\psi(n+1) = \lambda\psi(n), \tag{8.66}$$

the last one valid for $n \geq 2$. With the aim of rendering the equations symmetric, we define the quantities $\varphi(n)$ by

$$\psi(n) = r^n \varphi(n), \qquad r = \sqrt{a/b}. \tag{8.67}$$

The equations become

$$-(a+b)\varphi(1) + \sqrt{ab}\,\varphi(2) = \lambda\varphi(1), \tag{8.68}$$

$$\sqrt{ab}\varphi(n-1) - (a+b)\varphi(n) + \sqrt{ab}\,\varphi(n+1) = \lambda\varphi(n). \tag{8.69}$$

The solution is

$$\varphi_k(n) = \frac{2}{\sqrt{2\pi}} \sin kn, \qquad 0 < k < \pi, \tag{8.70}$$

what can be checked by inspection and which yields the eigenvalues

$$\lambda_k = -(a+b) + 2\sqrt{ab}\cos k, \tag{8.71}$$

and is orthonormalized in such a way that

$$\int_0^\pi \varphi_k(n)\varphi_k(n')dk = \delta_{nn'}. \tag{8.72}$$

In addition they obey the relation

$$\sum_{n=1}^{\infty} \varphi_k(n)\varphi_{k'}(n) = \delta(k - k').$$ (8.73)

The right eigenvectors, ψ_k, and the left eigenvalues, ϕ_k, are given by

$$\psi_k(n) = r^n \varphi_k(n), \qquad \phi_k(n) = r^{-n} \varphi_k(n), \qquad n = 1, 2, \ldots$$ (8.74)

Replacing the eigenvectors into Eq. (7.39), we reach the following result for the probability distribution

$$P(n, t) = \frac{2}{\pi} r^{n-n_0} \int_0^{\pi} e^{t\lambda_k} \sin kn \sin kn_0 \, dk,$$ (8.75)

valid for $n \neq 0$.

The probability $P(0, t)$, which is the probability of finding the particle in the absorbing state at time t, is obtained by means of (8.64). It can also be understood as the probability of the particle being absorbed up to time t. Therefore, the probability of permanence of the particle up to time t, which we denote by $\mathscr{P}(t)$, is equal to $\mathscr{P}(t) = 1 - P(0, t)$. Using (8.64) and the result (8.75) for $P(1, t)$ we get the following expression

$$\mathscr{P}(t) = 1 - \frac{2}{\pi}\sqrt{ab}\, r^{-n_0} \int_0^{\pi} \frac{1 - e^{t\lambda_k}}{|\lambda_k|} \sin k \sin kn_0 \, dk.$$ (8.76)

The total probability of permanence $\mathscr{P}^* = \lim_{t \to \infty} \mathscr{P}(t)$ is thus

$$\mathscr{P}^* = 1 - \frac{2}{\pi}\sqrt{ab}\, r^{-n_0} \int_0^{\pi} \frac{\sin k \sin kn_0}{a + b - 2\sqrt{ab} \cos k} \, dk.$$ (8.77)

Performing the integral, we get

$$\mathscr{P}^* = \begin{cases} 1 - \left(\frac{b}{a}\right)^{n_0}, & a > b, \\ 0, & a \leq b, \end{cases}$$ (8.78)

From the result (8.78), we can draw the following conclusions. When $a \leq b$ the particle is absorbed by the wall no matter what the initial position, that is, for times large enough the particle will be at position $n = 0$. When $a < b$, there is a probability that the particle will not be absorbed by the wall, that is, the particle will remain forever outside the point $n = 0$. We call this state, an active state. The probability of occurrence of the active state is nonzero and proportional to $a - b$ for a near b, $\mathscr{P}^* = n_0(b - a)/a$.

Another quantity that characterizes the random walk with an absorbing state is the probability of absorption at time t per unit time, which we denote by $\Phi(t)$. In other terms, $\Phi(t) = dP(0,t)/dt$ or yet $\Phi(t) = -d\mathscr{P}(t)/dt$. Deriving (8.76) with respect to time, we get

$$\Phi(t) = \frac{2}{\pi}\sqrt{ab}\; r^{-n_0} \int_0^\pi e^{t\lambda_k} \sin k \sin kn_0 \, dk, \qquad (8.79)$$

which can also be obtained directly from $R(t) = bP(1,t)$, which follows from (8.61) and $R(t) = dP(0,t)/dt$. Taking into account that $d\lambda_k/dk = -2\sqrt{ab}\sin k$, an integration by parts leads us to the result

$$\Phi(t) = \frac{n_0}{\pi t}\; r^{-n_0} \int_0^\pi e^{t\lambda_k} \cos kn_0 \, dk. \qquad (8.80)$$

For large times

$$\Phi(t) = \frac{n_0}{\pi t}\; r^{-n_0} e^{-(\sqrt{a}-\sqrt{b})^2 t} \int_0^\infty e^{-t\sqrt{ab}\,k^2} \cos kn_0 \, dk. \qquad (8.81)$$

Performing the integral

$$\Phi(t) = \frac{n_0 r^{-n_0}}{t\sqrt{4\pi t \sqrt{ab}}}\, e^{-(\sqrt{a}-\sqrt{b})^2 t}, \qquad (8.82)$$

which gives the probability of absorption of the particle at time t per unit time, being the particle at position $x = n_0$ at time $t = 0$.

Inserting the expression for λ_k, given by (8.71), into (8.75), we get the following result for large times

$$P(n,t) = \frac{2}{\pi} r^{n-n_0} e^{-(\sqrt{a}-\sqrt{b})^2 t} \int_0^\infty e^{-t\sqrt{ab}k^2} \sin kn_0 \sin kn \, dk. \qquad (8.83)$$

Performing the integral, we reach the expression

$$P(n,t) = r^{n-n_0} e^{-(\sqrt{a}-\sqrt{b})^2 t} \frac{1}{2\sqrt{\pi t \sqrt{ab}}} \left(e^{-(n-n_0)^2/4t\sqrt{ab}} - e^{(n+n_0)^2/4t\sqrt{ab}} \right) \qquad (8.84)$$

This formula is similar to (4.99) for the probability distribution related to the problem of a Brownian particle with absorbing wall, seen in Sect. 4.6.

8.6   Random Walk with a Reflecting Wall

Now we examine the random walk along the x-axis but with the restriction $x \geq 0$. A particle jumps to the right with rate a and to left with rate b,

$$W(n+1,n) = a, \qquad W(n-1,n) = b, \qquad n = 1,2,3,\ldots \quad (8.85)$$

except for $x = 0$. Since the motion is restrict to the positive semi-axis, $W(-1,0) = 0$. Unlike the absorbing case, the particle may leave the point $x = 0$ jumping to the right with rate b, that is, $W(1,0) = b$. We may say that there is a reflecting wall at $x = 0$. Figure 8.2 shows the corresponding transition diagram.

The probability $P(n,t)$ of finding the particle at position $x = n$ at time t obeys the master equation

$$\frac{d}{dt}P(0,t) = bP(1,t) - aP(0,t), \qquad (8.86)$$

$$\frac{d}{dt}P(n,t) = aP(n-1,t) + bP(n+1,t) - (a+b)P(n,t), \qquad (8.87)$$

valid for $n \geq 1$. The stationary probability $P_e(n)$ is such that $bP_e(n+1) = aP_e(n)$. The normalization is possible only if $a < b$. When $a \geq b$, there is no stationary probability distribution.

The eigenvalue equation reads

$$-a\psi(0) + b\psi(1) = \lambda\psi(0), \qquad (8.88)$$

$$a\psi(n-1) - (a+b)\psi(n) + b\psi(n+1) = \lambda\psi(n). \qquad (8.89)$$

for $n \geq 1$. To render the equations symmetric, we define the quantities $\varphi(n)$ by

$$\psi(n) = r^n\varphi(n), \qquad r = \sqrt{a/b}. \qquad (8.90)$$

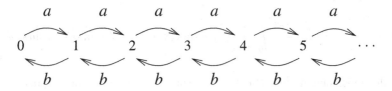

Fig. 8.2 Transition diagram for the random walk with a reflecting state. The rate of jump to right is a and to the left is b, for any state, except $n = 0$, which is a reflecting state. When $a < b$, the particle remains next to the reflecting state. When $a > b$, the particle moves to the right with a finite velocity. When $a = b$, the motion is purely diffusive

The eigenvalue equations become

$$-a\varphi(0) + \sqrt{ab}\varphi(1) = \lambda\varphi(0), \tag{8.91}$$

$$\sqrt{ab}\varphi(n-1) - (a+b)\varphi(n) + \sqrt{ab}\varphi(n+1) = \lambda\varphi(n). \tag{8.92}$$

for $n \geq 1$. The solutions are of the form

$$\varphi_k(n) = A_k(\sqrt{a}\sin k(n+1) - \sqrt{b}\sin kn), \qquad\qquad 0 < k < \pi, \tag{8.93}$$

which can be checked by substitution, giving the eigenvalues

$$\lambda_k = 2\sqrt{ab}\cos k - (a+b). \tag{8.94}$$

The orthogonalization

$$\int_0^\pi \varphi_k(n)\varphi_k(n') = \delta_{nn'} \tag{8.95}$$

leads us to the result $1/A_k^2 = \sqrt{\pi/2}\,(a + b - 2\sqrt{ab}\cos k) = \sqrt{\pi/2}\,|\lambda_k|$, which gives the normalization constant A_k. Thus

$$\varphi_k(n) = \frac{2}{\sqrt{2\pi|\lambda_k|}}(\sqrt{a}\sin k(n+1) - \sqrt{b}\sin kn), \qquad 0 < k < \pi, \tag{8.96}$$

The right eigenvectors, ψ_k, and left eigenvectors, ϕ_k, of W are given by

$$\psi_k(n) = r^n\varphi_k(n), \qquad\qquad \phi_k(n) = r^{-n}\varphi_k(n), \qquad\qquad n = 0, 1, 2, \ldots \tag{8.97}$$

In addition to these eigenvalues, there is also the eigenvalue corresponding to the stationary state $P_e(n) = (1 - r^2)r^{2n} = \psi_0(n)$, as long as $r < 1$, with zero eigenvalue. The left eigenvector is $\phi_0(n) = 1$.

Replacing the eigenvector into Eq. (7.39) and considering that at the initial time the particle is at $n = 0$, that is, $P(n, 0) = \delta_{n0}$, we reach the following result for the probability distribution

$$P(n, t) = P_e(n) + \frac{2r^n}{\pi}\int_0^\pi \frac{e^{\lambda_k t}}{|\lambda_k|}(\sqrt{a}\sin k(n+1) - \sqrt{b}\sin kn)\sqrt{a}\sin k \, dk, \tag{8.98}$$

with the understanding that $P_e(n) = 0$ when $a \geq b$. Inserting this expression for λ_k in this equation, we obtain, for large times, an expression similar to Eq. (4.115), obtained in Sect. 4.7 for the Brownian motion with a reflecting wall. In analogy to that case, we draw the following conclusions. When $a < b$, $P(n, t)$ approaches the stationary distribution, which means that the particle is found near the reflecting wall. When $a > b$, the particle moves away from the wall with finite velocity. The probability distribution has a bump that moves away from the reflecting wall. When $a = b$, the particle performs a pure diffusive movement.

8.7   Multidimensional Random Walk

We study here a random walk on a hypercubic lattice of dimension d. The possible states are denoted by the vector $n = (n_1, n_2, \ldots, n_d)$, where n_i takes the integer values. For convenience we define unit vectors e_i as the ones having all components zero except the i-th which is equal to one. Thus, the vector n can be written as $n = \sum_i n_i e_i$. The possible transition are those such that $n \to n \pm e_i$ with all rates equal, which we take to be equal to α. The master equation, which governs the time evolution of the probability distribution $P(n, t)$ of state n at time t, is given by

$$\frac{d}{dt} P(n, t) = \alpha \sum_i \{ P(n + e_i, t) + P(n - e_i, t) - 2P(n, t) \}, \qquad (8.99)$$

where the sum in i varies from 1 to d.

The characteristic function $G(k, t)$, where $k = (k_1, k_2, \ldots, k_d, t)$, is defined by

$$G(k, t) = \sum_n e^{ik \cdot n} P(n, t), \qquad (8.100)$$

and obeys the equation

$$\frac{d}{dt} G(k, t) = \gamma_k G(k, t), \qquad (8.101)$$

where

$$\gamma_k = -2\alpha \sum_i (1 - \cos k_i), \qquad (8.102)$$

and whose solution is given by

$$G(k, t) = e^{\gamma_k t} G(k, 0). \qquad (8.103)$$

Supposing that initially the particle that performs the random walk is at $n_i = m_i$, then $P(n, 0) = \prod_i \delta(n_i - m_i)$ and $G(k, 0) = e^{ik \cdot m}$, where $m = (m_1, \ldots, m_d)$. The characteristic function is thus

$$G(k, t) = e^{ik \cdot m + \gamma_k t} \qquad (8.104)$$

The probability distribution is obtained by means of the integral

$$P(n,t) = \frac{1}{(2\pi)^d} \int e^{-ik \cdot n} G(k,t) dk, \qquad (8.105)$$

where $dk = dk_1 dk_2 \dots dk_d$ and the integral is performed over the region of the space k defined by $-\pi \leq k_i \leq \pi$. Therefore, we get the solution of the master equation in the form

$$P(n,t) = \frac{1}{(2\pi)^d} \int e^{-ik \cdot (n-m) + \gamma_k t} dk. \qquad (8.106)$$

We see that the distribution is a product,

$$P(n,t) = \prod_j \frac{1}{2\pi} \int_{-\pi}^{\pi} e^{-ik_j (n_j - m_j) - 2\alpha(1 - \cos k_j)t} dk_j \qquad (8.107)$$

For large times, each integral is dominated by the region in k_j around $k_j = 0$. The solution for large times is obtained thus by expanding the exponent in the integrand up to order k_j^2 and extending the limits of the integral, that is,

$$P(n,t) = \prod_j \frac{1}{2\pi} \int_{-\infty}^{\infty} e^{ik_j (n_j - m_j) - \alpha t k_j^2} dk_j. \qquad (8.108)$$

Each integral is a Gaussian of mean m_j and variance $2\alpha t$,

$$P(n,t) = \prod_j \frac{e^{-(n_j - m_j)^2 / 4\alpha t}}{\sqrt{4\pi \alpha t}}. \qquad (8.109)$$

8.8 Multidimensional Walk with an Absorbing State

Now we consider a random walk such that the state $n = 0$ is absorbing. This means to say that the probability of occurrence of the state $n = 0$ at time t, which we denote by $P_0(t)$, obeys the equation

$$\frac{d}{dt} P_0(t) = \frac{1}{2d} \sum_i \{ P(e_i, t) + P(-e_i, t) \}. \qquad (8.110)$$

For convenience, we set $\alpha = 1/2d$ which can be done by rescaling the time. The probability $P(n,t)$ of state $n \neq 0$ at time t obeys the equation

$$\frac{d}{dt} P(n,t) = \frac{1}{2d} \sum_i \{ P(n + e_i, t) + P(n - e_i, t) - 2P(n,t) \}, \qquad (8.111)$$

which is valid only for $n \neq 0$ and with the proviso that $P(0,t) = 0$ at the right-hand side. We remark that the probability of occurrence of the state $n = 0$ is $P_0(t)$, which obeys (8.110), and not $P(0,t)$ which is set as being identically zero.

Since $P_0(t)$ does not enter into Eqs. (8.111), these can be solved without the use of Eq. (8.110). The solution can be obtained more easily by the introduction of the variable $\Phi(t)$, defined by

$$\Phi(t) = \frac{1}{2d} \sum_i \{P(e_i,t) + P(-e_i,t)\}. \tag{8.112}$$

With the introduction of this variable we can write the equations for the probabilities in the form

$$\frac{d}{dt}P(n,t) = \frac{1}{2d} \sum_i \{P(n+e_i,t) + P(n-e_i,t) - 2P(n,t)\} - \Phi(t)\delta(n), \tag{8.113}$$

which is valid for any value of n including $n = 0$. The solution of these equations depends on $\Phi(t)$, which must be chosen so that $P(0,t) = 0$, since in this case the equation for $n = 0$ becomes equivalent to Eq. (8.112).

Comparing (8.112) with (8.110), we see that the variable $\Phi(t)$ is related to the probability $P_0(t)$ of state $n = 0$ by

$$\frac{d}{dt}P_0(t) = \Phi(t) \tag{8.114}$$

so that $\Phi(t)$ is interpreted as the flux of probability to the absorbing state.

Using the master equation (8.113) we get the following equation for the Fourier transform of $P(n,t)$

$$\frac{d}{dt}G(k,t) = \gamma_k G(k,t) - \Phi(t), \tag{8.115}$$

where γ_k is given by

$$\gamma_k = -\frac{1}{d} \sum_i (1 - \cos k_i), \tag{8.116}$$

Next, defining the Laplace transform of $G(k,t)$,

$$\hat{G}(k,s) = \int_0^\infty G(k,t)e^{-st}dt, \tag{8.117}$$

we conclude that it obeys the equation

$$s\hat{G}(k,s) - G(k,0) = \gamma_k G(k,s) - \hat{\Phi}(s), \tag{8.118}$$

where

$$\hat{\Phi}(s) = \int_0^\infty \Phi(t)e^{-st}dt, \tag{8.119}$$

Assuming the initial condition $P(n,0) = \delta(n_1 - 1)\delta(n_2)\ldots\delta(n_d)$, then $G(k,0) = e^{ik_1}$ and

$$\hat{G}(k,s) = \frac{e^{ik_1} - \hat{\Phi}(s)}{s - \gamma_k}. \tag{8.120}$$

The initial condition we use is the one in which the walker is the closest possible to the absorbing state.

Now it is convenient to define the integral

$$\mathscr{G}(n,s) = \frac{1}{(2\pi)^d}\int \frac{e^{-ik\cdot n}}{s - \gamma_k}dk, \tag{8.121}$$

where the integral in the space k is over a region defined by $-\pi \le k_i \le \pi$ for each component. With the help of this definition we can write

$$\hat{P}(n,s) = \mathscr{G}(n - e_1, s) - \hat{\Phi}(s)\mathscr{G}(n,s), \tag{8.122}$$

where $\hat{P}(n,s)$ is the Laplace transform of $P(n,t)$,

$$\hat{P}(n,s) = \int_0^\infty P(n,t)e^{-st}dt. \tag{8.123}$$

We notice that $\hat{P}(n,s)$ is not determined by Eq. (8.122) because we do not yet know $\hat{\Phi}(s)$. To obtain this quantity, we recall that $\Phi(t)$ must be determined by requiring that $P(0,t) = 0$, what amounts to impose $P(0,s) = 0$. Inserting this condition into Eq. (8.122), we reach the desired result

$$\hat{\Phi}(s) = \frac{\mathscr{G}(-e_1, s)}{\mathscr{G}(0, s)}. \tag{8.124}$$

From the definition (8.121), we can check that the following relation is correct, $\mathscr{G}(-e_1, s) = (1 + s)\mathscr{G}(0, s) - 1$, so that

$$\hat{\Phi}(s) = 1 - \frac{1}{\mathscr{G}(0, s)} + s. \tag{8.125}$$

Integrating (8.61) and taking into account that $P_0(0) = 0$,

$$P_0(\infty) = \int_0^\infty \Phi(t)dt = \hat{\Phi}(0). \tag{8.126}$$

Therefore, to determine $P_0(\infty)$, which is the probability of finding the walker at $n = 0$ in any time, we should look to the behavior of $\mathscr{G}$ when $s \to 0$. For $d = 1$

$$\mathscr{G}(0, s) = \frac{1}{2\pi} \int_{-\pi}^{\pi} \frac{dk}{s + 1 - \cos k} = \frac{1}{\sqrt{2s + s^2}}, \tag{8.127}$$

and therefore $\mathscr{G}$ diverges when $s \to 0$ so that $\hat{\Phi}(0) = 1$ and we conclude that in $d = 1$ the walker will enter the absorbing state if we wait enough time. Explicitly,

$$\hat{\Phi}(s) = 1 + s - \sqrt{2s + s^2}, \tag{8.128}$$

which takes us to the following expression for $\Phi(t)$,

$$\Phi(t) = \frac{e^{-t}}{t\pi} \int_0^{\pi} e^{t \cos k} \cos k \, dk. \tag{8.129}$$

For large times,

$$\Phi(t) = \frac{1}{t\pi} \int_0^{\pi} e^{-tk^2/2} \cos k \, dk = \frac{1}{t\sqrt{2\pi t}}. \tag{8.130}$$

For $d = 2$, we should consider the integral

$$\mathscr{G}(n_1, n_2, s) = \frac{1}{(2\pi)^2} \int_{-\pi}^{\pi} \int_{-\pi}^{\pi} \frac{dk_1 dk_2}{s + 1 - (\cos k_1 + \cos k_2)/2}. \tag{8.131}$$

For small values of s, this integral behaves as $\mathscr{G} = |\ln s|/\pi$ and diverges when $s \to 0$. Therefore, $P_0(\infty) = \hat{\Phi}(0) = 1$ and we conclude that in $d = 2$ the walker will also enter the absorbing state if we wait enough time.

For $d \geq 3$, the integral $\mathscr{G}(n, 0)$ is finite and hence

$$P_0(\infty) = \hat{\Phi}(0) = 1 - \frac{1}{\mathscr{G}(0, 0)}. \tag{8.132}$$

We conclude that for $d \geq 3$ there is a probability that the walker will never reach the absorbing state. This probability is equal to $1 - P_0(\infty) = 1/\mathscr{G}(0, 0)$.

As a final comment to this section, we remember that the problem of first passage of a stochastic process is equivalent to the same stochastic process with an absorbing state, as seen in Sect. 7.5. Therefore the results obtained here can be restate in the following terms. In one and two dimensions the walker returns to the starting point with probability one. In three or more dimensions the probability of returning is strictly less than one, what means that the walker can never return. This result is the same seen in the random walk in discrete time, studied in Sect. 6.12.

Chapter 9
Phase Transitions and Criticality

9.1 Introduction

The water, when heated at constant pressure, boils at a well defined temperature, turning into steam. To each value of the pressure imposed on the water, there corresponds a boiling temperature. The water-steam transition temperature increases with pressure and, in a temperature-pressure diagram, it is represented by a line with a positive slope. On the transition line the liquid and vapor coexist in any proportion. However, the liquid and the vapor present well defined and distinct densities that depend only on the transition temperature. As we increase the temperature, along the coexistence line, the difference between the densities of the liquid and vapor becomes smaller and vanishes at a point characterized by well defined temperature and pressure. At this point, called critical point, the liquid and vapor becomes identical and the line of coexistence ends. Beyond this point there is no distinction between liquid and vapor.

Other types of phase transition occur in condensed matter physics. A ferromagnetic substance, when heated, loses its spontaneous magnetization at a well defined temperature, called Curie point, becoming paramagnetic. In the paramagnetic phase, the substance acquires a magnetization only when subject to a magnetic field. In the ferromagnetic phase, in contrast, the magnetization remains after the magnetic field is removed.

The zinc-copper alloy undergoes an order-disorder transition as one varies the temperature. Imagine the crystalline structure as being composed by two intertwined sublattices, which we call sublattice A and sublattice B. At low temperatures, the copper atoms are located preferentially in one of the sublattices, while the zinc atoms are located preferentially in the other sublattice. At high temperatures however a zinc atom can be found in any sublattice with equal probability. The same happens with a copper atom. Increasing the temperature, the alloy passes from an ordered phase to a disordered phase when a critical temperature is reached. Below

© Springer International Publishing Switzerland 2015
T. Tomé, M.J. de Oliveira, *Stochastic Dynamics
and Irreversibility*, Graduate Texts in Physics,
DOI 10.1007/978-3-319-11770-6_9

the critical temperature, there are two possible ordered states. In one of them the zinc atoms are located with greater probability in the sublattice A, and in the other, in B. Above the critical temperature, there is a single state such that a zinc atom has equal probability of being found in any sublattice. The symmetry is higher in the disordered state and lower in the ordered state. Thus, when the temperature is decreased, there is a spontaneous symmetry breaking at the critical temperature.

To describe the phase transitions, it is convenient to introduce a quantity called order parameter, whose most important property is to take the zero value at the disordered phase, which is the phase of higher symmetry, and a nonzero value at the ordered phase, which is the phase of lower symmetry. In the liquid-vapor transition, it is defined as the difference between the densities of the liquid and the vapor in coexistence. In ferromagnet systems, the order parameter is simply the magnetization, called spontaneous. In a binary alloy discussed above, we may define it as the difference between the concentrations of the zinc in the two sublattices. In all three cases discussed above, the order parameter vanishes at temperatures above the critical temperature.

9.2   Majority Model

To illustrate the transition between a disordered and a ordered phase with two coexistence states, we begin by examining a simple model which consists in an extension of the Ehrenfest model. In the Ehrenfest model, we imagine a system of N particles each of which can be found in state A or B. A configuration of the system is defined by the number of particles n in state A. The number of particles in state B is $N - n$. At each time interval, one particle is chosen at random and change its state. If initially all particles are in one state, after some time, the two states will have, in the average, the same number of particles. The stationary probability distribution of the number of particles in state A has a single peak, similar to a Gaussian distribution. In the extension of the Ehrenfest model that we will present shortly, the change of states is done with a certain probability defined a priori, which depends on the number of particles in each state. As we will see, this modification can result in a stationary probability distribution with two peaks and therefore qualitatively distinct from that corresponding to the Ehrenfest model.

The model we consider can be understood as a model that describes the system zinc-copper, if we interpret the number of particles in states A and B as the number of atoms of zinc in the sublattices A and B, respectively. When one particle passes from A to B, this means that a zinc atom passes from sublattice A to sublattice B and, as a consequence, a copper atom passes from sublattice B to A. The model can also be used to describe a ferromagnetic system in which each magnetic dipole takes only two values of opposite signs. The numbers of particles in A and B are interpreted as the numbers of dipoles in a certain direction and in the opposite direction, respectively. When a particle passes from one state to the other, it means that a dipole changes sign. From these associations, we see that the Ehrenfest model predicts only the disordered state for the binary alloy and only the paramagnetic state for the magnetic system. The model that we introduce next, on the other hand, can predict also the ordered state for the former and the ferromagnetic state for the latter.

In the Ehrenfest model, at each time interval one particle is chosen at random and changes state. Denoting by n the number of particles in state A and defining the transition rates $W_{n+1,n} = a_n$ from $n \to n+1$ and $W_{n-1,n} = b_n$ from $n \to n-1$, then the master equation, that governs the evolution of the probability $P_n(t)$ of n at time t, is given by

$$\frac{d}{dt} P_n = a_{n-1} P_{n-1} + b_{n+1} P_{n+1} - (a_n + b_n) P_n. \tag{9.1}$$

As seen before, the rates of the Ehrenfest model are

$$a_n = \alpha(N - n), \tag{9.2}$$

$$b_n = \alpha n, \tag{9.3}$$

where α is a positive constant and N is the total number of particles. The stationary probability is given by

$$P_n = \frac{1}{2^N} \binom{N}{n}. \tag{9.4}$$

This distribution has a maximum when the state A has half the particles. When N is large, the distribution approaches a Gaussian. The presence of a single peak in the distribution indicates that the system presents only one thermodynamic phase, in this case, a disordered phase.

Now we modify this model in such a way that the stationary distribution might have two peaks instead of just one. At regular intervals of time, we choose a particle at random and next we check in which state it is. If it is in a state with a smaller number of particles, it changes its state with probability q. If however it is in a state with a larger number of particles, it changes its state with probability $p = 1 - q$. If $q > p$, these rules favor the increase in the number of particles in the state with more particles. According to these rules, the transition rates $W_{n+1,n} = a_n$ from $n \to n+1$ and $W_{n-1,n} = b_n$ from $n \to n-1$ are given by

$$a_n = \alpha(N - n)p, \qquad n < N/2, \tag{9.5}$$

$$a_n = \alpha(N - n)q, \qquad n \geq N/2, \tag{9.6}$$

$$b_n = \alpha n q, \qquad n \leq N/2, \tag{9.7}$$

$$b_n = \alpha n p, \qquad n > N/2. \tag{9.8}$$

Otherwise, the transition rate is zero. The process is governed by the master equation (9.1), which must be solved with the boundary condition $P_{-1} = 0$ and $P_{N+1} = 0$.

It is possible to show that the stationary probability is

$$P_n = \frac{1}{Z}\binom{N}{n} \exp\{2C\,|n - \frac{N}{2}|\}, \tag{9.9}$$

where Z is a normalization constant and the constant C is related to p through

$$p = \frac{e^{-C}}{e^{C} + e^{-C}}. \tag{9.10}$$

Notice that the probability distribution has the symmetry $P_{N-n} = P_n$, that is, it is invariant under the transformation $n \to N - n$.

If $p = 1/2$, that is, $C = 0$, the stationary distribution has a single peak that occurs at $n = N/2$ because in this case we recover the Ehrenfest model. If $p > 1/2$, it also has a single peak at $n = N/2$. However, if $p < 1/2$, the probability distribution presents two peaks, what can be checked more easily if we examine the case in which N and n are very large. To this end, it is convenient to consider the probability distribution $\rho(x)$ of the variables $x = n/N$. Using Stirling formula and taking into account that $\rho(x) = NP_n$, we get

$$\rho(x) = \frac{1}{Z}\frac{\sqrt{N}}{\sqrt{2\pi x(1-x)}} e^{-NF(x)}, \tag{9.11}$$

where

$$F(x) = x \ln x + (1-x)\ln(1-x) - C|2x - 1|. \tag{9.12}$$

Figure 9.1 shows the function $F(x)$ and the probability distribution $\rho(x)$ for several values of C. When $p < 1/2$, this function has minima at $x = x_+$ and $x = x_-$, where

$$x_+ = 1 - p \qquad e \qquad x_- = p. \tag{9.13}$$

When $p \geq 1/2$, the function $F(x)$ presents just one minimum located at $x = 1/2$. The probability distribution $\rho(x)$ has maxima at those points as seen in Fig. 9.1.

Some features must be pointed out. First, the distance between the maxima, given by

$$x_+ - x_- = 1 - 2p, \tag{9.14}$$

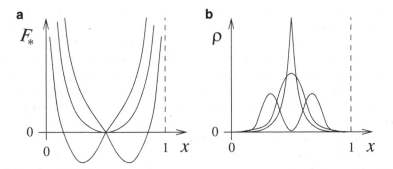

Fig. 9.1 Majority model. (**a**) Plot of $F_*(x) = F(x) + \ln 2$ versus x, where the function $F(x)$ is given by (9.12), for $p < p_c$, $p = p_c$ and $p > p_c$. (**b**) Probability density $\rho(x)$ versus x according to (9.11) for the same values of p. The maxima of ρ correspond to the minima of $F(x)$

vanishes when $p \rightarrow 1/2$. When $p = 1/2$, we have seen that the probability distribution has a single maximum at $n = N/2$ or $x = 1/2$. Therefore, there are two regimes: one characterize by two maxima, when $p < 1/2$, and the other by just one maximum, when $p \geq 1/2$. Around $x = x_+$ and $x = x_-$, the probability distribution $\rho(x)$ has a shape similar to a Gaussian with width proportional to $\sqrt{pq/N}$.

Second, in the limit $N \rightarrow \infty$ and for $p < 1/2$, the height of the maxima increases without bounds while the value of the probability distribution at $x = 1/2$ vanishes. Therefore, in this limit, the probability distribution of the variable x is characterized by two Dirac delta functions symmetric located at $x = x_+$ and $x = x_-$. This means that, in the limit $N \rightarrow \infty$, the system exhibits two stationary states. Depending on the initial condition, the system may reach one or the other state.

The order parameter m of this model is defined as the difference between the concentrations x_+ and x_- of the two states, that is,

$$m = 1 - 2p, \qquad (9.15)$$

valid for $p \leq 1/2$. Thus, in the ordered phase, $m \neq 0$, while in the disordered phase, $m = 0$.

Spontaneous symmetry breaking To characterize the spontaneous symmetry breaking, we consider a phase transition described by the model we have just presented. We have seen that the stationary distribution $\rho(x)$ passes from a regime with a single peak, which characterizes the disordered state, to one with two peaks, which characterizes the ordered state. If the two peaks are sharp so that around $x = 1/2$ the distribution is negligible, then it is possible to find the system in only one of the two states defined by each one of the two peaks.

The emergence of the two stationary states, characterized each one by one peak, indicates the occurrence of the spontaneous symmetry breaking. It is necessary thus to examine the conditions under which the stochastic process can have more than one stationary state. If the stochastic process, governed by a master equation, has a finite number of states and if any state can be reached from any other, there will be no spontaneous symmetry breaking. It suffices to apply the Perron-Frobenius theorem to the corresponding evolution matrix to conclude that there is just one stationary state. Thus, as long as the number of states is finite, there is just one stationary state and, therefore, no symmetry breaking. For the occurrence of a symmetry breaking, that is, the existence of more than one stationary state, it is necessary, but not sufficient, that the number of states is infinite.

To understand in which way the system will reach one of the two stationary states and therefore present a spontaneous symmetry breaking, let us examine the stochastic process defined above and consider N to be very large. Suppose that the stochastic process is simulated numerically from a initial condition such that all particles are in the state A, that is, initially $n = N$ and therefore $x = 1$. During a certain time interval, which we denote by τ, the fraction of particles x in A will fluctuate around x_+ with a dispersion proportional to $\sqrt{pq/N}$. The time interval τ will be the larger, the larger is N. When $N \to \infty$, the dispersion vanishes so that the state A will always have a larger number of particles than the state B. In this case $\tau \to \infty$ and we are faced to a spontaneous symmetry breaking.

In general the numerical simulations are performed with a finite number of particles so that τ is finite. If the observation time is much greater than τ, then the probability distribution which one obtains from simulation will be symmetric. However, if the observation time is smaller or of the same order of τ, the probability distribution might not be symmetric. If the observation time is much smaller than τ, then we will get a distribution with just one peak and a symmetry breaking.

9.3   Ferromagnetic Model

In the model introduced in the previous section, the probability of changing the state does note take into account the number of particles in each state. In the present model, if the particle is in a state with a smaller number of particles, it changes state with a probability which is the larger, the larger is the number of particle in the other state. If it is in a state with a larger number of particles, it changes state with a probability which is the smaller, the smaller is the number of particles in the other state. Various transitions can be set up from these considerations. Here we present one of them.

The transition rates $W_{n+1,n} = a_n$ from $n \to n + 1$ and $W_{n-1,n} = b_n$ from $n \to n - 1$, which should be inserted into the master equation (9.1), are written as

$$a_n = \alpha(N - n)p_n, \tag{9.16}$$

$$b_n = \alpha n q_n, \tag{9.17}$$

where p_n and q_n are given by

$$p_n = \frac{1}{2}\{1 + \tanh \frac{K}{N}(2n - N + 1)\}, \tag{9.18}$$

$$q_n = \frac{1}{2}\{1 - \tanh \frac{K}{N}(2n - N - 1)\}. \tag{9.19}$$

where K is a parameter. The simulation of the model can be performed as follows. At each time step, we choose a particle at random and check its state. If the particle is in state A, it changes to state B with probability q_n. If it is in state B, it changes to A with probability p_n.

The dependence of p_n and q_n on n was chosen so that the stationary probability P_n corresponding to the stochastic process defined by the transition rates above is

$$P_n = \frac{1}{Z}\binom{N}{n}\exp\{\frac{2K}{N}(n - \frac{N}{2})^2\}, \tag{9.20}$$

which is the probability distribution corresponding to the equilibrium model called Bragg-Williams, used to describe binary alloys, where the constant K is proportional to the inverse of the absolute temperature. The probability distribution (9.20) can also be interpreted as that corresponding to a ferromagnetic system whose energy is

$$E = -\frac{2J}{N}(n - \frac{N}{2})^2, \tag{9.21}$$

where J is a parameter. The constant K is related to J by $K = J/kT$, where k is the Boltzmann constant and T the absolute temperature. According to this interpretation, n is equal to the number of dipoles that point to a certain direction. We can verify that the expressions of p_n and q_n, given by (9.18) and (9.19), are such that the detailed balance $a_n P_n = b_{n+1} P_{n+1}$ is obeyed, guaranteeing that P_n is indeed the stationary distribution.

The stationary probability (9.20) has two maxima as long as the parameter K is larger that a certain critical value K_c and a single maxima otherwise. To determine the critical value K_c, we examine the probability distribution P_n for the case in which N is very large. Again, we consider the probability distribution $\rho(x)$ of the variable $x = n/N$. Using Stirling formula, we get

$$\rho(x) = \frac{1}{Z}\frac{\sqrt{N}}{\sqrt{2\pi x(1 - x)}}e^{-NF(x)}, \tag{9.22}$$

where

$$F(x) = x \ln x + (1 - x)\ln(1 - x) - 2K(x - \frac{1}{2})^2. \tag{9.23}$$

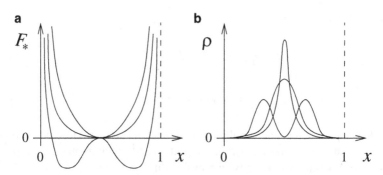

Fig. 9.2 Ferromagnetic model. (**a**) Plot of $F_*(x) = F(x) + \ln 2$ versus x, where the function $F(x)$ is given by (9.23), for $T > T_c$, $T = T_c$ and $T < T_c$. (**b**) Probability density ρ versus x according to (9.22) for the same values of T. The maxima of ρ correspond to the minima of $F(x)$

The two maxima of $\rho(x)$ occur at the minima of $F(x)$. The function $F(x)$ has the aspect shown in Fig. 9.2 and the probability density is shown in the same figure.

To analyze the behavior of the function $F(x)$, we determine its derivatives, given by

$$F'(x) = \ln x - \ln(1 - x) - 4K(x - \frac{1}{2}), \tag{9.24}$$

$$F''(x) = \frac{1}{x} + \frac{1}{1 - x} - 4K. \tag{9.25}$$

Since $F'(1/2) = 0$ and $F''(1/2) = 4(1 - K)$, then the function $F(x)$ has a minimum at $x = 1/2$ as long as $K < 1$, as shown in Fig. 9.2. When $K > 1$, the function $F(x)$ has a local maximum at $x = 1/2$ and develops two minima located symmetrically with respect to $x = 1/2$, as can be seen in Fig. 9.2. Therefore, the stationary probability distribution $\rho(x)$ has a single maximum for $K \leq K_c = 1$ or $T \geq T_c = J/k$, and two maxima for $K > K_c$ or $T < T_c$, as seen in Fig. 9.2.

These two minima are solutions of $F'(x) = 0$, which we write as

$$2K(2x - 1) = \ln \frac{x}{1 - x}. \tag{9.26}$$

When K approaches the critical value K_c, these minima are located near $x = 1/2$. Therefore, to determine the minima for values of K around the critical value, we expand the right-hand side of (9.26) around $x = 1/2$. Up to terms of cubic order in the deviation, we obtain the equation

$$(K - 1)(x - \frac{1}{2}) = \frac{4}{3}(x - \frac{1}{2})^3. \tag{9.27}$$

The minima occur for

$$x_\pm = \frac{1}{2} \pm \frac{1}{2}\sqrt{3(K-1)}. \qquad (9.28)$$

The order parameter m is defined in the same way as was done in the previous section so that $m = x_+ - x_-$, that is,

$$m = \sqrt{3(K-1)}, \qquad K > 1. \qquad (9.29)$$

For $K \leq 1$ the order parameter vanishes.

A system described by the ferromagnetic model presents, therefore, a phase transition when the external parameter K reaches its critical value $K_c = 1$. When $K \leq K_c$ there is a single stationary state corresponding to the disordered or paramagnetic phase. When $K > K_c$, there are two stationary states corresponding to the two ordered or ferromagnetic phases.

Order parameter To determine the time dependence of $\rho(x,t)$, we start by writing the evolution equation for $\rho(x,t)$. From the master equation (9.1) and remembering that $x = n/N$, we get

$$\frac{\partial}{\partial t}\rho(x,t) = \frac{1}{\varepsilon}\{a(x-\varepsilon)\rho(x-\varepsilon,t) + b(x+\varepsilon)\rho(x+\varepsilon,t) - [a(x)+b(x)]\rho(x,t)\},$$
$$(9.30)$$

where $\varepsilon = 1/N$ and

$$a(x) = \frac{\alpha}{2}(1-x)\{1 + \tanh K(2x-1)\}, \qquad (9.31)$$

$$b(x) = \frac{\alpha}{2}x\{1 - \tanh K(2x-1)\}. \qquad (9.32)$$

For small values of ε,

$$\frac{\partial}{\partial t}\rho(x,t) = -\frac{\partial}{\partial x}[a(x)-b(x)]\rho(x,t) + \frac{\varepsilon}{2}\frac{\partial^2}{\partial x^2}[a(x)+b(x)]\rho(x,t). \qquad (9.33)$$

We see thus that the master equation is reduced, for sufficiently large N, to a Fokker-Planck equation, which gives the time evolution of $\rho(x,t)$.

For convenience we use from now on the variable s defined by $s = 2x - 1$. The probability distribution $\rho(s,t)$ in terms of this variable obeys the Fokker-Planck equation

$$\frac{\partial}{\partial t}\rho(s,t) = -\frac{\partial}{\partial s}f(s)\rho(s,t) + \varepsilon\frac{\partial^2}{\partial s^2}g(s)\rho(s,t), \qquad (9.34)$$

where $f = 2(a - b)$ and $g = 2(a + b)$ or

$$f(s) = \alpha(-s + \tanh Ks), \tag{9.35}$$

$$g(s) = \alpha(1 - s \tanh Ks). \tag{9.36}$$

It is worth to note that to the Fokker-Planck (9.34) we can associate the following Langevin equation

$$\frac{ds}{dt} = f(s) + \zeta(s, t), \tag{9.37}$$

where $\zeta(s, t)$ is a multiplicative noise with the properties

$$\langle \zeta(s, t) \rangle = 0, \tag{9.38}$$

$$\langle \zeta(s, t)\zeta(s, t') \rangle = 2\varepsilon g(s)\, \delta(t - t'). \tag{9.39}$$

From the Fokker-Planck equation, we get the equation that gives the time evolution of the average $\langle s \rangle$,

$$\frac{d}{dt}\langle s \rangle = \langle f(s) \rangle, \tag{9.40}$$

a result that is obtained by an integration by parts and taking into account that ρ and $d\rho/ds$ vanish in the limits of integration. Taking into account that ε is small (N very large), we presume that the probability distribution $\rho(s, t)$ becomes very sharp when s takes the value $m = \langle s \rangle$ so that, in the limit $\varepsilon \to 0$, it becomes a Dirac delta function in $s = m$. In this limit, $\langle f(s) \rangle \to f(m)$, and we arrive at the following equation

$$\frac{dm}{dt} = f(m), \tag{9.41}$$

which gives the time evolution of the order parameter. In explicit form,

$$\frac{dm}{dt} = \alpha(-m + \tanh Km). \tag{9.42}$$

Before searching for a time-dependent solution, we notice that the solution for large times obeys the equation

$$m = \tanh Km, \tag{9.43}$$

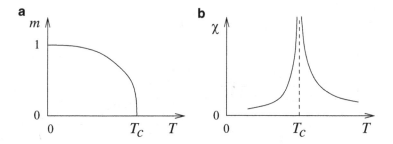

Fig. 9.3 Ferromagnetic model. (**a**) Order parameter m and (**b**) variance χ as functions of the temperature T

which can be written as

$$2Km = \ln \frac{1+m}{1-m}, \tag{9.44}$$

which is equivalent to Eq. (9.26). It suffices to use the relation $m = 2x - 1$. Therefore, the stationary values of m coincide with those obtained from the location of the peaks of the stationary probability distribution. In addition to the trivial solution $m = 0$, Eq. (9.44) also has, for $K > K_c = 1$ or $T < T_c$, the nonzero solutions $m = m^*$ and $m = -m^*$. The solution $m = m^*$ is shown in Fig. 9.3. When $t \to \infty$, we expect that $m(t)$ approaches one of these solutions.

The time-dependent solution of Eq. (9.42) can be obtained explicitly when m may be considered sufficiently small so that the right-hand side of (9.42) can be approximated by the first terms of its expansion in m. Indeed, this can be done as long as K is close to K_c. Up to terms of cubic order in m,

$$\frac{d}{dt}m = -am - bm^3, \tag{9.45}$$

where $a = \alpha(1 - K)$ and $b = \alpha/3 > 0$. Multiplying both sides of Eq. (9.45) by m we arrive at the equation

$$\frac{1}{2}\frac{d}{dt}m^2 = -am^2 - bm^4, \tag{9.46}$$

which can be understood as a differential equation in m^2. The solution of (9.46) for the case $a \neq 0$ is shown in Fig. 9.4 and is given by

$$m^2 = \frac{a}{ce^{2at} - b}, \tag{9.47}$$

where c is a constant which should be determined from the initial condition.

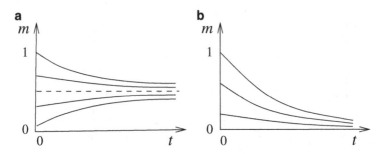

Fig. 9.4 Order parameter m of the ferromagnetic model as a function of time t for several initial conditions. (**a**) $T < T_c$, in this case m approaches a nonzero value when $t \to \infty$. (**b**) $T > T_c$, in this case $m \to 0$ in this limit. In both cases the decay to the final value is exponential

When $t \to \infty$ we get distinct solutions depending on the sign of the parameter a. If $a > 0$ ($K < 1$), then $m \to 0$, which corresponds to the disordered state. If $a < 0$ ($K > 1$), then $m \to \pm m^*$, where $m^* = \sqrt{|a|/b} = \sqrt{3(K-1)}$, which corresponds to the ordered state and is identified with the solution (9.29).

When $K < K_c = 1 (a > 0)$ we see, from the solution (9.47), that, for large times, m decay exponentially to its zero value. Writing

$$m = m_0 e^{-t/\tau}, \tag{9.48}$$

then the relaxation time $\tau = 1/a$. Hence τ behaves as

$$\tau \sim |K - 1|^{-1}, \tag{9.49}$$

and diverges at the critical point.

Similarly when $K > K_c = 1$ ($a < 0$), m decays exponentially to its stationary nonzero value, for large times. Indeed, taking the limit $t \to \infty$ in (9.47) and taking into account that $a < 0$, we obtain $m^2 \to |a|/b$ a hence $m \to m^*$ or $m \to -m^*$ depending on the initial condition. The relaxation time also behaves according to (9.49).

When $K = K_c = 1$ ($a = 0$), the decay is no longer exponential and becomes algebraic, From equation

$$\frac{dm}{dt} = -bm^3, \tag{9.50}$$

valid for $a = 0$, we see that the solution is $m = 1/\sqrt{2bt + c}$. The asymptotic behavior of m for large times is thus algebraic

$$m = \left(\frac{2\alpha t}{3}\right)^{-1/2}, \tag{9.51}$$

remembering that $b = \alpha/3$.

The order parameter decays exponentially out of the critical point with a characteristic time, the relaxation time τ, whose behavior is given by (9.49). Close to the critical point, the relaxation time increases without bounds and eventually diverges at the critical point. At the critical point, the time behavior of the order parameter is no longer exponential and becomes algebraic as given by expression (9.51).

Variance Here we determine the time evolution of the variance χ, defined by

$$\chi = N[\langle s^2 \rangle - \langle s \rangle^2]. \tag{9.52}$$

From the Fokker-Planck equation (9.34),

$$\frac{d}{dt}\langle s^2 \rangle = 2\langle sf(s) \rangle + 2\varepsilon\langle g(s) \rangle, \tag{9.53}$$

a result obtained by integrating by parts and taking into account that ρ and $d\rho/ds$ vanish in the limits of integration. Next we use the result (9.40) to obtain

$$\frac{d\chi}{dt} = \frac{2}{\varepsilon}[\langle sf \rangle - \langle s \rangle\langle f \rangle] + 2\langle g \rangle. \tag{9.54}$$

When ε is very small (N very large), we assume that the probability distribution $\rho(s,t)$ becomes sharp at s equal to $\langle s \rangle = m$ so that the variance is of the order of ε and that the other cumulants are of higher order. In the limit $\varepsilon \to 0$ it becomes a Dirac delta function in $s = m$. In this limit $\langle g(s) \rangle \to g(m)$ and in addition $\langle sf \rangle - \langle s \rangle\langle f \rangle \to f'(m)[\langle s^2 \rangle - \langle s \rangle^2] = f'(m)\varepsilon\chi$. With these results we get the equation for χ,

$$\frac{d\chi}{dt} = 2f'(m)\chi + 2g(m). \tag{9.55}$$

Before searching for the time-dependent solution, we determine the stationary solution, given by

$$\chi = -\frac{g(m)}{f'(m)} = \frac{1 - m^2}{1 - K(1 - m^2)}, \tag{9.56}$$

and shown in Fig. 9.3. The second equality is obtained recalling that $m = \tanh Km$ at the stationary state. When $K < 1$, we have seen that $m = 0$ so that

$$\chi = \frac{1}{1 - K}, \qquad K < 1. \tag{9.57}$$

When $K > 1$, we have seen that around $K = 1$, $m^2 = 3(K - 1)$ and hence

$$\chi = \frac{1}{2(K - 1)}, \qquad K > 1. \tag{9.58}$$

To determine χ as a function of time we use the time dependence of m. At the critical point, $K = 1$, we have seen that $m = (2\alpha t/3)^{-1/2}$ for large times. Since m is small, then, at the critical point, $f'(m) = -\alpha m^2$ and $g(m) = \alpha$ and the equation for χ becomes

$$\frac{d\chi}{dt} = -\frac{3\chi}{t} + 2\alpha. \tag{9.59}$$

whose solution is

$$\chi = \frac{\alpha t}{2}, \tag{9.60}$$

and χ diverges at the critical point when $t \to \infty$.

Time correlation The behavior of a stochastic variable at a certain instant of time may related to the same variable at an earlier time. The relation is measured by the time correlation, defined for the variable s as

$$\langle s(t)s(0)\rangle - \langle s(t)\rangle\langle s(0)\rangle. \tag{9.61}$$

We choose arbitrarily the previous instant of time as being zero.

The correlation $\langle s(t)s(0)\rangle$ is defined by

$$\langle s(t)s(0)\rangle = \int s' K(s', s, t)s\rho(s)ds'ds, \tag{9.62}$$

where $K(s', s, t)$ is the conditional probability density of the occurrence of s' at time t given the occurrence of s at time zero and $\rho(s)$ is the probability distribution at time zero. The distribution $K(s', s, t)$ is the solution of the Fokker-Planck equation (9.34), with the initial condition such that at $t = 0$ it reduces to a $\delta(s' - s)$. Therefore, $K(s', s, t)$ is a Gaussian whose mean $m(t)$ is the solution of (9.41) with the initial condition $m(0) = s$. To explicit this property, we write $m(s, t)$ as a function of s and t. Taking into account that $K(s', s, t)$ is a Gaussian in s', we integrate (9.62) in s' to obtain

$$\langle s(t)s(0)\rangle = \int m(s, t)s\rho(s)ds = \langle ms\rangle. \tag{9.63}$$

Moreover,

$$\langle s(t)\rangle = \int s K(s', s, t)\rho(s)ds'ds = \int m(s, t)\rho(s)ds = \langle m\rangle, \tag{9.64}$$

$$\langle s(0)\rangle = \int s\rho(s)ds = \langle s\rangle, \tag{9.65}$$

so that the correlation (9.61) is given by

$$\langle ms \rangle - \langle m \rangle \langle s \rangle = \langle (m - \langle m \rangle)(s - \langle s \rangle) \rangle. \tag{9.66}$$

In this equation, we should understand that the averages are taken using the initial distribution $\rho(s)$ and that m is a function of s.

Next we suppose that $\rho(s)$ has variance χ/N so that for N large enough the right-hand side of (9.66) becomes $\langle (s - \langle s \rangle)^2 \rangle \partial m/\partial s$, where the derivative is taken at $s = \langle s \rangle$. Defining the time correlation function as

$$C(t) = \frac{\langle ms \rangle - \langle m \rangle \langle s \rangle}{\langle s^2 \rangle - \langle s \rangle \langle s \rangle}, \tag{9.67}$$

we reach the following result

$$C(t) = \frac{\partial m}{\partial s}, \tag{9.68}$$

where the derivative should be taken at $s = \langle s \rangle$, which is the average of s taken by using the initial distribution $\rho(s)$.

The correlation function can be determined deriving both sides of (9.41) with respect to s. Carrying out the derivation, we obtain the following equation for the correlation function

$$\frac{dC}{dt} = f'(m)C. \tag{9.69}$$

Knowing m as a function of time, which is the solution of (9.41), we can solve Eq. (9.69) and get $C(t)$.

For large times, we replace m in $f'(m)$ by the equilibrium value, which is independent of time. The solution of (9.69) in this regime is

$$C(t) \sim e^{-t/\tau}, \tag{9.70}$$

where $1/\tau = f'(m) = \alpha(1 - K(1 - m^2))$. For $K < K_c = 1$, $m = 0$ and we get the result $\tau = 1/\alpha(K - K_c)$. For $K > K_c$ close to K_c we have seen that $m^2 = 3(K - K_c)$ so that $\tau = 1/2\alpha(K_c - K)$. In both cases

$$\tau \sim |K - K_c|^{-1}. \tag{9.71}$$

When $K = K_c = 1$, we have seen that, for large times, $m^2 = 3/2\alpha t$ so that the equation for C becomes

$$\frac{dC}{dt} = -\frac{3}{2t}C, \tag{9.72}$$

whose solution is

$$C \sim t^{-3/2}. \tag{9.73}$$

9.4   Model with Absorbing State

We consider now another modification of the Ehrenfest model such that one of the possible configuration is an absorbing state. Here we choose the absorbing state as the configuration in which all particles are in the state B or equivalently that the state A has no particles. At regular times, a particle is chosen at random. If it is in state A, it passes to state B with probability q. If it is in state B, it passes to state A with probability $p = 1 - q$ times the fraction n/N of particles in state A. Therefore, if there is no particle in state A, the process ends. According to these rules, the transition rates $W_{n+1,n} = a_n$ from $n \to n+1$ and $W_{n-1,n} = b_n$ from $n \to n-1$ are

$$a_n = \alpha'(N - n)\frac{n}{N}p, \tag{9.74}$$

$$b_n = \alpha' n q, \tag{9.75}$$

which would be inserted into the master equation (9.1). For convenience we choose $\alpha' q = 1$ so that

$$a_n = (N - n)\lambda \frac{n}{N}, \tag{9.76}$$

$$b_n = n, \tag{9.77}$$

where $\lambda = p/q$. We see clearly that the state $n = 0$ is absorbing.

It is possible to show by inspection that the distribution

$$P_n = \frac{K N^{N-n} \lambda^n}{(N - n)! n}, \tag{9.78}$$

for $n \neq 0$ and $P_0 = 0$, where K is a normalization constant, is stationary. Another stationary distribution is obviously $P_n = 0$ for $n \neq 0$ and $P_0 = 1$. With the purpose of examining the properties of the stationary probability distribution (9.78), we will consider a large value of N and focus on the probability density $\rho(x)$ of the variable $x = n/N$. Using Stirling formula, we get

$$\rho(x) = \frac{K}{x\sqrt{2\pi N x}} e^{-NF(x)}, \tag{9.79}$$

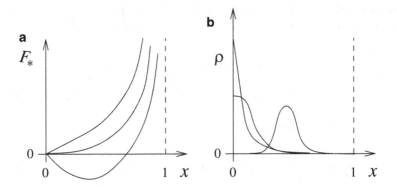

Fig. 9.5 Model with an absorbing state. (a) Plot of $F_*(x) = F(x) + \ln 2$ versus x, where the function $F(x)$ is given by (9.80), for three different values of $\lambda < \lambda_c$, $\lambda = \lambda_c$ and $\lambda > \lambda_c$. (b) Probability density $\rho(x)$ versus x according to (9.79) for the same values of λ. The maxima of ρ correspond to the minima of $F(x)$

where

$$F(x) = -x \ln \lambda - (1 - x) + (1 - x)\ln(1 - x). \tag{9.80}$$

The function (9.80) has the aspect shown in Fig. 9.5. The probability density (9.80) is shown in the same figure. The maxima of ρ occur at the minima of F.

To analyze the behavior of the function $F(x)$, we determined its derivatives, given by

$$F'(x) = -\ln \lambda - \ln(1 - x), \tag{9.81}$$

$$F''(x) = \frac{1}{1 - x}. \tag{9.82}$$

The function $F(x)$ has a single maximum, which we denote by x_0. For $\lambda > 1$, it is obtained from $F'(x_0) = 0$ and gives us $x_0 = (\lambda - 1)/\lambda$. Notice that $F''(x_0) = \lambda > 0$ and therefore x_0 corresponds indeed to a minimum. When $\lambda \leq 1$, the function $F(x)$ is an increasing monotonic and therefore the minimum occurs at $x = 0$ as seen in Fig. 9.5.

We see thus that the model has two behaviors. When $\lambda \leq 1$, the probability density has a peak located at $x = 0$. When $\lambda > 1$, it has a peak located at $x = x_0 > 0$, which moves as λ varies. Therefore we identify $\lambda_c = 1$ as the critical value of λ, which separates the two behaviors. We identify x_0 as the order parameter, which varies around the critical point according to

$$x_0 = \lambda - \lambda_c. \tag{9.83}$$

When $\lambda < \lambda_c$, the order parameter vanishes.

Time correlation From the master equation (9.1) for $P_n(t)$, we get the equation that gives the time evolution of the distribution $\rho(x,t)$ of the variable $x = n/N$,

$$\frac{\partial}{\partial t}\rho(x,t) = \frac{1}{\varepsilon}\{\lambda(1 - x + \varepsilon)(x - \varepsilon)\rho(x - \varepsilon, t) + (x + \varepsilon)\rho(x + \varepsilon, t)$$

$$-[\lambda x(1 - x) + x]\rho(x,t)\}, \tag{9.84}$$

where $\varepsilon = 1/N$. For small values of ε,

$$\frac{\partial}{\partial t}\rho(x,t) = -\frac{\partial}{\partial x}f(x)\rho(x,t) + \frac{\varepsilon}{2}\frac{\partial^2}{\partial x^2}g(x)\rho(x,t), \tag{9.85}$$

where

$$f(x) = \lambda x(1 - x) - x, \tag{9.86}$$

$$g(x) = \lambda x(1 - x) + x. \tag{9.87}$$

We see thus that the master equation is reduced to a Fokker-Planck equation.

It is worth to note that the Fokker-Planck equation (9.85) is associated to the Langevin equation

$$\frac{dx}{dt} = f(x) + \zeta(x,t), \tag{9.88}$$

where $\zeta(x,t)$ is a multiplicative noise with the properties

$$\langle\zeta(x,t)\rangle = 0, \tag{9.89}$$

$$\langle\zeta(x,t)\zeta(x,t')\rangle = \varepsilon g(x)\delta(t - t'). \tag{9.90}$$

From the Fokker-Planck equation we find the equation which gives the time evolution of the average $\langle x \rangle$,

$$\frac{d}{dt}\langle x \rangle = \langle f(x) \rangle, \tag{9.91}$$

a result that is obtained by means of integration by parts and taking into account that ρ and $d\rho/dx$ vanish in the limits of integration. Considering that ε is very small (N very large) we presume that the probability distribution $\rho(x,t)$ will become very sharp when x takes the value $r = \langle x \rangle$ so that, in the limit $\varepsilon \to \infty$, it becomes a Dirac delta function in $x = r$. This allows us to say that in this limit $\langle f(x) \rangle \to f(r)$ and to arrive at the following equation for r

$$t\frac{dr}{dt} = f(r), \tag{9.92}$$

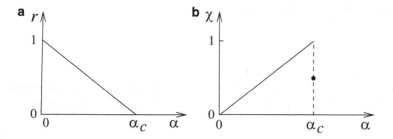

Fig. 9.6 Model with an absorbing state. (**a**) Order parameter r and (**b**) variance χ as functions of $\alpha = 1/\lambda$

or yet

$$\frac{dr}{dt} = \lambda r(1 - r) - r. \tag{9.93}$$

Before determining the time-dependent solutions of (9.93), we calculate r for large times, which is the solution of

$$r = \lambda r(1 - r). \tag{9.94}$$

This equation admits one trivial solution $r = 0$, which we identify as the absorbing state, and a non trivial solution $r = (\lambda - 1)/\lambda$, that corresponds to the state called active. As the parameter λ varies, there occurs a phase transition at $\lambda = \lambda_c = 1$ between the active state, characterized by $r \neq 0$, and the absorbing state, characterized by $r = 0$. The quantity r, which is the fraction of particles in state A, is the order parameter, shown in Fig. 9.6, and behaves close to the critical point as

$$r = (\lambda - \lambda_c). \tag{9.95}$$

Next we move on to the time-dependent solution of (9.93). When $\lambda \neq 1$, this equation is equivalent to

$$(\lambda - 1)\frac{dt}{dx} = \frac{1}{x} - \frac{1}{x - c}, \tag{9.96}$$

where $c = (\lambda - 1)/\lambda$. Integrating, we get

$$r = \frac{c}{1 + be^{-(\lambda-1)t}}, \tag{9.97}$$

where b is determined by the initial condition. We see thus that, if $\lambda < 1$ we reach the trivial solution $r = 0$ when $t \to \infty$. On the other hand, if $\lambda > 1$, we reach the non trivial solution $r = c \neq 0$. In both cases, the decay to the stationary value is exponential with a relaxation time τ which behaves as

$$\tau \sim |\lambda - \lambda_c|^{-1}, \tag{9.98}$$

and hence diverges at the critical point $\lambda_c = 1$.

If $\lambda = 1$, Eq. (9.93) is reduced to equation $dr/dt = -r^2$ whose solution is $r = 1/(c + t)$. The behavior for large times is thus algebraic and given by

$$r = t^{-1}. \tag{9.99}$$

Variance Here we determine the time evolution of the variance χ defined by

$$\chi = N[\langle x^2 \rangle - \langle x \rangle^2]. \tag{9.100}$$

Using the Fokker-Planck equation (9.85),

$$\frac{d}{dt}\langle x^2 \rangle = 2\langle x f(x) \rangle + \varepsilon \langle g(x) \rangle, \tag{9.101}$$

a result that is obtained by integrating by parts and taking into account that ρ and $d\rho/dx$ vanish in the limits of integration. Next, we use the result (9.91) to get

$$\frac{d\chi}{dt} = \frac{2}{\varepsilon}[\langle xf \rangle - \langle x \rangle \langle f \rangle] + \langle g \rangle. \tag{9.102}$$

When ε is very small (N very large), we assume that the probability distribution $\rho(x, t)$ becomes sharp in x equal to $\langle x \rangle = r$ so that the variance is of the order ε and that the other cumulants are of higher order. In the limit $\varepsilon \to 0$ it becomes a Dirac delta function in $x = r$. In this limit $\langle g(x) \rangle \to g(r)$ and in addition $\langle xf \rangle - \langle x \rangle \langle f \rangle \to f'(r)[\langle x^2 \rangle - \langle x \rangle^2]$. With these results, we get the equation for χ,

$$\frac{d\chi}{dt} = 2f'(r)\chi + g(r). \tag{9.103}$$

Before searching for the time-dependent solution, we determine the stationary solution which is given by

$$\chi = -\frac{g(r)}{2f'(r)}. \tag{9.104}$$

When $\lambda < 1$, $r = 0$ and therefore $\chi = 0$. When $\lambda = 1$, $r = 0$ and we get $\chi = 1/2$. When $\lambda > 1$, $r = (\lambda - 1)/\lambda$ and we obtain the result

$$\chi = \frac{1}{\lambda}, \qquad \lambda > 1. \tag{9.105}$$

Therefore, the variance is finite and has a jump at the critical point, as seen in Fig. 9.6.

To find χ as a function of time, we use the time dependence of r. At the critical point, $\lambda = 1$, $f'(r) = -2r$ and for large times, we have seen that $r = 1/t$. Since r is small, then, at the critical point, $g(r) = 2r$ and the equation for χ becomes

$$\frac{d\chi}{dt} = -\frac{4\chi}{t} + \frac{2}{t},\qquad(9.106)$$

whose solution is

$$\chi = \frac{1}{2} + at^{-4},\qquad(9.107)$$

where a is an integration constant.

Time correlation The time correlation function $C(t)$ is obtained by the same procedure used before in the case of the ferromagnetic model. The equation for $C(t)$ is given by $dC/dt = f'(r)C$ or

$$\frac{dC}{dt} = (\lambda - 1 - 2\lambda r)C,\qquad(9.108)$$

which we will solve for large times. When $\lambda > 1$, we have seen that r approaches its stationary value $r = (\lambda - 1)/\lambda$ so that

$$\frac{dC}{dt} = -(\lambda - 1)C,\qquad(9.109)$$

whose solution is $C \sim e^{-t/\tau}$ with

$$\tau = \frac{1}{\lambda - 1}.\qquad(9.110)$$

When $\lambda < 1$ we have seen that r approaches the zero value so that

$$\frac{dC}{dt} = (\lambda - 1)C,\qquad(9.111)$$

whose solution is $C \sim e^{-t/\tau}$ with

$$\tau = \frac{1}{1 - \lambda}.\qquad(9.112)$$

When $\lambda = 1$, we have seen that $r = 1/t$ for large times so that in this case

$$\frac{dC}{dt} = -\frac{2}{t}C,\qquad(9.113)$$

whose solution is

$$C \sim t^{-2}, \tag{9.114}$$

and therefore the correlation function decays algebraically.

Exercises

1. Perform a simulation of the majority model defined in Sect. 9.2 for several values of p and N. Construct the histogram $h(n)$ of the number n of particles in state A and make a normalized plot of h versus n/N. Determine also the average and the variance of n. Make the plots of $\langle n \rangle/N$ and $[\langle n^2 \rangle - \langle n \rangle^2]/N$ versus p.

2. Perform a simulation of the ferromagnetic model defined in Sect. 9.3 for several values of K and N. Construct the histogram $h(n)$ of the number n of particles in state A and make a normalized plot of h versus n/N. Determine also the average and the variance of n. Make the plots of $\langle n \rangle/N$ and $[\langle n^2 \rangle - \langle n \rangle^2]/N$ versus $T = 1/K$.

3. Perform the simulation of the model with an absorbing state defined in Sect. 9.4 for several values of λ and N. Construct the histogram $h(n)$ of the number n of particles in state A and make a normalized plot of h versus n/N. Determine also the average and variance of n. Make the plots of $\langle n \rangle/N$ and $[\langle n^2 \rangle - \langle n \rangle^2]/N$ versus $\alpha = 1/\lambda$.

Chapter 10
Reactive Systems

10.1 Systems with One Degree of Freedom

Extension of reaction Here we study the kinetics of chemical reactions from the stochastic point of view. We consider a system comprised by a vessel inside which several chemical species transform into each other according to certain chemical reactions. The thermochemical process caused by the chemical reactions is understood as a stochastic process. The stochastic approach which we use here corresponds to the description level in which the chemical state is defined by the numbers of molecules of the various chemical species. The reaction rates, understood as the transition rates between the chemical states, are considered to depend only on the numbers of molecules of each species involved in the reactions. A microscopic approach, that takes into account the individuality of the constitutive units and not only the number of them, will be the object of study later on.

We start our study by considering the occurrence of a single reaction and its reverse, involving q chemical species and represented by the chemical equation

$$\sum_{i=1}^{q} v_i' A_i \rightleftharpoons \sum_{i=1}^{q} v_i'' A_i, \tag{10.1}$$

where A_i are the chemical formula of the chemical species, $v_i' \geq 0$ are the stoichiometric coefficients of the reactants and $v_i'' \geq 0$ are the stoichiometric coefficients of products. If a chemical species does not appear among the reactants, the corresponding stoichiometric coefficients v_i' vanishes. Similarly, if a chemical species does not appear among the products, the corresponding stoichiometric coefficients v_i'' vanishes. We remark that if a certain chemical species, which we call catalytic, may appear as among the reactants as among the products. In this case, both stoichiometric coefficients v_i' and v_i'' of the chemical species are nonzero.

© Springer International Publishing Switzerland 2015
T. Tomé, M.J. de Oliveira, *Stochastic Dynamics
and Irreversibility*, Graduate Texts in Physics,
DOI 10.1007/978-3-319-11770-6_10

Equation (10.1) tell us that in the forward reaction, v_i' molecules of species A_i disappear and v_i'' molecules of species A_i are created so that the number of the molecules of type A_i increases by a value equal to

$$v_i = v_i'' - v_i'. \tag{10.2}$$

We denote by n_i the number of molecules of the i-th chemical species, considered to be a stochastic variable. Due to the chemical reactions, the quantities n_i do not remain constant along the thermochemical process. However, their variations are not arbitrary but must be in accordance with the variations (10.2), dictated by the chemical reaction represented by (10.1).

With the purpose of describing the changes in n_i prescribed by the chemical reaction (10.1), we introduce the quantity ℓ, called extension of reaction, and defined as follows. (i) The quantity ℓ takes only integer values. (ii) When ℓ increases by one unit, there occurs a forward reaction, that is, n_i undergoes a variation equal to v_i. (iii) When ℓ decreases by one unit, there occurs the backward reaction that is, n_i undergoes a variation equal to $-v_i$. (iv) When the variables n_i takes pre-determined values, which we denote by N_i, the extension of reaction takes the zero value. We see thus that n_i is related with ℓ by

$$n_i = N_i + v_i \ell. \tag{10.3}$$

Since the variations in n_i must be in accordance with (10.3), then the chemical state of the system in consideration is determined by the variable ℓ only and the system has only one degree of freedom.

Reaction rate The stochastic process that we consider is that in which ℓ varies by just one unit. The rate of the transition $\ell \to \ell + 1$, which we denote by a_ℓ, is identified with the rate of the forward reaction (from reactants to the products) and the rate of the transition $\ell \to \ell - 1$, which we denote by b_ℓ, is identified with the backward reaction (from the products to the reactants). The probability $P_\ell(t)$ of the occurrence of the state ℓ at time t obeys the master equation

$$\frac{d}{dt} P_\ell = a_{\ell-1} P_{\ell-1} + b_{\ell+1} P_{\ell+1} - (a_\ell + b_\ell) P_\ell. \tag{10.4}$$

From the master equation we obtain the time evolution of the average of ℓ

$$\frac{d}{dt} \langle \ell \rangle = \langle a_\ell \rangle - \langle b_\ell \rangle, \tag{10.5}$$

and of the variance of ℓ

$$\frac{d}{dt} (\langle \ell^2 \rangle - \langle \ell \rangle^2) = 2 \langle (\ell - \langle \ell \rangle)(a_\ell - b_\ell) \rangle + \langle a_\ell \rangle + \langle b_\ell \rangle. \tag{10.6}$$

To establish the reaction rates, related to the reactions (10.1), we assume that they obey detailed balance

$$\frac{a_\ell}{b_{\ell+1}} = \frac{P_{\ell+1}^e}{P_\ell^e}, \tag{10.7}$$

with respect to the probability distribution P_ℓ^e that describes the thermodynamic equilibrium. Assuming that the system is weakly interacting, then, according to statistical mechanics of equilibrium systems, the probability distribution has the form

$$P_\ell^e = C \prod_i \frac{(z_i V)^{n_i}}{n_i!}, \tag{10.8}$$

where C is a normalization constant, V is the volume and z_i is a parameter, called activity, that depends only on temperature and on the characteristics of the i-th chemical species. Recalling that n_i is connected to the extension of reaction ℓ by means of (10.3), then

$$\frac{P_{\ell+1}}{P_\ell} = \prod_i \frac{n_i!(z_i V)^{\nu_i}}{(n_i + \nu_i)!}. \tag{10.9}$$

The rates of forward and backward reactions, a_ℓ and b_ℓ, respectively, are assumed to be

$$a_\ell = k^+ V \prod_i \frac{n_i!}{(n_i - \nu_i')! V^{\nu_i'}}, \tag{10.10}$$

$$b_\ell = k^- V \prod_i \frac{n_i!}{(n_i - \nu_i'')! V^{\nu_i''}}, \tag{10.11}$$

where k^+ and k^- are parameters called constants of reaction or specific transition rates. As we will see below they satisfy the detailed balance as long as the constants of reaction are connected to the activities by a certain relation to be found. Notice that the transition rates depend on the extension of reaction because the variables n_i are related to ℓ according to (10.3). We point out that if a chemical species does not belong to the reactants, then ν_j' vanishes and the corresponding term on the product of (10.10) is absent. Similarly, if a chemical species does not belong to the products of the reaction, then ν_j'' vanishes and the corresponding term on the product of (10.11) is absent.

To determined $b_{\ell+1}$ from b_ℓ we should remember that the variation $\ell \to \ell + 1$ entails the variation $n_i \to n_i + \nu_i$. Thus the expression for $b_{\ell+1}$ is that obtained from (10.11) by the substitution of n_i by $n_i + v_i$. Taking into account that $v_i'' - v_i = v_i'$, we get

$$\frac{a_\ell}{b_{\ell+1}} = \frac{k^+}{k^-} \prod_i \frac{n_i! V^{v_i}}{(n_i + v_i)!}.$$

(10.12)

Comparing with (10.9), we see that the detailed balance (10.7) is obeyed if

$$\frac{k^+}{k^-} = \prod_i z_i^{v_i}.$$

(10.13)

The right-hand side of this equation $\prod_i z_i^{v_i} = K$ is called equilibrium constant. The rates of reactions are adopted as being those given by (10.10) and (10.11) and such that the constant of reactions satisfy (10.13).

When n_i is large enough, we may approximate

$$\frac{(n_i + v_i)!}{n_i!} = (n_1 + 1)(n_i + 2) \ldots (n_i + v_1)$$

(10.14)

by $n_i^{v_i}$ so that the reaction rates in this regime are given by

$$a_\ell = k^+ V \prod_i \left(\frac{n_i}{V}\right)^{v_i'},$$

(10.15)

$$b_\ell = k^- V \prod_i \left(\frac{n_i}{V}\right)^{v_i''}.$$

(10.16)

Thermodynamic limit The thermochemical properties are obtained, according to the stochastic approach that we use, from the master equation (10.4) and, in particular, from Eqs. (10.5) and (10.6). The appropriate properties in the description of a thermochemical system are those obtained considering the number of particles and the volume as macroscopic quantities. This means to say that both quantities, number of particles and volume, must be sufficiently large. More precisely, we should take the thermodynamic limit, defined as the limit where the volume and the numbers of particles of each species grow without limit in such a way that the concentrations $n_i/V = x_i$ remain finite. As a consequence, the ratios $\ell/V = \xi$ and $N_i/V = a_i$ also remain finite because from (10.3) the concentrations x_i and the extension of reaction ξ are related by

$$x_i = a_i + v_i \xi.$$

(10.17)

In the thermodynamic limit the quantities $a_\ell/V = \lambda$ and $b_\ell/V = \mu$, which are the reaction rates per unit volume, are finite and given by

$$\lambda(\xi) = k^+ \prod_i x_i^{v_i'}, \qquad \mu(\xi) = k^- \prod_i x_i^{v_i''}. \qquad (10.18)$$

Notice that $\lambda(\xi)$ and $\mu(\xi)$ must be considered functions of ξ due to the relation (10.17).

When the volume is large, we assume that the probability distribution $P_\ell(t)$, solution of the master equation (10.4) with transition rates given by (10.10) and (10.11), or equivalently by (10.15) and (10.16), becomes a Gaussian distribution such that the average of the variable ℓ/V and the variance of the variable $\ell/\sqrt{V}$ are finite in the thermodynamic limit. In other terms, we assume that $\langle \ell \rangle/V$ and $\chi = [\langle \ell^2 \rangle - \langle \ell \rangle^2]/V$ are finite in the thermodynamic limit.

To proceed in our analysis it is convenient to use the stochastic variable $\xi = \ell/V$ and focus on the probability distribution $\rho(\xi, t)$ of this variable. The average of ξ we denote by $\bar{\xi}$, that is, $\bar{\xi} = \langle \ell \rangle/V$. According to the assumptions above $\bar{\xi}$ is finite in the thermodynamic limit. Moreover, the variance of ξ, which is given by $\langle \xi^2 \rangle - \langle \xi \rangle^2 = \chi/V$, is inversely proportional to V and vanishes in the thermodynamic limit because χ is finite in this limit. We may conclude therefore that for V large enough, $\rho(\xi, t)$, is a Gaussian with mean $\bar{\xi}$ finite and variance χ/V, which becomes thus a Dirac delta function in the limit $V \to \infty$.

Using Eqs. (10.5) and (10.6), we see that the time evolutions of $\langle \xi \rangle$ and χ obey the equations

$$\frac{d}{dt}\langle \xi \rangle = \langle \lambda(\xi) \rangle - \langle \mu(\xi) \rangle, \qquad (10.19)$$

$$\frac{d}{dt}\chi = 2V \langle (\xi - \langle \xi \rangle)(\lambda(\xi) - \mu(\xi)) \rangle + \langle \lambda(\xi) \rangle + \langle \mu(\xi) \rangle, \qquad (10.20)$$

where $\lambda(\xi)$ and $\mu(\xi)$ are given by (10.18).

Taking into account that the probability distributions $\rho(\xi, t)$ becomes a Dirac delta function when $V \to \infty$, then we may conclude that in this limit the average $\langle f(\xi) \rangle$ of any analytical function of ξ approaches $f(\bar{\xi})$. From this result we get

$$\frac{d\bar{\xi}}{dt} = \lambda(\bar{\xi}) - \mu(\bar{\xi}), \qquad (10.21)$$

which gives the time evolution of $\bar{\xi}$. In this limit, the following result holds

$$\lim_{V\to\infty} V \langle (\xi - \langle \xi \rangle) f(\xi) \rangle = f'(\langle \xi \rangle)\chi, \qquad (10.22)$$

for any analytical function of $f(\xi)$. Using this result, Eq. (10.20) becomes the equation

$$\frac{d}{dt}\chi = 2[\lambda'(\bar\xi) - \mu'(\bar\xi)]\chi + \lambda(\bar\xi) + \mu(\bar\xi),\tag{10.23}$$

which gives the time evolution of χ. From $\bar\xi$, the concentrations are determined by means of

$$\bar x_i = a_i + v_i\bar\xi.\tag{10.24}$$

Stationary state In the stationary state $\lambda(\bar\xi) = \mu(\bar\xi)$, what leads us to the result

$$k^+\prod_i \bar x_i{}^{v_i'} = k^-\prod_i \bar x_i{}^{v_i''},\tag{10.25}$$

which can be written in the form

$$\prod_i \bar x_i{}^{v_i} = \frac{k^+}{k^-},\tag{10.26}$$

bearing in mind that $v_i = v_i'' - v_i'$. We assume that the reaction constants k^+ and k^- are both nonzero. This expression constitutes the Guldberg-Waage law.

To calculate the variance, we should first determine $\lambda'(\bar\xi)$ and $\mu'(\bar\xi)$. From (10.18) and using (10.24), we get

$$\lambda'(\bar\xi) = \lambda(\bar\xi)\sum_i \frac{v_i' v_i}{\bar x_i}, \qquad \mu'(\bar\xi) = \mu(\bar\xi)\sum_i \frac{v_i'' v_i}{\bar x_i}.\tag{10.27}$$

Taking into account that in the stationary state $\lambda(\bar\xi) = \mu(\bar\xi)$, then, from (10.23), we reach the following expression for the variance in the stationary state

$$\frac{1}{\chi} = \sum_i (v_i'' - v_i')\frac{v_i}{\bar x_i}.\tag{10.28}$$

Since $v_i = v_i'' - v_i'$, then Eq. (10.28) takes the form

$$\frac{1}{\chi} = \sum_i \frac{v_i^2}{\bar x_i}.\tag{10.29}$$

To determine the variance $\chi_i = V(\langle x_i^2\rangle - \langle x_i\rangle^2)$, we use the relation (10.17) to reach the result

$$\chi_i = v_i^2\chi.\tag{10.30}$$

Fokker-Planck equation   The Gaussian form of $\rho(\xi,t)$, assumed above when V is large enough, can be obtained from the transformation of the master equation into a Fokker-Planck equation. For this, we start by writing the master equation (10.4) in terms of the probability distribution $\rho(\xi,t)$,

$$\frac{\partial}{\partial t}\rho(\xi,t) = \frac{1}{\varepsilon}\{\lambda(\xi-\varepsilon)\rho(\xi-\varepsilon,t) + \mu(\xi+\varepsilon)\rho(\xi+\varepsilon,t)$$

$$-\lambda(\xi)\rho(\xi,t) - \mu(\xi)\rho(\xi,t)\}, \tag{10.31}$$

where $\epsilon = 1/V$. For small values of ε

$$\frac{\partial}{\partial t}\rho(\xi,t) = -\frac{\partial}{\partial\xi}[\lambda(\xi)-\mu(\xi)]\rho(\xi,t) + \frac{\varepsilon}{2}\frac{\partial^2}{\partial\xi^2}[\lambda(\xi)+\mu(\xi)]\rho(\xi,t), \tag{10.32}$$

which is a Fokker-Planck equation. From this equation we see that the variance of ξ is proportional to ε and that the other cumulants are of order higher than ε. Therefore, for ε small enough, $\rho(\xi,t)$ behaves as a Gaussian with variance proportional to ε. From the Fokker-Planck equation we get immediately Eqs. (10.19) and (10.20) by integration by parts and eventually Eqs. (10.21) and (10.23) by using the properties we have just mentioned.

It worth to notice yet that the Fokker-Planck equation has as the associate Langevin equation the following equation

$$\frac{d\xi}{dt} = \lambda(\xi) - \mu(\xi) + \{\varepsilon[\lambda(\xi) + \mu(\xi)]\}^{1/2}\zeta(t), \tag{10.33}$$

where $\zeta(t)$ has the properties $\langle\zeta(t)\rangle = 0$ and $\langle\zeta(t)\zeta(t')\rangle = \delta(t-t)$.

10.2   Reactions with One Degree of Freedom

Next, we apply the results above to a system consisting of one or two chemical reactions involving a certain number of chemical species. We assume that the system is large enough in such a way that it can be characterized only by the quantities $\bar{\xi}$ and χ whose evolution equations are given by (10.21) and (10.23). From $\bar{\xi}$, we obtain the concentrations $\bar{x}_i$ by means of (10.24). To simplify notation, from here on we write ξ in the place of $\bar{\xi}$ and x_i in the place of $\bar{x}_i$. Thus the equation for the evolution of the extension of reaction are given by

$$\frac{d}{dt}\xi = \lambda - \mu, \tag{10.34}$$

where

$$\lambda = k^+ \prod_i x_i^{v_i'}, \qquad \mu = k^- \prod_i x_i^{v_i''}, \qquad (10.35)$$

and x_i is related to ξ by

$$x_i = a_i + v_i \xi. \qquad (10.36)$$

It is possible to determine x_i directly from their evolution equations. They are obtained from (10.34) using (10.36) and are given by

$$\frac{d}{dt} x_i = v_i (\lambda - \mu), \qquad (10.37)$$

$i = 1, 2, \ldots, q$. Notice however that these q equations are not independent because the concentrations x_i are not independent but are connected by (10.36).

If the constants of reaction k^+ and k^- are both nonzero, then, in the stationary state $\lambda(\xi) = \mu(\xi)$, equation that determines ξ in the stationary state, and the variance χ is given by (10.29), that is,

$$\frac{1}{\chi} = \sum_i \frac{v_i^2}{x_i}. \qquad (10.38)$$

In the following, we will solve Eq. (10.34) or Eqs. (10.37) for reactive systems consisting of two or more chemical species. When convenient, we will use the notation x, y and z in the place of x_1, x_2 and x_3.

Example 1

We start by examining the chemical reactions represented by the chemical equation

$$A \rightleftharpoons B, \qquad (10.39)$$

with forward and backward constants of reaction k_1 and k_2. We denote by x and y, the concentrations of the chemical species A and B, respectively, and by ξ the extension of reaction. As initial condition we set $x = 0$ and $y = a$ so that, according to (10.36), $x = -\xi$, $y = a + \xi$, from which we get the constraint $x + y = a$. The forward and backward rates are $\lambda = k_1 x$ and $\mu = k_2 y$. The evolution equation of x, according to (10.37), is given by

$$\frac{dx}{dt} = -k_1 x + k_2 (a - x) = k_2 a - kx, \qquad (10.40)$$

where $k = k_1 + k_2$, whose solution with initial condition $x = 0$ is

$$x = \frac{ak_2}{k} (1 - e^{-kt}), \qquad (10.41)$$

from which we get

$$y = \frac{a}{k}(k_1 + k_2 e^{-kt}).$$ (10.42)

When $t \to \infty$, we get the equilibrium concentrations

$$x = \frac{ak_2}{k}, \qquad\qquad y = \frac{ak_1}{k}.$$ (10.43)

Therefore, in equilibrium, $y/x = k_1/k_2$. Still in equilibrium, the variance χ is obtained from (10.38) and is

$$\chi = a\frac{k_1 k_2}{k^2}.$$ (10.44)

Example 2

Now, we examine the reaction

$$2A \rightleftharpoons B,$$ (10.45)

with forward and backward reaction constants k_1 and k_2. We denote by x and y the concentrations of the species A and B, respectively, and by ξ the extension of reaction. As initial condition we set $x = 0$ and $y = a$ so that $x = -2\xi$ and $y = a + \xi$, from which we get the constraint $x + 2y = a$. The forward and backward rates are $\lambda = k_1 x^2$ and $\mu = k_2 y$. The time evolution of x, according to (10.37), is given by

$$\frac{dx}{dt} = -2k_1 x^2 + 2k_2(a - 2x),$$ (10.46)

which can be written in the form

$$\frac{dx}{dt} = -2k_1(x - c_0)(x - c_1),$$ (10.47)

where c_0 and c_1 are the roots of the right-hand side of (10.46), given by

$$c_0 = \frac{1}{k_1}\left(-k_2 + \sqrt{k_2^2 + k_1 k_2 a}\right),$$ (10.48)

$$c_1 = \frac{1}{k_1}\left(-k_2 - \sqrt{k_2^2 + k_1 k_2 a}\right).$$ (10.49)

Equation (10.47) can then be written in the equivalent form

$$-\alpha \frac{dt}{dx} = \frac{1}{x - c_0} - \frac{1}{x - c_1}, \tag{10.50}$$

where $\alpha = 2k_1(c_0 - c_1)$. Integrating, we get

$$-\alpha t = \ln \frac{(x - c_0)c_1}{(x - c_1)c_0}, \tag{10.51}$$

where the integration constant was chosen according to the initial condition, $x = 0$ at $t = 0$, from which we reach the result

$$x = c_0 c_1 \frac{1 - e^{-\alpha t}}{c_1 - c_0 e^{-\alpha t}}. \tag{10.52}$$

When $t \to \infty$, $y \to c_0$ since $\alpha > 0$. The decay to the stationary value is exponential. The concentrations of the two species are

$$x = a - 2c_0, \qquad\qquad y = c_0, \tag{10.53}$$

which are the equilibrium result and are in accordance with the equilibrium condition, which is $y/x^2 = k_1/k_2$. In equilibrium, the variance χ is obtained from (10.38) and is

$$\chi = \frac{c_0(a - 2c_0)}{3c_0 + a}. \tag{10.54}$$

Example 3

We examine now the reaction

$$A + B \rightleftharpoons C. \tag{10.55}$$

The rate of forward and backward reaction are k_1 and k_2. We denote by x, y and z the concentrations of the species A, B and C, respectively, and the extension of reaction by ξ. As initial conditions we set $x = 0$, $y = 0$ and $z = a$ so that $x = y = -\xi$ and $z = a + \xi$. The concentrations obey therefore the restrictions $z = a - x$ and $y = x$. The rate of forward and backward reactions are given by $\lambda = k_1 xy = k_1(a - z)^2$ and $\mu = k_2 z$ and the equation for x, according to (10.37), is given by

$$\frac{dx}{dt} = -k_1 x^2 + k_2(a - x), \tag{10.56}$$

which can be written in the form

$$\frac{dx}{dt} = -k_1(x - c_0)(x - c_1), \tag{10.57}$$

where c_0 and c_1 are the roots of the right-hand side of Eq. (10.56), given by

$$c_0 = \frac{1}{2k_1}\left(-k_2 + \sqrt{k_2^2 + 4k_1k_2a}\right), \tag{10.58}$$

$$c_1 = \frac{1}{2k_1}\left(-k_2 - \sqrt{k_2^2 + 4k_1k_2a}\right). \tag{10.59}$$

Employing the same method used previously, we get the solution

$$x = c_0c_1\frac{1 - e^{-\alpha t}}{c_1 - c_0e^{-\alpha t}}, \tag{10.60}$$

where $\alpha = k_1(c_0 - c_1)$ and the integration constant was chosen according to the initial condition, $x = 0$ at $t = 0$. The decay to stationary value is exponential.

When $t \to \infty$, $x \to c_0$ because $\alpha > 0$, and the concentrations of the three species are

$$x = y = c_0, \qquad\qquad z = a - c_0, \tag{10.61}$$

which are the equilibrium values and are such that $z/xy = k_1/k_2$. In equilibrium, the variance χ is obtained from (10.38) and is

$$\chi = \frac{c_0(a - c_0)}{2a + c_0}. \tag{10.62}$$

Example 4

We examine now the reaction

$$A + B \rightleftharpoons 2C. \tag{10.63}$$

The constants of the forward and backward reactions are, respectively, k_1 and k_2. We denote by x, y and z the concentrations of species A, B and C, respectively, and the extension of reaction by ξ. As initial condition we set $x = y = 0$ and $z = a$ so that $x = y = -\xi$ and $z = a + 2\xi$. The concentrations are restricted to $z = a - 2x$ and $y = x$. The rates of forward and backward reactions are $\lambda = k_1xy$ and $\mu = k_2z^2$.

The time evolution of x, according to (10.37), is given by

$$\frac{dx}{dt} = -k_1x^2 + k_2(a - 2x)^2, \tag{10.64}$$

which we write as

$$\frac{dx}{dt} = -(k_1 - 4k_2)(x - c_0)(x - c_1), \tag{10.65}$$

where

$$c_0 = \frac{a\sqrt{k_2}}{2\sqrt{k_2} + \sqrt{k_1}}, \qquad c_1 = \frac{a\sqrt{k_2}}{2\sqrt{k_2} - \sqrt{k_1}}. \tag{10.66}$$

Using the same approach used previously, we obtain the following solution

$$x = c_0 c_1 \frac{1 - e^{-\alpha t}}{c_1 - c_0 e^{-\alpha t}}, \tag{10.67}$$

where $\alpha = (k_1 - 4k_2)(c_0 - c_1)$.

When $t \to \infty$, $x \to c_0$ because $\alpha > 0$. The decay to the stationary value is exponential. The concentrations of the three species are

$$y = a - 2c_0, \qquad x = z = c_0, \tag{10.68}$$

which are the results for equilibrium and are in accordance with $z^2/xy = k_1/k_2$. In equilibrium, the variance χ is obtained from (10.38) and is

$$\chi = \frac{c_0(a - 2c_0)}{5a - 9c_0}. \tag{10.69}$$

Example 5

We consider now an autocatalytic reaction,

$$A + B \rightleftharpoons 2A. \tag{10.70}$$

We denote by x and y the concentrations of A and B, respectively. The equation for x, according to (10.37), is given by

$$\frac{dx}{dt} = k_1 xy - k_2 x^2. \tag{10.71}$$

Taking into account that $x + y = a$ is constant, then we can write

$$\frac{dx}{dt} = -kx(x - c), \tag{10.72}$$

where $k = k_1 + k_2$ and $c = k_1 a/k$. Writing this equation in the form

$$-ck\frac{dt}{dx} = \frac{1}{x - c} - \frac{1}{x}, \tag{10.73}$$

we can integrate to get

$$x = \frac{ac}{a - (a - c)e^{-ckt}}, \tag{10.74}$$

where the integration constant was chosen so that $x = a$ at $t = 0$.

When $t \to \infty$, $x = c$ and $y = a - c$, that is,

$$x = \frac{ak_1}{k}, \qquad\qquad y = \frac{ak_2}{k}, \qquad (10.75)$$

which fulfills the condition $x/y = k_1/k_2$, and therefore the equilibrium solution is the same solution obtained for the reaction $B \rightleftharpoons A$. We conclude that the catalyst does not modify the thermochemical equilibrium. However, it can accelerate the process of approaching equilibrium. To perceive this effect it suffices to compare the relaxation times given by Eq. (10.74) with those given by Eq. (10.41). We notice in addition that the variance is the same as that corresponding to the reaction $B \rightleftharpoons A$, given by (10.44).

Example 6

Now we examine an example in which the reverse reaction is forbidden,

$$\nu A \to B, \qquad (10.76)$$

where ν can be 1, 2, ... We denote by x and y the concentrations of A and B, respectively, and by ξ the extension of reaction. As initial condition we choose $x = a$ and $y = 0$ so that, according to (10.36), $x = a - \nu\xi$ and $y = \xi$, from which we get the constraint $x + \nu y = a$. Denoting by k the reaction constant, then the rate of reaction is $\lambda = kx^{\nu}$ and the equation for x, according to (10.37), is given by

$$\frac{dx}{dt} = -\nu kx^{\nu}. \qquad (10.77)$$

Initially, we examine the reaction represented by the equation A→B, corresponding to $\nu = 1$. The reaction rate is $\lambda = kx$ and the equation for x is given by

$$\frac{dx}{dt} = -kx, \qquad (10.78)$$

whose solution is $x = ae^{-kt}$. Therefore,

$$y = a(1 - e^{-kt}). \qquad (10.79)$$

We see that the decay to the solution $x = a$ and $y = 0$ is exponential.

Next, we examine the reaction 2A→B, corresponding to $\nu = 2$. The reaction rate is $\lambda = kx^2$ and

$$\frac{dx}{dt} = -2kx^2, \qquad (10.80)$$

whose solution is

$$x = \frac{a}{2kat + 1}, \qquad (10.81)$$

valid for the initial condition $x = a$. We see that the decay becomes algebraic. For large times $x \sim t^{-1}$.

The solution of (10.77) for $\nu > 1$ is given by

$$x = a[\nu(\nu - 1)ka^{\nu-1}t + 1]^{-1/(\nu-1)}, \tag{10.82}$$

and the decay is algebraic. For large times, $x \sim t^{-1/(\nu-1)}$.

10.3   Systems with More that One Degree of Freedom

Reaction rate   The results obtained previously for reactive systems with one degree of freedom can be generalized to reactive systems with more than one degree of freedom. Generically, we suppose that a vessel contains q chemical species that react among them according to r chemical reactions. The chemical reactions are represented by the chemical equations

$$\sum_{i=1}^{q} v'_{ij} A_i \;\rightleftharpoons\; \sum_{i=1}^{q} v''_{ij} A_i, \tag{10.83}$$

$j = 1, 2, \ldots, r$, with $r \le q$, where $v'_{ij} \ge 0$ are the stoichiometric coefficients of the reactants and $v''_{ij} \ge 0$ are the stoichiometric coefficients of the products of the reaction. Equation (10.83) tell us that in the j-th forward reaction, v'_{ij} molecules of species A_i disappear and v''_{ij} molecules of species A_i are created so that the number of molecules of type A_i increase by an amount

$$v_{ij} = v''_{ij} - v'_{ij}. \tag{10.84}$$

For each chemical reaction and its reverse, described by the chemical equation (10.83), we define a extension of reaction ℓ_j such that the number of molecules n_i of the several chemical species are given by

$$n_i = N_i + \sum_{j=1}^{r} v_{ij}\ell_j, \tag{10.85}$$

where N_i are predetermined values of n_i, which we may consider as those occurring at a certain instant of time. Notice that the number of variables ℓ_j is equal to the number r of chemical reactions and the number of variables n_i is equal to the number q of chemical species. Since $r \le q$, the variables n_i are not independent but have constraints whose number is $q - r$.

The r chemical reactions taking place inside the vessel among the q chemical species are understood as a stochastic process whose state space we consider to

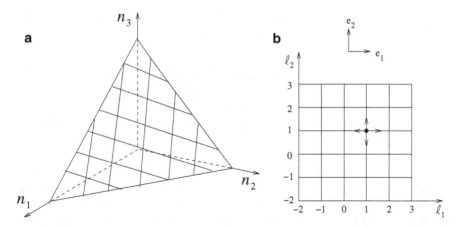

Fig. 10.1 Illustration of the possible states of a system with two chemical reactions involving three chemical species. (**a**) Space of the number of molecules n_1, n_2 and n_3 of the chemical species. The transitions occur along the straight lines belonging to the plane that cuts the three axis. (**b**) Space of the extensions of reaction ℓ_1 and ℓ_2. The *arrows* represent the possible transitions from a state represented by a *full circle*. The occurrence of the reaction j correspond to a displacement e_j and its reverse to a displacement $-e_j$

be the vector $\ell = (\ell_1, \ell_2, \ldots, \ell_r)$, as shown in Fig. 10.1. From ℓ_j we obtain n_i by means of (10.85). The stochastic process that we adopt here is defined by a transition rate consisting by various rates, each one corresponding to a chemical reaction. We assume that the transitions in space ℓ are such that only one component of ℓ changes its value, which means that the chemical reactions occur independent of each other. In other terms, the allowed changes of the vector ℓ are those parallel to the axis in space ℓ, as seen in Fig. 10.1. In analogy to the case of one degree of freedom, the allowed transitions are such that the variations in ℓ_j equals plus or minus one, that is, $\ell_j \to \ell_j \pm 1$, as seen in Fig. 10.1.

Here it is convenient to define the unit vector e_j that has all components equal to zero except the j-th which is 1. Thus we write

$$\ell = \sum_{j=1}^{r} \ell_j e_j. \tag{10.86}$$

In terms of the unit vectors e_j, the possible transitions are those such that $\ell \to \ell \pm e_j$. Denoting by a_ℓ^j the transition rate $\ell_j \to \ell_j + 1$ and by b_ℓ^j the transition rate $\ell_j \to \ell_j - 1$, then the time evolution of $P_\ell(t)$ is governed by the master equation

$$\frac{d}{dt} P_\ell = \sum_j \{ a_{\ell-e_j}^j P_{\ell-e_j} + b_{\ell+e_j}^j P_{\ell+e_j} - (a_\ell^j + b_\ell^j) P_\ell \}. \tag{10.87}$$

The rate of the forward reaction a_ℓ^j and the rate of backward reaction b_ℓ^j are adopted in a way similar to that done previously,

$$a_\ell^j = k_j^+ V \prod_i \frac{n_i!}{(n_i - v_{ij}')! V^{v_{ij}'}}, \tag{10.88}$$

$$b_\ell^j = k_j^- V \prod_i \frac{n_i!}{(n_i - v_{ij}'')! V^{v_{ij}''}}, \tag{10.89}$$

where k_j^+ and k_j^- are the constant of reaction corresponding to the j-th reaction. Notice that the rate of reactions depend on the extension of reaction because the variable n_i are related with ℓ_j according to (10.85). The constants of reaction are adopted in such a way that the detailed balance is obeyed with relation to the probability distribution (10.8), what results in the following relations

$$\frac{k_j^+}{k_j^-} = \prod_i z_i^{v_{ij}}. \tag{10.90}$$

These relations can always be fulfilled because the number of reactions is smaller or equal to the number of chemical species.

When V is large enough, the rates are given with good approximation by

$$a_\ell^j = k_j^+ V \prod_i \left(\frac{n_i}{V}\right)^{v_{ij}'}, \tag{10.91}$$

$$b_\ell^j = k_j^- V \prod_i \left(\frac{n_i}{V}\right)^{v_{ij}''}, \tag{10.92}$$

Thermodynamic limit   In the thermodynamic limit the ratio $\ell_j = \xi_j / V, n_i / V = x_i$ and $N_i / V = a_i$ are finite and, according to (10.85), are related by

$$x_i = a_i + \sum_{j=1}^{r} v_{ij} \xi_j. \tag{10.93}$$

The ratios $\lambda_j = a_n^j / V$ and $\mu_j = b_n^j / V$ are also finite in the thermodynamic limit and are given by

$$\lambda_j(\xi) = k_j^+ \prod_i x_i^{v_{ij}'}, \qquad \mu_j(\xi) = k_j^- \prod_i x_i^{v_{ij}''}, \tag{10.94}$$

where we denote by ξ the vector $\xi = (\xi_1, \xi_2 \ldots, \xi_r)$. Notice that $\lambda_j(\xi)$ and $\mu_j(\xi)$ must be considered functions of ξ in view of the relation (10.93).

Now we must consider the probability distribution $\rho(\xi, t)$ of the variable ξ. When V is large enough, the master equation is transformed in the Fokker-Planck equation

$$\frac{\partial}{\partial t}\rho(\xi, t) = -\sum_j \frac{\partial}{\partial \xi_j}[\lambda_j(\xi) - \mu_j(\xi)]\rho(\xi, t) + \frac{\varepsilon}{2}\sum_j \frac{\partial^2}{\partial \xi_j^2}[\lambda_j(\xi) + \mu_j(\xi)]\rho(\xi, t),$$
(10.95)

where $\varepsilon = 1/V$ and $\lambda_j(\xi)$ and $\mu_j(\xi)$ are given by (10.94).

It is worth to note that the Fokker-Planck equation is associated to the following Langevin equation

$$\frac{d\xi_j}{dt} = \lambda_j(\xi) - \mu_j(\xi) + \{\varepsilon[\lambda_j(\xi) + \mu_j(\xi)]\}^{1/2}\zeta_j(t),$$
(10.96)

where $\zeta_j(t)$ has the properties $\langle \zeta_j(t) \rangle = 0$ and $\langle \zeta_j(t)\zeta_{j'}(t') \rangle = \delta_{jj'}\delta(t - t')$.

From the Fokker-Planck equation we obtain the equations for $\langle \xi_j \rangle$,

$$\frac{d}{dt}\langle \xi_j \rangle = \langle \lambda_j(\xi) \rangle - \langle \mu_j(\xi) \rangle,$$
(10.97)

and for $\chi_j = V[\langle \xi_j^2 \rangle - \langle \xi_j \rangle^2]$,

$$\frac{d}{dt}\chi_j = 2V\langle (\xi_j - \langle \xi_j \rangle)(\lambda_j(\xi) - \mu_j(\xi)) \rangle + \langle \lambda_j(\xi) \rangle + \langle \mu_j(\xi) \rangle.$$
(10.98)

In the thermodynamic limit $V \to \infty$, using the notation $\bar{\xi}_j = \langle \xi_j \rangle$,

$$\frac{d}{dt}\bar{\xi}_j = \lambda_j(\bar{\xi}) - \mu_j(\bar{\xi}),$$
(10.99)

$$\frac{d}{dt}\chi_j = 2\left(\frac{\partial}{\partial \bar{\xi}_j}\lambda_j(\bar{\xi}) - \frac{\partial}{\partial \bar{\xi}_j}\mu_j(\bar{\xi})\right)\chi_j + \lambda_j(\bar{\xi}) + \mu_j(\bar{\xi}).$$
(10.100)

Notice that the averages $\bar{x}_i$ of the concentrations are obtained from $\bar{\xi}_j$ by

$$\bar{x}_i = a_i + \sum_{j=1}^{r} v_{ij}\bar{\xi}_j.$$
(10.101)

In the stationary state $\lambda_j(\bar{\xi}) = \mu_j(\bar{\xi})$ so that from (10.94),

$$\prod_i x_i^{v_{ij}} = \frac{k_j^+}{k_j^-},$$
(10.102)

where we have taken into account that $v_{ij} = v''_{ij} - v'_{ij}$. The right-hand side of this equation is the equilibrium constant related to the j-th reaction.

To determine the variance χ_j in the stationary state, we proceed as we have done in the case of one degree of freedom to obtain the following result

$$\frac{1}{\chi_j} = \sum_i \frac{v_{ij}^2}{\bar{x}_i}. \tag{10.103}$$

To determine the variance $\chi_i^* = V(\langle x_i^2 \rangle - \langle x_i \rangle^2)$, we use the relation (10.93) to reach the result

$$\chi_i^* = \sum_j v_{ij}^2 \chi_j. \tag{10.104}$$

10.4   Reactions with More than One Degree of Freedom

Example 1

We start by the set of reactions

$$A \rightleftharpoons B, \qquad\qquad B \rightleftharpoons C. \tag{10.105}$$

We denote by x, y and z the concentrations A, B, and C, respectively. The extension of the first reaction we denote by ξ and the second by η. Suppose that at the initial instant of time $x = a$ and $y = z = 0$, then according to (10.93), $x = a - \xi$, $y = \xi - \eta$ and $z = \eta$. The equation for the extension of reaction ξ and z are

$$\frac{d\xi}{dt} = k_1(a - \xi) - k_2\xi, \tag{10.106}$$

$$\frac{dz}{dt} = k_3(\xi - z) - k_4z. \tag{10.107}$$

The first equation is independent of the second and its solution for the initial condition $\xi = 0$ at $t = 0$ is given by

$$\xi = \frac{ak_1}{k}(1 - e^{-kt}), \tag{10.108}$$

where $k = k_1 + k_2$. Replacing in the second

$$\frac{dz}{dt} = \frac{ak_1k_3}{k}(1 - e^{-kt}) - k'z, \tag{10.109}$$

where $k' = k_3 + k_4$, whose solution for the initial condition $z = 0$ is

$$z = \frac{ak_1k_3}{k}e^{-k't}\int_0^t (1 - e^{-kt'})e^{k't'}dt',$$

(10.110)

or, performing the integral,

$$z = \frac{ak_1k_3}{k}\left(\frac{1 - e^{-k't}}{k'} - \frac{e^{-kt} - e^{-k't}}{k' - k}\right).$$

(10.111)

When $t \to \infty$, we get the solutions $\xi = ak_1/k$ and $z = ak_1k_3/kk'$, from which the following equilibrium concentrations follow,

$$x = \frac{ak_2}{k}, \qquad y = \frac{ak_1k_4}{kk'}, \qquad z = \frac{ak_1k_3}{kk'}.$$

(10.112)

In equilibrium, the variances χ_j are obtained from (10.103) and are

$$\chi_1 = \frac{ak_1k_2k_4}{k(k_1k_4 + k'k_2)}, \qquad \chi_2 = \frac{ak_1k_3k_4}{k(k')^2}.$$

(10.113)

Example 2

Next we examine the reactions

$$A \to B, \qquad\qquad B \to C,$$

(10.114)

which can be understood as the reactions (10.105), with $k_2 = 0$ and $k_4 = 0$, so that the solutions for ξ and z are reduced to

$$\xi = a(1 - e^{-k_1t}),$$

(10.115)

$$z = a - \frac{a}{k_1 - k_3}(k_1e^{-k_3t} - k_3e^{-k_1t}),$$

(10.116)

from which we obtain

$$x = ae^{-k_1t},$$

(10.117)

$$y = \frac{ak_1}{k_1 - k_3}(e^{-k_3t} - e^{-k_1t}).$$

(10.118)

In Fig. 10.2 we show the concentrations x, y and z as functions of time for $k_1 = 1$ and $k_3 = 0.5$ and for initial conditions $x = 1$, $y = 0$ and $z = 0$. The concentrations x and y of species A and C have monotonic behavior. The concentration of A decreases and eventually vanishes. The concentration of C increases and eventually reaches the saturation value. The concentration y of species B, on the other hand, increases at the beginning, reaches a maximum value and eventually decreases to its zero value. The maximum of y occurs at $t_{max} = \ln(k_3/k_1)/(k_3 - k_1)$.

Fig. 10.2 Concentrations x, y and z of the chemical species A, B and C as functions of time t corresponding to the reactions (10.114) with constants of reactions $k_1 = 1$ and $k_3 = 0.5$. The initial conditions are $x = 1$ and $y = z = 0$

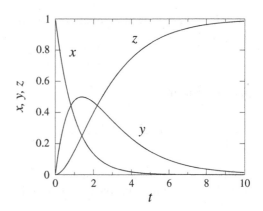

Catalytic oxidation of the carbon monoxide Here we examine the reaction of oxidation of the carbon monoxide on a catalytic surface. In the presence of the catalytic surface, the carbon monoxide and the oxygen are adsorbed and react forming the carbon dioxide, which leaves the surface. The reaction mechanism presumes that the catalytic surface is composed by sites, which can be occupied either by an oxygen atom or by a carbon monoxide molecule. The reactions are the following

$$C \rightarrow A, \qquad 2C \rightarrow 2B, \qquad A + B \rightarrow 2C. \tag{10.119}$$

In the first reaction, of constant k_1, an empty site (C), is occupied by a carbon monoxide (A) molecule. In the second reaction, of constant k_2, two empty sites are occupied by two oxygen (B) atoms, coming from the dissociation of an oxygen molecule. In the third reaction, of constant k_3, an oxygen atom reacts with a carbon monoxide molecule forming the carbon dioxide, leaving two sites empty. We assume further that the carbon monoxide can leave the catalytic surface according to the reaction

$$A \rightarrow C, \tag{10.120}$$

with constant k_4, which is identified with the reverse of the first reaction. Denoting by x, y and z the concentrations of A, B and C, respectively, then

$$\frac{dx}{dt} = k_1 z - k_3 xy - k_4 x, \tag{10.121}$$

$$\frac{dy}{dt} = 2k_2 z^2 - k_3 xy, \tag{10.122}$$

$$\frac{dz}{dt} = -k_1 z - 2k_2 z^2 + 2k_3 xy + k_4 x. \tag{10.123}$$

These equations are not independent because $x + y + z = a$, a constant.

Next we analyze the stationary state, given by the equations

$$k_1(a - x - y) = k_3xy + k_4x, \qquad (10.124)$$

$$2k_2(a - x - y)^2 = k_3xy, \qquad (10.125)$$

and treat the case in which there is no desorption, $k_4 = 0$. In this case, a solution is $x = a$, $y = 0$ and $z = 0$. The other is $z = r/2$,

$$x = \frac{1}{2}\left(a - \frac{r}{2} + \sqrt{(a - \frac{r}{2})^2 - \frac{2k_2r^2}{k_3}} \right), \qquad (10.126)$$

$$y = \frac{1}{2}\left(a - \frac{r}{2} - \sqrt{(a - \frac{r}{2})^2 - \frac{2k_2r^2}{k_3}} \right), \qquad (10.127)$$

where $r = k_1/k_2$, valid for $r < r_0$, where

$$r_0 = a \left(\sqrt{2k_2/k_3} + \frac{1}{2} \right)^{-1}. \qquad (10.128)$$

Figure 10.3 shows x and y as functions of k_1/k_2. As we see the concentrations has a jump at $r = r_0$. Above r_0, the concentration $x = a$ and $y = 0$, what means that the catalytic surface is saturated by the carbon monoxide.

When $k_4 \neq 0$, the stationary solutions for x and y are obtained by solving numerically the evolution equations starting from the initial condition such that $z = 0$, y is very small and $x = a - y$. The results are shown in Fig. 10.3. For

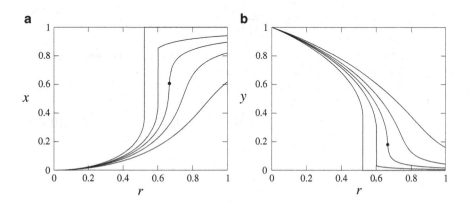

Fig. 10.3 Concentrations x, y of the chemical species A, B as functions of $r = k_1/k_2$ corresponding to the reactions (10.119) of the catalytic oxidation of the carbon monoxide for $k_3/k_2 = 1$, $a = 1$ and for the following values of k_4/k_2, from left to right: 0, 0.05, 0.08, 0.12 and 0.2. The *full circle* indicates the critical point

small values of k_4/k_2, the concentrations have a jump at a certain value of r, characterizing a discontinuous phase transition. At a certain value of k_4/k_2 this jump vanishes characterizing the appearance of a critical point. Above this value of k_4/k_2 there is no phase transition.

10.5   Open Systems

The contact with a reservoir of particles of type A_i may be understood as the occurrence of the reaction $O \rightleftharpoons A_i$, where O represents the absence of particles. According to this idea, the rate of addition of a particle, $n_i \rightarrow n_i + 1$, is adopted as $kz_i V$ and of removal, $n_i \rightarrow n_i - 1$, as kn_i. The transition rates per unit volume corresponding to the addition and removal of particles are given by

$$\alpha_i = kz_i, \tag{10.129}$$

$$\beta_i = kx_i, \tag{10.130}$$

where $x_i = n_i/V$. The average flux of particles of species i, per unit volume, is thus

$$\phi_i = k(z_i - \bar{x}_i). \tag{10.131}$$

Here we consider a system that consists of a vessel where a set of reaction take place and that particles are exchange with the environment. Thus, the variation in the number of particles is not only due to the chemical reaction but also to the flux from the environment. The equation for the time evolution of x_i is given by

$$\frac{dx_i}{dt} = \sum_{j=1}^{r} v_{ij} R_j + \phi_i, \tag{10.132}$$

$i = 1, 2, \ldots, q$, where $R_j = \lambda^j - \mu^j$ and ϕ_i is the flux per unit volume of particles of type i from the environment to inside the vessel.
   The entropy flux Φ is given by

$$\Phi = \sum_{j} R_j \ln \frac{\lambda_j}{\mu_j} + \sum_{i} \phi_i \ln \frac{z_i}{\bar{x}_i}. \tag{10.133}$$

In the stationary state, x_i is constant so that

$$\phi_i = -\sum_{j=1}^{r} v_{ij} R_j, \tag{10.134}$$

valid for $i = 1, 2, \ldots, q$.

When the number of chemical species is larger than the number of reactions, $q > r$, the fluxes cannot be arbitrary but must hold a relation between them. Let a_{ij}, $i, j = 1, 2, \ldots, r$, the elements of a matrix which is the inverse of the matrix whose entries are the stoichiometric coefficients ν_{ij}, $i, j = 1, 2, \ldots, r$, that is,

$$\sum_{i=1}^{r} a_{ki}\nu_{ij} = \delta_{kj}. \tag{10.135}$$

Using this relation we get

$$R_j = -\sum_{i=1}^{r} a_{ji}\phi_i. \tag{10.136}$$

valid only for $j = 1, 2, \ldots, r$, which replaced in (10.134) gives $q - r$ constraints among the fluxes, given by

$$\phi_i = \sum_{j=1}^{r}\sum_{k=1}^{r} \nu_{ij}a_{jk}\phi_k, \tag{10.137}$$

valid for $i = r + 1, \ldots, q$.

Example 1

We start by examining the reactions

$$A \rightleftharpoons B. \tag{10.138}$$

Denoting by x and y the concentrations of A and B, respectively, the evolution equations are

$$\frac{dx}{dt} = -k_1 x + k_2 y + \phi_x, \tag{10.139}$$

$$\frac{dy}{dt} = k_1 x - k_2 y + \phi_y, \tag{10.140}$$

where

$$\phi_x = \alpha(z_1 - x), \qquad \phi_y = \beta(z_2 - y). \tag{10.141}$$

In the stationary state $\phi_x = -\phi_y = k_1 x - k_2 y$, from which we obtain the solution

$$x = \frac{\beta k_2 z_2 + \alpha k_2 z_1 - \alpha\beta z_1}{\beta k_1 + \alpha k_2 - \alpha\beta}, \tag{10.142}$$

$$y = \frac{\alpha k_1 z_1 + \beta k_1 z_2 - \alpha\beta z_2}{\beta k_1 + \alpha k_2 - \alpha\beta}. \tag{10.143}$$

The fluxes ϕ_x and ϕ_y are given by

$$\phi_x = -\phi_y = \alpha\beta \frac{k_1 z_1 - k_2 z_2}{\beta k_1 + \alpha k_2 - \alpha\beta}. \tag{10.144}$$

The entropy flux, which is identified with the entropy production rate,

$$\Phi = R \ln \frac{k_1 x}{k_2 y} + \phi_x \ln \frac{z_1}{x} + \phi_y \ln \frac{z_2}{y}, \tag{10.145}$$

where $R = k_1 x - k_2 y$, is

$$\Phi = \frac{\alpha\beta(k_1 z_1 - k_2 z_2)}{\beta k_1 + \alpha k_2 - \alpha\beta} \ln \frac{z_1 k_1}{z_2 k_2}. \tag{10.146}$$

In equilibrium the fluxes vanish so that $z_1 k_1 = z_2 k_2$, which is identified with the equilibrium condition (10.13) and the entropy flux vanishes. In this case, $x = z_1$ and $y = z_2$.

Lindemann mechanism The reactions

$$2A \rightleftharpoons A + B, \qquad\qquad B \to C, \tag{10.147}$$

correspond to the transformation of molecules of type A in molecules of type C by means of an intermediate process in which A is transformed into B, which is interpreted as an activated A molecule, which in turn is transformed into C. This transformation of A in C is called Lindemann mechanism.

We denote by x, y and z the concentrations of A, B, and C, respectively. We assume the existence of flux of particles of type A and C so that the equations for the concentrations are

$$\frac{dx}{dt} = -k_1 x^2 + k_2 xy + \phi_x, \tag{10.148}$$

$$\frac{dy}{dt} = k_1 x^2 - k_2 xy - k_3 y, \tag{10.149}$$

$$\frac{dz}{dt} = k_3 y + \phi_z. \tag{10.150}$$

In the stationary state

$$\phi_x = -\phi_z = k_3 y, \qquad\qquad y = \frac{k_1 x^2}{k_2 x + k_3}. \tag{10.151}$$

Eliminating y, we reach the following relation between the flux ϕ_x and the concentration x,

$$\phi_x = kx, \qquad k = \frac{k_1 k_3 x}{k_2 x + k_3}. \tag{10.152}$$

If $k_2 x \gg k_3$, then $k = k_1 k_3 / k_2$ and ϕ_x is proportional to x. If $k_2 x \ll k_3$, then $k = k_1 x$ and ϕ_x is proportional to x^2.

Michaelis-Menten mechanism   In the reactions

$$A + B \rightleftharpoons C, \qquad C \rightarrow B + D, \tag{10.153}$$

molecules of type A are transformed in D by means of the Michaelis-Menten mechanism, in which A is transformed into a complex C by means of a catalyst (enzyme) B. The complex can be broken, which corresponds to the backward of the first reaction, or yield the product D, which is the second reaction. We denote by x, y, z and u the concentrations of species A, B, C and D, and we assume the existence of flux of particles of type A and D. The equations for the concentrations are therefore

$$\frac{dx}{dt} = -k_1 xy + k_2 z + \phi_x, \tag{10.154}$$

$$\frac{dy}{dt} = -k_1 xy + k_2 z + k_3 z, \tag{10.155}$$

$$\frac{dz}{dt} = k_1 xy - k_2 z - k_3 z, \tag{10.156}$$

$$\frac{du}{dt} = k_3 z + \phi_u. \tag{10.157}$$

Notice that $y + z = y_0$ is a constant.

In the stationary state,

$$\phi_x = -\phi_u = k_3 z, \qquad z = \frac{xy}{K_m}, \tag{10.158}$$

where K_m is the Michaelis constant, given by

$$K_m = \frac{k_2 + k_3}{k_1}. \tag{10.159}$$

Since $y = y_0 - z$, then, using (10.158), we get the following relation between z and x,

$$z = \frac{y_0 x}{K_m + x},$$

(10.160)

and therefore

$$\phi_x = \frac{k_3 y_0 x}{K_m + x},$$

(10.161)

which is the Michaelis-Menten equation. The saturation flux (obtained when $x \to \infty$) is $\phi_m = k_3 y_0$ so that the Michaelis-Menten equation is written in the equivalent form

$$\frac{\phi_x}{\phi_m} = \frac{x}{K_m + x}.$$

(10.162)

Schlögl first model The Schlögl first model is defined by the set of reactions

$$A + B \rightleftharpoons 2A, \qquad A \rightleftharpoons C,$$

(10.163)

with the reaction rates k_1, k_2, k_3 and k_4, respectively. The first is a catalytic reaction and the second is a spontaneous annihilation and spontaneous creation. We denote by x, y and z the concentrations of the species A, B and C, respectively. The equation for x is given by

$$\frac{dx}{dt} = k_1 xy - k_2 x^2 - k_3 x + k_4 z,$$

(10.164)

$$\frac{dy}{dt} = -k_1 xy + k_2 x^2 + \phi_y,$$

(10.165)

$$\frac{dz}{dt} = k_3 x - k_4 z + \phi_z.$$

(10.166)

In the stationary state,

$$\phi_x = -\phi_z = k_1 xy - k_2 x^2 = k_3 x - k_4 z.$$

(10.167)

Denoting by a and b the solutions of these equations for y and z, respectively, then the solution for x is the root of

$$k_2 x^2 - (k_1 a - k_3)x - k_4 b = 0,$$

(10.168)

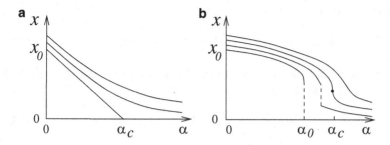

Fig. 10.4 Concentration x of chemical species A as a function of $\alpha = k_3/k_2$ corresponding to the Schlögl first (**a**) and second (**b**) model. In both cases, $x_0 = k_1 a/k_2$. The *full circle* represents the critical point

and is given by

$$x = \frac{1}{2k_2} \left(k_1 a - k_3 + \sqrt{(k_1 a - k_3)^2 + 4k_2 k_4} \right). \tag{10.169}$$

In Fig. 10.4, we show x as a function of $\alpha = k_3/k_2$ for several values of k_4. When $k_4 = 0$, the solution is $x = 0$ or

$$x = \frac{k_1 a - k_3}{k_2}, \tag{10.170}$$

as shown in Fig. 10.4. Therefore, there is a phase transition, that occurs at $k_3 = k_1 a$, or at $\alpha = \alpha_c = k_1 a/k_2$, between a state characterized by $x \neq 0$ to a state characterized by $x = 0$. Notice that the transition is continuous because x vanishes continuously when k_3 approaches $k_1 a$.

Schlögl second model   The Schlögl second model consists of the equations

$$2A + B \rightleftharpoons 3A, \qquad A \rightleftharpoons C. \tag{10.171}$$

Using the same notation as in the case of the Schlögl first model, now we have the following equations for the concentrations

$$\frac{dx}{dt} = k_1 x^2 y - k_2 x^3 - k_3 x + k_4 z, \tag{10.172}$$

$$\frac{dy}{dt} = -k_1 x^2 y + k_2 x^3 + \phi_y, \tag{10.173}$$

$$\frac{dz}{dt} = k_3 x - k_4 z + \phi_z. \tag{10.174}$$

In the stationary state,

$$\phi_x = -\phi_z = k_1 x^2 y - k_2 x^3 = k_3 x - k_4 z. \tag{10.175}$$

Denoting by a and b the solutions of these equations for y and z, respectively, then the solution for x is the root of

$$k_2 x^3 - k_1 x^2 a + k_3 x - k_4 b = 0, \tag{10.176}$$

whose solution is shown in Fig. 10.4 as a function of $\alpha = k_3/k_2$ for several values of k_4.

When $k_4 = 0$, one solution is $x = 0$. The other solution, that occurs for $k_3 > k_3^* = (k_1 a)^2/4k_2$, is characterized by $x \neq 0$. Therefore, there is a phase transition that occurs at $k_3 = k_3^*$. When k_3 approaches k_3^*, x has a jump, characterized by a discontinuous phase transition. The transition occurs at $\alpha = \alpha_0 = (k_1 a/2k_2)^2$. When $k_4 \neq 0$ the concentration x has a jump as shown in Fig. 10.4 for small values of k_4. When $k_4 = (k_1 a)^3/27k_2^2$ this jump vanishes characterizing a continuous phase transition that occurs at $k_3 = (k_1 a)^2/3k_2$ or $\alpha_c = (k_1 a/k_2)^2/3$.

Chapter 11
Glauber Model

11.1 Introduction

In this chapter we consider systems with stochastic dynamics, governed by master equations and defined on a lattice. The system to be studied comprises a lattice with N sites. To each site one associates a stochastic variable that takes discrete values. Here we consider the simplest example in which the variables take only two values: $+1$ or -1. We denote by σ_i the variable associated to the site i, so that $\sigma_i = \pm 1$. The total number of configurations or states of the system is 2^N. A possible state of the system is a vector

$$\sigma = (\sigma_1, \sigma_2, \ldots, \sigma_i, \ldots, \sigma_N), \tag{11.1}$$

whose i-th component is the state σ_i of site i.

The time evolution of the system is governed by the master equation

$$\frac{d}{dt} P(\sigma, t) = \sum_{\sigma'(\neq\sigma)} \{W(\sigma, \sigma') P(\sigma', t) - W(\sigma', \sigma) P(\sigma, t)\}, \tag{11.2}$$

where $P(\sigma, t)$ is the probability of occurrence of configuration σ at time t, and $W(\sigma, \sigma')$ is the transition rate from σ' to σ. We can say also that $W(\sigma, \sigma')\Delta t$ is the conditional probability of transition from σ' to σ in a small time interval Δt.

Similarly to what we have seen previously, we may here also define the $2^N \times 2^N$ matrix W, whose nondiagonal entries are $W(\sigma, \sigma')$ and the diagonal entries $W(\sigma, \sigma)$ are defined by

$$W(\sigma, \sigma) = -\sum_{\sigma'(\neq\sigma)} W(\sigma, \sigma'). \tag{11.3}$$

© Springer International Publishing Switzerland 2015
T. Tomé, M.J. de Oliveira, *Stochastic Dynamics*
and Irreversibility, Graduate Texts in Physics,
DOI 10.1007/978-3-319-11770-6_11

Thus the master equation can also be written in the form

$$\frac{d}{dt}P(\sigma,t) = \sum_{\sigma'} W(\sigma,\sigma')P(\sigma',t), \tag{11.4}$$

in which the sum is unrestricted, or yet, in the matrix form

$$\frac{d}{dt}P(t) = WP(t), \tag{11.5}$$

where $P(t)$ is the column matrix whose elements are $P(\sigma,t)$. The solution of this equation is

$$P(t) = e^{tW}P(0). \tag{11.6}$$

In this and in the next chapter we will treat only the case in which the transitions occur between configurations that differ only by the state of one site. Accordingly, we write

$$W(\sigma',\sigma) = \sum_{i} \delta(\sigma_1',\sigma_1)\delta(\sigma_2',\sigma_2)\ldots\delta(\sigma_i',-\sigma_i)\ldots\delta(\sigma_N',\sigma_N)w_i(\sigma), \tag{11.7}$$

where $\delta(\sigma_j,\sigma_j')$ is the Kronecker delta, that takes the value 1 or 0 according to whether σ_j and σ_j' are equal or different, respectively. The factor $w_i(\sigma)$ is interpreted as the rate of changing the sign of the state of the i-th site, that is the rate of σ_i to $-\sigma_i$. Thus the master equation reads

$$\frac{d}{dt}P(\sigma,t) = \sum_{i=1}^{N} \{w_i(\sigma^i)P(\sigma^i,t) - w_i(\sigma)P(\sigma,t)\}, \tag{11.8}$$

where the configuration σ^i is that configuration obtained from configuration σ by changing σ_i to $-\sigma_i$, that is, it is defined by

$$\sigma^i = (\sigma_1,\sigma_2,\ldots,-\sigma_i,\ldots,\sigma_N). \tag{11.9}$$

The time evolution of the average $\langle f(\sigma)\rangle$ of a state function $f(\sigma)$, defined by

$$\langle f(\sigma)\rangle = \sum_{\sigma} f(\sigma)P(\sigma,t), \tag{11.10}$$

is obtained by multiplying both sides of (11.7) by $f(\sigma)$ and summing over σ,

$$\frac{d}{dt}\langle f(\sigma)\rangle = \sum_{i=1}^{N} \langle\{f(\sigma^i) - f(\sigma)\}w_i(\sigma)\rangle. \tag{11.11}$$

In particular, the time evolution of the average $\langle \sigma_j \rangle$ is

$$\frac{d}{dt}\langle \sigma_j \rangle = -2\langle \sigma_j w_j(\sigma) \rangle, \tag{11.12}$$

and the two point correlation $\langle \sigma_j \sigma_k \rangle$, $j \neq k$, is

$$\frac{d}{dt}\langle \sigma_j \sigma_k \rangle = -2\langle \sigma_j \sigma_k w_j(\sigma) \rangle - 2\langle \sigma_j \sigma_k w_k(\sigma) \rangle. \tag{11.13}$$

Other correlations can be obtained similarly.

A quantity particularly important in the characterization of the system we are studying is the variance χ, defined by

$$\chi = \frac{1}{N}\{\langle \mathcal{M}^2 \rangle - \langle \mathcal{M} \rangle^2\}, \tag{11.14}$$

where $\mathcal{M} = \sum_i \sigma_i$. When $\langle \mathcal{M} \rangle = 0$ and assuming that the correlations are invariant by translation, it can be written as

$$\chi = \sum_i \langle \sigma_0 \sigma_i \rangle. \tag{11.15}$$

11.2   One-Dimensional Glauber Model

We begin by studying the one-dimensional model introduced originally by Glauber and defined by the inversion rate

$$w_\ell(\sigma) = \frac{\alpha}{2}\{1 - \frac{1}{2}\gamma\sigma_\ell(\sigma_{\ell-1} + \sigma_{\ell+1})\}, \tag{11.16}$$

with periodic boundary condition, that is, $\sigma_{N+\ell} = \sigma_\ell$, where N is the number of sites in the linear chain. The Glauber rate has two parameters, α and γ. The first defines the time scale. The second will be considered later.

The Glauber rate fulfills the detailed balance

$$\frac{w_\ell(\sigma^\ell)}{w_\ell(\sigma)} = \frac{P(\sigma)}{P(\sigma^\ell)}, \tag{11.17}$$

for the stationary distribution

$$P(\sigma) = \frac{1}{Z}e^{K\sum_\ell \sigma_\ell \sigma_{\ell+1}}, \tag{11.18}$$

provided $\gamma = \tanh 2K$. Indeed, replacing $P(\sigma)$ and $w_\ell(\sigma)$ in the detailed balance condition, we see that it is obeyed when $\sigma_{\ell+1} = -\sigma_{\ell-1}$. It suffices to check the cases in which $\sigma_{\ell+1} = \sigma_{\ell-1} = \pm\sigma_\ell$. In both cases we get the relation

$$\frac{1+\gamma}{1-\gamma} = e^{4K}, \tag{11.19}$$

which is equivalent to $\gamma = \tanh 2K$.

The stationary distribution $P(\sigma)$ is the Boltzmann-Gibbs distribution corresponding to the one-dimensional Ising model with interaction between neighboring sites, whose energy is given by

$$E(\sigma) = -J \sum_\ell \sigma_\ell \sigma_{\ell+1}, \tag{11.20}$$

where J is a parameter, and that is found in equilibrium at a temperature T such that $K = J/kT$, where k is the Boltzmann constant. Therefore, the Glauber dynamics may be interpreted as a dynamic for the one-dimensional Ising model.

Using Eq. (11.12), we write the time evolution of the magnetization $m_\ell = \langle \sigma_\ell \rangle$ as

$$\frac{d}{dt}m_\ell = -\alpha m_\ell + \frac{\alpha}{2}\gamma(m_{\ell-1} + m_{\ell+1}). \tag{11.21}$$

Assuming that at $t = 0$ the magnetization is the same for any site, that is, $m_\ell = m^0$, then the solution is homogeneous, $m_\ell = m$, and

$$\frac{d}{dt}m = -\alpha(1-\gamma)m, \tag{11.22}$$

whose solution is

$$m = m^0 e^{-\alpha(1-\gamma)t}. \tag{11.23}$$

We see thus that the magnetization m vanishes when $t \to \infty$, as long as $\gamma \neq 1$ ($T \neq 0$). The relaxation time τ is given by $\tau^{-1} = \alpha(1-\gamma)$ and diverges when $\gamma \to 1$ ($T \to 0$) according to

$$\tau \sim (1-\gamma)^{-1}. \tag{11.24}$$

The solution of Eq. (11.21) for a general condition is obtained as follows. We define

$$g(k) = \sum_\ell e^{ik\ell} m_\ell, \tag{11.25}$$

from which one obtains m_ℓ by means of the Fourier transform

$$m_\ell = \frac{1}{2\pi} \int_{-\pi}^{\pi} g(k)e^{-ik\ell}dk. \tag{11.26}$$

Deriving $g(k)$ with respect to time and using (11.21),

$$\frac{d}{dt}g(k) = -\alpha(1 - \gamma \cos k)g(k), \tag{11.27}$$

whose solution is

$$g(k) = g_0(k)e^{-\alpha(1-\gamma \cos k)t}, \tag{11.28}$$

where $g_0(k)$ must be determined by the initial conditions. Considering the generic initial condition m_ℓ^0, then

$$g_0(k) = \sum_\ell e^{ik\ell} m_\ell^0. \tag{11.29}$$

Replacing $g(k)$ into the expression for m_ℓ, we get

$$m_\ell = \sum_{\ell'} A_{\ell,\ell'} m_{\ell'}^0, \tag{11.30}$$

where

$$A_{\ell,\ell'} = \frac{1}{2\pi} \int_{-\pi}^{\pi} e^{-\alpha(1-\gamma \cos k)t} \cos k(\ell - \ell')dk, \tag{11.31}$$

an expression which can be written in the form

$$A_{\ell,\ell'} = e^{-\alpha t} I_{|\ell-\ell'|}(\alpha \gamma t), \tag{11.32}$$

where $I_\ell(z)$ is the modified Bessel function of order ℓ, defined by

$$I_\ell(z) = \frac{1}{2\pi} \int_{-\pi}^{\pi} e^{z \cos k} \cos k\ell \, dk. \tag{11.33}$$

For large times, that is, for large values of the argument, the Bessel function behaves as $I_\ell(z) = e^z/\sqrt{2\pi z}$. Therefore, if $\gamma \neq 1$, the quantity $A_{\ell,\ell'}$ vanishes exponentially in the limit $t \to \infty$ and the same occurs with m_ℓ for any initial condition.

Static pair correlation Using Eq. (11.13), we may write the evolution equation for the correlation $\phi_\ell = \langle \sigma_0 \sigma_\ell \rangle$ between the site 0 and the site ℓ. Assuming that the correlations have translational invariance, $\langle \sigma_{\ell'} \sigma_{\ell'+\ell} \rangle = \langle \sigma_0 \sigma_\ell \rangle$, which is possible if the initial conditions have this property, then

$$\frac{d}{dt}\phi_\ell = -2\alpha\phi_\ell + \alpha\gamma(\phi_{\ell-1} + \phi_{\ell+1}). \tag{11.34}$$

which is valid for $\ell = \pm 1, \pm 2, \pm 3, \ldots$ with the condition that $\phi_0 = 1$ at any time.

In the stationary state the correlation ϕ_ℓ obeys the equation

$$- 2\phi_\ell + \gamma(\phi_{\ell-1} + \phi_{\ell+1}) = 0, \tag{11.35}$$

whose solution is $\phi_\ell = \Lambda^{|\ell|}$ with

$$\Lambda = \frac{1}{\gamma}\{1 - \sqrt{1 - \gamma^2}\} = \tanh K. \tag{11.36}$$

Notice that the correlation decays exponentially with distance. Indeed, writing

$$\phi_\ell = e^{-|\ell|/\xi}, \tag{11.37}$$

we see that $\xi = 1/|\ln \Lambda|$, so that, for values of γ close to 1, we get

$$\xi \sim (1 - \gamma)^{-1/2}. \tag{11.38}$$

The variance χ is obtained from (11.15), that is,

$$\chi = \sum_\ell \phi_\ell = 1 + 2\sum_{\ell=1}^{\infty} \Lambda^\ell = \frac{1 + \Lambda}{1 - \Lambda} = \sqrt{\frac{1 + \gamma}{1 - \gamma}}, \tag{11.39}$$

from which we conclude that χ behaves near $\gamma = 1$ as

$$\chi \sim (1 - \gamma)^{-1/2}. \tag{11.40}$$

Replacing the result $\Lambda = \tanh K$ in the expression for the variance, we conclude that $\chi = e^{2K}$, which is the result one obtains directly from the probability distribution of the one-dimensional Ising model.

Time pair correlation Next we obtain the general solution of (11.34), that is, the time-dependent solution. However, instead of looking for solution to this equation, we solve the equation for $\phi'_\ell = d\phi_\ell/dt$, given by

$$\frac{d}{dt}\phi'_\ell = -2\alpha\phi'_\ell + \gamma\alpha(\phi'_{\ell-1} + \phi'_{\ell+1}), \tag{11.41}$$

valid for $\ell = 1, 2, 3, \ldots$, with the condition $\phi'_0 = 0$ for any time. Defining

$$a_k = \sum_{\ell=1}^{\infty} \phi'_\ell \sin k\ell, \tag{11.42}$$

then y_ℓ is obtained from a_k by the use of the following Fourier transform

$$\phi'_\ell = \frac{2}{\pi} \int_0^{\pi} a_k \sin k\ell \, dk, \tag{11.43}$$

and we see that $\phi'_0 = 0$ is satisfied. The coefficients a_k depend on time and fulfills the equation

$$\frac{d}{dt} a_k = -\lambda_k a_k, \tag{11.44}$$

where

$$\lambda_k = 2\alpha(1 - \gamma \cos k), \tag{11.45}$$

which is obtained by deriving (11.42) and using (11.41). The solution is

$$a_k(t) = b_k e^{-\lambda_k t}. \tag{11.46}$$

The coefficients b_k must be determined by the initial conditions. Here we consider an uncorrelated initial condition, that is, such that $\phi_\ell(0) = 0$. Equation (11.34) gives us $\phi'_\ell = -2\alpha\phi_\ell + \alpha\gamma(\phi_{\ell-1} + \phi_{\ell+1})$, so that $\phi'_\ell(0) = 0$, except $\phi'_1(0) = \alpha\gamma$ since $\phi_0 = 1$. Replacing in (11.42), and bearing in mind that b_k is the value of a_k at $t = 0$, we see that $b_k = \alpha\gamma \sin k$ and therefore

$$a_k(t) = \alpha\gamma \sin k \, e^{-\lambda_k t}, \tag{11.47}$$

which replaced in (11.41) yields

$$\phi'_\ell = \frac{2\alpha\gamma}{\pi} \int_0^{\pi} e^{-\lambda_k t} \sin k \sin k\ell \, dk, \tag{11.48}$$

The correlation ϕ_ℓ is obtained by integrating this equation in time.

The time derivative of the variance is given by

$$\frac{d}{dt} \chi = 2 \sum_{\ell=1}^{\infty} \phi'_\ell, \tag{11.49}$$

from which we get

$$\frac{d}{dt}\chi = \frac{2\alpha\gamma}{\pi} \int_0^\pi (1 + \cos k)e^{-\lambda_k t}dk. \tag{11.50}$$

The variance χ is obtained by integrating this expression in time.

To get the asymptotic behavior of χ for large times, if suffices to bear in mind that, in this regime, the integrand will be dominated by small values of k, so that we may replace λ_k by $2\alpha(1 - \gamma) + \alpha\gamma k^2$ and extend the integral over the whole axis to get

$$\frac{d}{dt}\chi = \frac{2\alpha\gamma}{\pi}e^{-2\alpha(1-\gamma)t} \int_{-\infty}^{\infty} e^{-\alpha\gamma k^2 t}dk = \sqrt{\frac{4\alpha\gamma}{\pi}}e^{-2\alpha(1-\gamma)t}t^{-1/2}. \tag{11.51}$$

We should now consider two cases: (a) $\gamma \neq 1$ and (b) $\gamma = 1$. In the first, we see that $d\chi/dt$ decays exponentially. It is clear that χ also approaches its stationary value with the same exponential behavior. In the second case, $\gamma = 1$, the behavior ceases to be exponential and becomes algebraic, that is, $d\chi/dt \sim t^{-1/2}$. Therefore, integrating in t, we reach the following asymptotic behavior for the variance

$$\chi = \frac{4}{\sqrt{\pi}}(\alpha t)^{1/2}. \tag{11.52}$$

For $\gamma = 1$ ($T = 0$) the variance diverges in the limit $t \to \infty$ as $\chi \sim t^{1/2}$.

11.3 Linear Glauber Model

The one-dimensional Glauber model can be extended to two or more dimensions in various ways. One of them is done in such a way that the transition rate obeys detailed balance with respect to a stationary distribution which is identified with the Boltzmann-Gibbs distribution corresponding to the Ising model. This extension will be examined in the next chapter. Here we do a generalization which we call linear Glauber model, defined in a hypercubic lattice of dimension d (linear chain for $d = 1$, square lattice for $d = 2$, cubic lattice for $d = 3$, etc.) The inversion rate $w_\mathbf{r}(\sigma)$, corresponding to a site $\mathbf{r}$ of the hypercubic lattice, is defined by

$$w_\mathbf{r}(\sigma) = \frac{\alpha}{2}\{1 - \frac{\gamma}{2d}\sigma_\mathbf{r} \sum_\delta \sigma_{\mathbf{r}+\delta}\}, \tag{11.53}$$

where the sum is performed over the $2d$ neighbors of site $\mathbf{r}$, α is a parameter that defined the time scale and γ is a positive parameter restricted to the interval $0 \leq \gamma \leq 1$. The case $\gamma = 1$ corresponds to the voter model which will be seen later on. The linear Glauber model is irreversible in two or more dimensions. In one dimension we recover the model seen in Sect. 11.2 and is therefore reversible. The model has the inversion symmetry because the rate (11.53) remains invariant under the transformation $\sigma_{\mathbf{r}} \rightarrow -\sigma_{\mathbf{r}}$.

From Eq. (11.12) and using the rate (11.53), we obtain the following equation for the magnetization $m_{\mathbf{r}} = \langle \sigma_{\mathbf{r}} \rangle$,

$$\frac{d}{dt}m_{\mathbf{r}} = \alpha\{-m_{\mathbf{r}} + \frac{\gamma}{2d}\sum_{\delta}m_{\mathbf{r}+\delta}\}. \tag{11.54}$$

For the case of homogeneous solutions, such that $m_{\mathbf{r}} = m$, we get

$$\frac{d}{dt}m = -\alpha(1-\gamma)m, \tag{11.55}$$

whose solution is

$$m = m^0 e^{-\alpha(1-\gamma)t}. \tag{11.56}$$

Therefore, the magnetization decays exponentially to zero with a relaxation time τ given by $\tau^{-1} = \alpha(1-\gamma)$ similar to the one-dimensional case.

Next, we study the correlation $\langle \sigma_{\mathbf{r'}} \sigma_{\mathbf{r}} \rangle$ between two sites. To this end, we use Eq. (11.13) to derive an evolution equation for the correlation $\phi_{\mathbf{r}} = \langle \sigma_0 \sigma_{\mathbf{r}} \rangle$ between site 0 and site $\mathbf{r}$. Taking into account the translational invariance, we get the equation

$$\frac{d}{dt}\phi_{\mathbf{r}} = 2\alpha\{-\phi_{\mathbf{r}} + \frac{\gamma}{2d}\sum_{\delta}\phi_{\mathbf{r}+\delta}\}, \tag{11.57}$$

which is valid for any $\mathbf{r}$ except $\mathbf{r} = 0$ and with the condition $\phi_0 = 1$, which must be obeyed at any time.

The stationary solution obeys the equation

$$\frac{1}{2d}\sum_{\delta}(\phi_{\mathbf{r}+\delta} - \phi_{\mathbf{r}}) - \varepsilon\phi_{\mathbf{r}} = 0, \tag{11.58}$$

which is valid for $\mathbf{r} \neq 0$, with the condition $\phi_0 = 1$, where $\varepsilon = (1-\gamma)/\gamma$. To solve the difference equation (11.58), we start by introducing a parameter a such that

$$\frac{1}{2d}\sum_{\delta}(\phi_{\delta} - \phi_0) - \varepsilon\phi_0 = -a. \tag{11.59}$$

With this proviso we may write an equation which is valid for any site $\mathbf{r}$ of the lattice,

$$\frac{1}{2d}\sum_{\delta}(\phi_{\mathbf{r}+\delta} - \phi_{\mathbf{r}}) - \varepsilon\phi_{\mathbf{r}} = -a\delta_{\mathbf{r}0}, \tag{11.60}$$

where the parameter a must be chosen in such a way that $\phi_0 = 1$.

We define the Fourier transform of $\phi_{\mathbf{r}}$ by

$$\hat{\phi}_{\mathbf{k}} = \sum_{\mathbf{r}} \phi_{\mathbf{r}} e^{i\mathbf{k}\cdot\mathbf{r}}, \tag{11.61}$$

where the sum extends over the sites of a hypercubic lattice. Multiplying Eq. (11.60) by $e^{i\mathbf{k}\cdot\mathbf{r}}$ and summing over $\mathbf{r}$ we get

$$\Lambda_{\mathbf{k}} \hat{\phi}_{\mathbf{k}} = a, \tag{11.62}$$

where

$$\Lambda_{\mathbf{k}} = \varepsilon + \frac{1}{2d} \sum_{\boldsymbol{\delta}} (1 - \cos \mathbf{k} \cdot \boldsymbol{\delta}). \tag{11.63}$$

The inverse Fourier transform is

$$\phi_{\mathbf{r}} = \int \hat{\phi}_{\mathbf{k}} e^{i\mathbf{k}\cdot\mathbf{r}} \frac{d^{dk}}{(2\pi)^d}, \tag{11.64}$$

where the integral is performed on the region of space k defined by $-\pi \le k_i \le \pi$ for each component and therefore

$$\phi_{\mathbf{r}} = a \int_{Bz} \frac{e^{i\mathbf{k}\cdot\mathbf{r}}}{\Lambda_{\mathbf{k}}} \frac{d^{dk}}{(2\pi)^d}. \tag{11.65}$$

The parameter a is obtained by setting the condition $\phi_0 = 1$. Hence

$$\frac{1}{a} = \int \frac{1}{\Lambda_{\mathbf{k}}} \frac{d^{dk}}{(2\pi)^d}. \tag{11.66}$$

Defining the function $G_{\mathbf{r}}(\varepsilon)$ by

$$G_{\mathbf{r}}(\varepsilon) = \int \frac{e^{i\mathbf{k}\cdot\mathbf{r}}}{\varepsilon + (2d)^{-1} \sum_{\boldsymbol{\delta}} (1 - \cos \mathbf{k} \cdot \boldsymbol{\delta})} \frac{d^{dk}}{(2\pi)^d}, \tag{11.67}$$

then the pair correlation in the stationary state is given by

$$\phi_{\mathbf{r}} = \frac{G_{\mathbf{r}}(\varepsilon)}{G_0(\varepsilon)}, \tag{11.68}$$

since $a = 1/G_0(\varepsilon)$.

Summing each side of Eq. (11.60) over $\mathbf{r}$ and using expression (11.15) for the variance,

$$\chi = \sum_{\mathbf{r}} \phi_{\mathbf{r}}, \tag{11.69}$$

we find the following relation $\chi = a/\varepsilon$ from which we reach the result

$$\chi = \frac{1}{\varepsilon G_0(\varepsilon)}. \tag{11.70}$$

The function $G_r(\varepsilon)$, given by (11.67), is finite for $\varepsilon > 0$. In the limit $\varepsilon \to 0$ it diverges for $d \le 2$, but is finite for $d > 2$. Formula (11.70) implies therefore that the variance diverges, in this limit, in any dimension. For small values of ε, we use the following results

$$G_0(\varepsilon) = A\varepsilon^{-1/2}, \qquad d = 1, \tag{11.71}$$

$$G_0(\varepsilon) = -\frac{1}{\pi}\ln\varepsilon, \qquad d = 2, \tag{11.72}$$

where A is a constant. Replacing in (11.70), we obtain the asymptotic behavior of the variance

$$\chi \sim \varepsilon^{-1/2}, \qquad d = 1, \tag{11.73}$$

$$\chi = \pi\frac{\varepsilon^{-1}}{|\ln\varepsilon|}, \qquad d = 2. \tag{11.74}$$

Since $G_0(0)$ is finite for $d > 2$, we get

$$\chi \sim \varepsilon^{-1}, \qquad d > 2. \tag{11.75}$$

Therefore, the variance diverges as $\chi \sim \varepsilon^{-d/2}$ for $d \le 2$, with logarithmic correction in $d = 2$, and as $\chi \sim \varepsilon^{-1}$ for $d \ge 2$.

From the solution (11.68) it is possible to get the asymptotic behavior of ϕ_r for large values of r. However, we use here another approach to get such behavior. The approach consists in assuming that the variable ϕ_i changes little with distance. With this assumption, we substitute the difference equation (11.58) by the differential equation

$$-\varepsilon\phi + \frac{1}{2d}\sum_{\nu=1}^{d}\frac{\partial^2\phi}{\partial x_\nu^2} = 0, \tag{11.76}$$

where x_ν are the components of the vector $\mathbf{r}$. Moreover, we choose solutions that are invariant by rotations, that is, such that $\phi(r, t)$ depends on x_ν through the variable $r = (x_1^2 + x_2^2 + \ldots + x_d^2)^{1/2}$, so that

$$-2d\varepsilon\phi + \frac{\partial^2\phi}{\partial r^2} + \frac{(d-1)}{r}\frac{\partial\phi}{\partial r} = 0. \tag{11.77}$$

In $d = 1$, the solution is

$$\phi(r) = A_1 e^{-r/\xi}, \tag{11.78}$$

where $\xi = \sqrt{1/2d\varepsilon}$ and A_1 is a constant. For $d > 1$, the solution for large values of r is given by

$$\phi(r) = A_d \frac{e^{-r/\xi}}{r^{(d-1)/2}}, \tag{11.79}$$

where A_d is a constant that depends of dimension. We see thus that the pair correlation decays exponentially with the distance with a correlation length that behaves near $\varepsilon = 0$ according to

$$\xi \sim \varepsilon^{-1/2}. \tag{11.80}$$

It is worth to note that expressions (11.79) are valid only for $\varepsilon \neq 0$. When $\varepsilon = 0$ the linear Glauber model reduces to the voter model that we will study in the next section.

11.4 Voter Model

Imagine a community of individuals in which each one holds an opinion about a certain issue, being in favor or against it. With the passage of time, they change their opinion according to the opinion of the individuals in their neighborhood. A certain individual chooses at random one of the neighbors and takes his opinion. In an equivalent way, we may say that the individual takes the favorable or the contrary opinion with probabilities proportional to the numbers of neighbor in favor or against the opinion, respectively.

To set up a model that describes such a community, we consider that the individuals are located at the sites of a lattice. We imagine this lattice as being a hypercubic lattice of dimension d. To each site $\mathbf{r}$ we associate a variable $\sigma_{\mathbf{r}}$ that takes the value 1 if the individual located at $\mathbf{r}$ if he has a favorable opinion and the value -1 if he has a contrary opinion. The time evolution of this system is governed by the master equation (11.8) with the following inversion rate

$$w_{\mathbf{r}}(\sigma) = \frac{\alpha}{2d} \sum_{\delta} \frac{1}{2}(1 - \sigma_{\mathbf{r}}\sigma_{\mathbf{r}+\delta}), \tag{11.81}$$

where α is a parameter that defines the time scale and the sum extends over all the $2d$ neighbors of site $\mathbf{r}$. Notice that each term in the summation vanishes if the neighbor has the same sign as the central site and equals one if the neighbor has opposite sign. The rate of inversion can also be written in the form

$$w_r(\sigma) = \frac{\alpha}{2}\{1 - \frac{1}{2d}\sigma_r \sum_\delta \sigma_{r+\delta}\},$$ (11.82)

from which we conclude that the linear Glauber model, defined by the inversion rate (11.53), reduces to the voter model when $\gamma = 1$.

The linear Glauber model can be interpreted as the voter model with a noise. At each time step a certain individual chooses at random a neighbor and takes his opinion with a probability $p = (1 + \gamma)/2$ and the contrary opinion with probability $q = (1 - \gamma)/2$. The parameter q can then be interpreted as a noise. When $q = 0$, $\gamma = 1$, the linear Glauber noise reduces to the voter model.

The voter model has two fundamental properties. The first is that the model has inversion symmetry. Changing the signs of all variables, that is, performing the transformation $\sigma_r \rightarrow -\sigma_r$, we see that the rate of inversion (11.82) remains invariant. The second is that the rate (11.82) has two absorbing states. One of them is the state $\sigma_r = +1$ for all sites. The other is the state $\sigma_r = -1$ for all sites. An absorbing state is the one that can be reach from other states, but the transition from it to any other state is forbidden. Once the system has reached an absorbing state, it remains forever in this state. In the present case, if all variables are equal to $+1$, the rate of inversion (11.82) vanish. The same is true when all variables are equal to -1.

From Eq. (11.12), and using the rate given by Eq. (11.82), we get the following equation for the magnetization $m_r = \langle \sigma_r \rangle$,

$$\frac{d}{dt}m_r = \alpha\{-m_r + \frac{1}{2d}\sum_\delta m_{r+\delta}\},$$ (11.83)

which can be written in the form

$$\frac{d}{dt}m_r = \frac{\alpha}{2d}\sum_\delta (m_{r+\delta} - m_r).$$ (11.84)

The average number of favorable individuals is equal to $\sum_r (1 + m_r)/2$.

In the stationary regime a solution is given by $m_r = 1$ and another is $m_r = -1$, which correspond to the two absorbing states. We show below that in one and two dimensions these are the only stationary states. In three or more dimensions however there might exist other stationary states such that $m_r = m$ can take any value. In this case the stationary value depends on the initial condition.

For each one of the absorbing states, the correlation $\langle \sigma_r \sigma_{r'} \rangle$ between any two sites takes the value 1. If this correlation takes values distinct from one, than other states might exist. Thus, we will determine the correlation $\langle \sigma_r \sigma_{r'} \rangle$ in the stationary state. From Eq. (11.13), using the rate given by Eq. (11.82) and taking into account the translational invariance, we get the following equation for the correlation $\phi_r =$

$\langle \sigma_0 \sigma_r \rangle$ between site 0 and site **r**,

$$\frac{d}{dt}\phi_r = \frac{\alpha}{d}\sum_\delta (\phi_{r+\delta} - \phi_r),$$ (11.85)

which is valid for any **r**, except **r** = 0, and with the condition $\phi_0 = 1$ at any time.

In the stationary regime we get

$$\sum_\delta (\phi_{r+\delta} - \phi_r) = 0.$$ (11.86)

We remark that these equations are homogeneous, except those for which **r** corresponds to a site which is a neighbor of the origin. In those cases, the following equation is valid

$$\sum_{\delta(r+\delta \neq 0)} (\phi_{r+\delta} - \phi_r) + (1 - \phi_r) = 0,$$ (11.87)

since $\phi_0 = 1$.

It is clear that a solution is $\phi_r = 1$ for any **r**. The other possible solutions must be inhomogeneous since, otherwise, that is, if they are of the type $\phi = \text{const}$, this constant must be equal to 1 in order to satisfy Eq. (11.87) so that we get the solution $\phi_r = 1$.

Assuming that for sites far away from the site **r** = 0 the variable ϕ_r changes little with distance, we may replace the difference equation above by a differential equation, as we did before. Taking into account that the linear Glauber model reduces to the voter mode when $\varepsilon = 0$, we may use the result (11.77). Setting $\varepsilon = 0$ on this equation, we get the differential equation for the correlation,

$$\frac{\partial^2 \phi}{\partial r^2} + \frac{(d-1)}{r}\frac{\partial \phi}{\partial r} = 0.$$ (11.88)

valid in the stationary regime and for ϕ invariant by rotation. Defining $\phi' = d\phi/dr$, this equation is equivalent to

$$\frac{d\phi'}{dr} + \frac{d-1}{r}\phi' = 0,$$ (11.89)

whose solution is $\phi' = A$, for $d = 1$, and $\phi' = Ar^{-(d-1)}$, for $d > 1$.

Therefore, for $d = 1$, the solution is $\phi(r) = Ar + B$. Since $\phi(r)$ must remain bounded in the limit $r \to \infty$, we conclude that the constant A must vanish so that $\phi = \text{const}$. Hence, the only stationary states are the two absorbing states for which $\phi = 1$.

For $d = 2$ we get $\phi' = A/r$, from which we obtain $\phi(r) = A\ln r + B$. In same way, since $\phi(r)$ must remain bounded in the limit $r \to \infty$, we conclude that the constant A must vanish so that $\phi = $ const. Again, the only stationary states are the two absorbing states for which $\phi = 1$.

For $d \geq 3$, we get

$$\phi(r) = B + \frac{C}{r^{d-2}}, \tag{11.90}$$

which is a non-homogeneous solution. Thus, for $d \geq 3$ there is another stationary solution in addition to the two absorbing states corresponding to $\phi = 1$.

Recalling that the consensus (all individuals with the same opinion) corresponds to an absorbing state, it will occur with certainty in one and two dimensions. In three or more dimensions the consensus may not occur since in these cases there are other stationary states in addition to the absorbing ones.

11.5 Critical Exponents

We have seen in the previous section that the linear Glauber model becomes critical when the parameter $\gamma \rightarrow 1$, when the model is converted into the voter model. Around this point some quantities have singular behavior of the type ε^{μ}, where μ is a critical exponent and $\varepsilon = 1 - \gamma$. We have seen, for example, that the variance χ behaves as

$$\chi \sim \varepsilon^{-\gamma}, \tag{11.91}$$

with $\gamma = 1/2$ for $d = 1$ and $\gamma = 1$ for $d \geq 2$ with logarithmic corrections in $d = 2$. The magnetization m vanishes except at the critical point when it takes a nonzero value. Writing

$$m \sim \varepsilon^{\beta}, \tag{11.92}$$

we see that we should assign to β a zero value.

Other critical exponents may be defined. The spatial correlation length behaves as

$$\xi \sim |\varepsilon|^{-\nu_{\perp}}, \tag{11.93}$$

with $\nu_{\perp} = 1/2$. The relaxation time, or time correlation length, behaves as

$$\tau \sim |\varepsilon|^{-\nu_{\parallel}}, \tag{11.94}$$

with $\nu_{\parallel} = 1$. Defining the dynamic exponent z by the relation between the spatial and time correlation lengths

$$\tau \sim \xi^{z}, \tag{11.95}$$

Table 11.1 Critical exponent
of the linear Glauber model

d	β	γ	$v_\perp$	$v_\parallel$	z	η	ζ
1	0	1/2	1/2	1	2	1	1/2
≥ 2	0	1	1/2	1	2	0	1

we see that $z = v_\parallel/v_\perp$ so that $z = 2$.

At the critical point, the pair correlation decays according to

$$\phi(r) \sim \frac{1}{r^{d-2+\eta}},\tag{11.96}$$

for r large enough. From the results obtained for the voter model, we see that $\eta = 0$ for $d \geq 2$ and $\eta = 1$ for $d = 1$. Still at the critical point, the variance grows according to the algebraic behavior

$$\chi \sim t^\zeta.\tag{11.97}$$

with $\zeta = 1/2$ in $d = 1$ and $\zeta = 1$ in $d > 2$. In $d = 2$, $\zeta = 1$, but the behavior of χ has logarithm corrections.

The values of the critical exponents of the linear Glauber model are shown in Table 11.1.

Chapter 12
Systems with Inversion Symmetry

12.1 Introduction

In this chapter we study systems governed by master equations and defined in a lattice. To each site of lattice there is a stochastic variable that takes only two values which we choose to be $+1$ or -1. The system have inversion symmetry, that is, they are governed by dynamics that are invariant by the inversion of the signs of all variables. In this sense they are similar to the models studied in the previous chapter. However, unlike them, the models studied here present in the stationary state a phase transition between two thermodynamic phases called ferromagnetic and paramagnetic. The transition from the paramagnetic to the ferromagnetic is due to a spontaneous symmetry breaking. We focus our study in two models of this type: the Glauber-Ising model, which in the stationary state is reversible or, equivalently, obeys the detailed balance condition, and the majority vote model which is irreversible.

We analyze here systems that have symmetry of inversion with inversion rates that involve only the sites in a small neighborhood of a central site. Such systems, reversible or irreversible, have the same behavior around the critical point and form therefore a universality class concerning the critical behavior. The principal component of this class is the Glauber-Ising model, that we will treat shortly. The universality of the critical behavior means that the critical exponents of such systems are the same and those of the Glauber-Ising model.

We denote by σ_i the stochastic variable associated to site i so that $\sigma_i = \pm 1$. The global configuration of the system is denoted by

$$\sigma = (\sigma_1, \sigma_2, \ldots, \sigma_N), \tag{12.1}$$

where N is the total number of sites. We consider a dynamic such that, at any time step, only one site has its state modified. Thus the time evolution of the probability

© Springer International Publishing Switzerland 2015
T. Tomé, M.J. de Oliveira, *Stochastic Dynamics and Irreversibility*, Graduate Texts in Physics, DOI 10.1007/978-3-319-11770-6_12

$P(\sigma, t)$ of state σ at time t is governed by the master equation

$$\frac{d}{dt} P(\sigma, t) = \sum_{i=1}^{N} \{ w_i(\sigma^i) P(\sigma^i, t) - w_i(\sigma) P(\sigma, t) \}, \qquad (12.2)$$

where

$$\sigma^i = (\sigma_1, \sigma_2, \ldots, -\sigma_i, \ldots, \sigma_N). \qquad (12.3)$$

and $w_i(\sigma)$ is the rate of inversion of the sign of site i. Unlike the rates used in the preceding chapter now they are not necessarily linear in σ_i.

For later reference in this chapter, we write below the evolution equations for the average

$$\frac{d}{dt} \langle \sigma_i \rangle = -2 \langle \sigma_i w_i(\sigma) \rangle, \qquad (12.4)$$

and for the pair correlation

$$\frac{d}{dt} \langle \sigma_i \sigma_j \rangle = -2 \langle \sigma_i \sigma_j w_i(\sigma) \rangle - 2 \langle \sigma_i \sigma_j w_j(\sigma) \rangle, \qquad (12.5)$$

which are obtained from the master equation (12.2).

The numerical simulation of the process defined by the master equation (12.2) is carried out by transforming the continuous time Markovian process into a Markov chain through a time discretization. To this end, we discretize the time in intervals Δt and write the transition rate in the form

$$w_i(\sigma) = \alpha \, p_i(\sigma), \qquad (12.6)$$

where α is chosen so that $0 \leq p_i \leq 1$, which is always possible to do. The conditional probability $T(\sigma^i, \sigma) = w_i(\sigma)/\Delta t = (\alpha/\Delta t) p_i(\sigma)$ related to the transition $\sigma \to \sigma^i$ is constructed by choosing $\Delta t = (\alpha N)^{-1}$, what leads us to the formula

$$T(\sigma^i, \sigma) = \frac{1}{N} \, p_i(\sigma). \qquad (12.7)$$

The Markovian process is performed according to this formula, interpreted as the product of two probabilities corresponding to two independent processes, as follows. (a) At each time step, we choose a site at random, what occurs with probability $1/N$. (b) The chosen site has its state modified with probability p_i.

In numerical simulations it is convenient to define a unit of time, which we call Monte Carlo step, as being equal to the interval of time elapsed during N time steps. Therefore, a Monte Carlo step is equal to $N\Delta t = 1/\alpha$. This definition is appropriate when we compare results coming from simulations performed in systems of distinct sizes because from this definition it follows that, in n Monte Carlo steps, each site will be chosen n times, in the average, independently of the size of the system.

12.2 Ising Model

Certain magnetic materials, known as ferromagnetic, have a natural magnetization that disappears when heated above a critical temperature, called Curie temperature. At low temperatures, the system is found in the ferromagnetic phase and, at high temperatures, in the paramagnetic phase. The simplest description of the ferromagnetic state and the ferromagnetic-paramagnetic phase transition is given by the Ising model.

Consider a lattice of sites and suppose that at each site there is a magnetic atom. The state of the magnetic atom is characterized by the direction of the magnetic dipole moment. In the Ising model the magnetic dipole moment can be found in only two state with respect to a certain z axis: in the direction or in the opposite direction of the z axis. Thus, the magnetic dipole moment μ_i of the i-th atom is given by $\mu_i = \gamma \sigma_i$, where γ is a constant and the variable σ_i takes the values $+1$ in case the dipole moment is in the direction $+z$ and -1 in the direction $-z$.

Consider two neighbor sites i and j. They can be in four configurations, in two of them the dipoles are parallel with each other and in the other two they are antiparallel. To favor the ferromagnetic ordering, we assume that the situation with the least energy is the one in which the dipole are parallel. Hence, the interaction energy between these two atoms is given by $-J\sigma_i\sigma_j$, where $J > 0$ is a constant that represents the interaction between the magnetic dipoles. The total energy $E(\sigma)$ corresponding to the configuration σ is thus

$$E = -J \sum_{(ij)} \sigma_i\sigma_j - h \sum_i \sigma_i, \tag{12.8}$$

where the first summation extends over all pairs of neighboring sites and the second summation describes the interaction with a field. In addition to the energy, another state function particularly relevant is the total magnetization

$$\mathcal{M} = \sum_i \sigma_i. \tag{12.9}$$

In thermodynamic equilibrium at a temperature T, the probability $P(\sigma)$ of finding the system in configuration σ is

$$P(\sigma) = \frac{1}{Z} e^{-\beta E(\sigma)}, \tag{12.10}$$

where $\beta = 1/kT$, k is the Boltzmann constant and Z is the partition function given by

$$Z = \sum_{\sigma} e^{-\beta E(\sigma)}, \tag{12.11}$$

where the sum extends over all the 2^N configurations.

Among the thermodynamic properties that we wish to calculate, we include:

(a) The average magnetization,

$$m = \frac{1}{N}\langle \mathcal{M} \rangle, \tag{12.12}$$

(b) The variance of the magnetization

$$\chi = \frac{1}{N}\{\langle \mathcal{M}^2 \rangle - \langle \mathcal{M} \rangle^2\}, \tag{12.13}$$

(c) The average energy per site,

$$u = \frac{1}{N}\langle E \rangle, \tag{12.14}$$

(d) The variance of energy per site

$$c = \frac{1}{N}\{\langle E^2 \rangle - \langle E \rangle^2\}. \tag{12.15}$$

From the equilibrium distribution (12.10) and the energy (12.8) we can show that the susceptibility $\chi^* = \partial m/\partial h$ is related with the variance of magnetization by means of $\chi^* = \chi/kT$, and that the specific heat $c^* = \partial u/\partial T$ is related to the variance of energy by $c^* = c/kT^2$.

Metropolis algorithm Next, we present the Monte Carlo method to generate configurations according to the probability given by (12.10). To this end, we use the stochastic dynamics called Metropolis algorithm. We start from an initial arbitrary configuration. From it, other configurations are generated. Suppose that, at a certain instant, the configuration is σ. The next configuration is chosen as follows.

A site of the lattice is chose at random, say, the i-th site. Next, we calculate the difference in energy $\Delta E = E(\sigma^i) - E(\sigma)$, which is given by

$$\Delta E = 2\sigma_i\{J \sum_{\delta} \sigma_{i+\delta} + h\}, \tag{12.16}$$

where the summation extends over the neighbors of site i.

(a) If $\Delta E \leq 0$, then the variable σ_i changes sign and the new configuration is σ^i.
(b) If $\Delta E > 0$, then we calculate $p = e^{-\beta \Delta E}$ and we generate a random number ξ uniformly distributed in the interval $[0, 1]$.

   (b1) If $\xi \leq p$, then σ_i has its sign changed and the new configuration is σ^i.
   (b2) If $\xi > p$, the variable σ_i does not change and the configuration remains the same, that is, it will be σ.

Using the procedure, we generate a sequence of configurations. For each configuration we calculate the desired state functions, such as, $\mathscr{M}$, $\mathscr{M}^2$, E and E^2. The estimate of the averages are obtained from the arithmetic means of these quantities, after discarding the first configurations

The Metropolis algorithm defined above is equivalent to a stochastic dynamics whose rate of inversion $w_i(\sigma)$ is given by

$$w_i(\sigma) = \begin{cases} \alpha e^{-\beta[E(\sigma^i) - E(\sigma)]}, & E(\sigma^i) > E(\sigma), \\ \alpha, & E(\sigma^i) \leq E(\sigma). \end{cases} \tag{12.17}$$

It is straightforward to show that this rate obey detailed balance

$$\frac{w_i(\sigma^i)}{w_i(\sigma)} = \frac{P(\sigma)}{P(\sigma^i)}, \tag{12.18}$$

with respect to the equilibrium probability (12.10) of the Ising model. The dynamics so defined can be interpreted as the one describing the contact of the system with a heat reservoir. At zero field, $h = 0$, the rate has the inversion symmetry, $w_i(-\sigma) = w_i(\sigma)$.

12.3 Glauber-Ising Dynamics

The Glauber-Ising stochastic dynamics is defined by the Glauber rate

$$w_i(\sigma) = \frac{\alpha}{2}\{1 - \sigma_i \tanh[K \sum_\delta \sigma_{i+\delta} + H]\}, \tag{12.19}$$

where α is a parameter that defines the time scale and the sum extends over all neighbors of site i. The sites comprise a regular lattice of dimension d. The Glauber rate, as the Metropolis rate, describes the interaction of the Ising system with a heat reservoir. In other terms, the Glauber rate obeys detailed balance (12.18) with respect to the probability distribution of the Ising model

$$P(\sigma) = \frac{1}{Z} \exp\{K \sum_{(ij)} \sigma_i \sigma_j + H \sum_i \sigma_i\}, \tag{12.20}$$

where the sum extends over the pairs of neighboring sites. To show detailed balance it suffices to write the Glauber rate in the equivalent form

$$w_i(\sigma) = \frac{\alpha e^{-\sigma_i E_i}}{e^{E_i} + e^{-E_i}}, \tag{12.21}$$

where $E_i = K \sum_\delta \sigma_{i+\delta} + H$ and take into account that $P(\sigma)/P(\sigma^i) = e^{2\sigma_i E_i}$. Therefore, the Glauber-Ising dynamics describes, in the stationary state, an Ising system whose energy is given by (12.8), in equilibrium at a temperature T such that $K = J/kT$ and $H = h/kT$. Notice that at zero field, $H = 0$, the rate has an inversion symmetry, $w_i(-\sigma) = w_i(\sigma)$.

To set up an algorithm for numerical simulation of the Glauber-Ising model, we proceed as follows. We start with an arbitrary initial configuration. Given a configuration σ, the next one is obtained as follows. A site of the lattice is chosen at random, say, the i-th site. Next we calculate

$$p = \frac{1}{2}\{1 - \sigma_i \tanh[K \sum_\delta \sigma_{i+\delta} + H]\}. \tag{12.22}$$

and generate a random number ξ uniformly distributed in the interval $[0, 1]$. If $\xi \leq p$, then the variable σ_i has its sign changes and the new configuration is σ^i. Otherwise, that is, if $\xi > p$, then the variable σ_i remains unchanged and the new configuration is σ.

An alternative algorithm, but equivalent, consists in choosing the the new sign of the variable σ_i independently of its present sign as follows. If $\xi \leq p$, then $\sigma_i = -1$, otherwise, $\sigma_i = +1$. In this case, the algorithm is called heat bath.

The algorithm for the simulation of the Glauber-Ising model can also be understood as a Monte Carlo method for the simulation of the equilibrium Ising model. For large times, the configurations are generated according to the equilibrium probability (12.20).

Figures 12.1 and 12.2 show the results obtained by the simulation of the Ising model defined in a square lattice with $N = L \times L$ sites and periodic boundary conditions using the Glauber rate at zero field, $H = 0$. The critical temperature is known exactly and is $kT_c/J = 2/\ln(1 + \sqrt{2}) = 2.269185\ldots$. We use a number of Monte Carlo steps equal to 10^6 and discarding the first 10^5 steps. Instead of the average magnetization defined above, we calculated the average of the absolute value of the magnetization,

$$m = \frac{1}{N}\langle|\mathcal{M}|\rangle. \tag{12.23}$$

Similarly, we determine the variance defined by

$$\chi = \frac{1}{N}\langle\mathcal{M}^2\rangle - \langle|\mathcal{M}|\rangle^2. \tag{12.24}$$

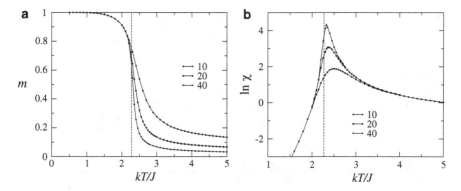

Fig. 12.1 (a) Magnetization m and (b) variance χ versus temperature T for the Ising model in a square lattice obtained by the Glauber-Ising dynamics for various values of L indicated. The susceptibility $\chi^* = \partial m/\partial h$ is related to χ by means of $\chi^* = \chi/kT$. The *dashed line* indicates the critical temperature, $kT_c/J = 2.269\ldots$

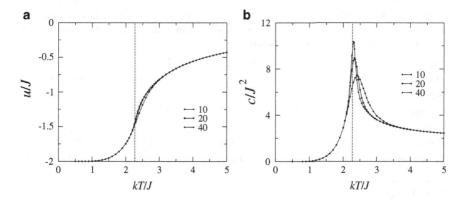

Fig. 12.2 (a) Energy per site u and (b) variance of energy per site c versus temperature T for the Ising model in the square lattice obtained by the Glauber-Ising dynamics for several values of L indicated. The specific heat $c^* = \partial u/\partial T$ is related to c by $c^* = c/kT^2$. The *dashed line* indicates the critical temperature, $kT_c/J = 2.269\ldots$

Figure 12.1 shows m and χ as functions of the temperature for several sizes L of the system. As we can see, the magnetization is finite for any value of T, but for $T \geq T_c$, $m \to 0$ when L increases without bounds. The variance is also finite, but at $T = T_c$ it diverges when $L \to \infty$. Figure 12.2 shows the average energy per site u and the variance of energy per site c as functions of temperature also for several values of the size of the system. The specific heat is finite, but diverges at $T = T_c$ when $L \to \infty$. In this same limit the slope of u at $T = T_c$ increases without bounds when $L \to \infty$ because du/dT is proportional to c.

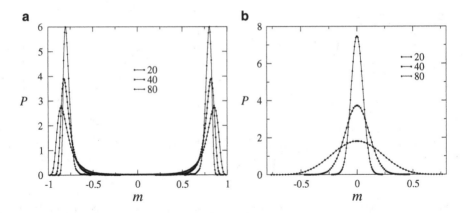

Fig. 12.3 Probability density $P(m)$ of the magnetization for the Ising model in the square lattice obtained by the Glauber-Ising dynamics for several values of L indicated. (**a**) $kT/J = 2.2$, (**b**) $kT/J = 2.8$

Table 12.1 Critical parameters T_c and $K_c = J/kT_c$ of the Ising model for various lattices of dimension d and number of neighbors z. For the two-dimensional lattices shown, the values are exact. For the cubic and hypercubic lattice, the errors are in the last digit

Lattice	d	z	kT_c/J	K_c	kT_c/Jz
Honeycomb	2	3	1.51865[a]	0.658478	0.506217
Square	2	4	2.26918[b]	0.440686	0.567296
Triangular	2	6	3.64095[c]	0.274653	0.606826
Diamond	3	4	2.7040	0.3698	0.676
Cubic	3	6	4.5116	0.22165	0.7519
Hypercubic	4	8	6.682	0.1497	0.8352

[a] $2/\ln(2 + \sqrt{3})$
[b] $2/\ln(1 + \sqrt{2})$
[c] $2/\ln\sqrt{3}$

Figure 12.3 shows the histograms of the magnetization for two temperatures, below and above the critical temperature. Above the critical temperature, there are two peaks located symmetrically, which characterize the ferromagnetic state. When L increases without bounds the same occurs with the height of the peaks.

The critical temperature of the Ising model with nearest neighbor interaction obtained by several methods is shown in Table 12.1 for several regular lattices.

12.4 Mean-Field Theory

An exact solution of the Glauber-Ising model can be obtained when the number of neighbors is very large and the interaction is weak. More precisely, we assume that the summation in (12.20) extends over all pair of sites and make the replacement $K \to K/N$. The Glauber rate (12.19) becomes

$$w_i(\sigma) = \frac{\alpha}{2}\{1 - \sigma_i \tanh(Ks + H)\}, \qquad (12.25)$$

where s is the stochastic variable defined by

$$s = \frac{1}{N}\sum_j \sigma_j. \qquad (12.26)$$

For N large enough we assume that s is distributed according to a Gaussian with mean m and variance χ/N,

$$P(s) = \frac{\sqrt{N}}{\sqrt{2\pi\chi}}e^{-N(s-m)^2/2\chi}, \qquad (12.27)$$

where m and χ depend on t. Notice that

$$\chi = N\{\langle s^2 \rangle - \langle s \rangle^2\}, \qquad (12.28)$$

and hence χ coincides with the definition (12.13) since $\mathcal{M} = Ns$.

Next, we use Eq. (12.4) which gives the time evolution of the magnetization. Replacing the inversion Glauber rate, given by (12.25), in (12.4), we get

$$\frac{1}{\alpha}\frac{d}{dt}\langle \sigma_i \rangle = -\langle \sigma_i \rangle + \langle \tanh(Ks + H) \rangle. \qquad (12.29)$$

In the limit $N \to \infty$, the distribution $P(s)$ becomes a Dirac delta function and $\langle \tanh(Ks + H) \rangle \to \tanh(Km + H)$ and we reach the following equation for the average magnetization

$$\frac{1}{\alpha}\frac{dm}{dt} = -m + \tanh(Km + H), \qquad (12.30)$$

where we have taken into account that $\langle \sigma_i \rangle = m$, independently of i. At zero field, when there is symmetry of inversion, it reduces to

$$\frac{1}{\alpha}\frac{dm}{dt} = -m + \tanh Km. \qquad (12.31)$$

The stationary solution is given by

$$m = \tanh Km. \tag{12.32}$$

which is the mean-field equation for the magnetization of the Ising model. For $K \leq 1$, the only solution is $m = 0$, which corresponds to the paramagnetic state. For $K < 1$, there is another solution $m \neq 0$, corresponding to the ferromagnetic state. Thus when $K = K_c = 1$, the system undergoes a phase transition. Since $K = J/kT$, the critical temperature T_c is given by $T_c = J/k$.

For large times and temperatures close to the critical temperature, the magnetization m is small so that we may approximate the right-hand side of Eq. (12.31) by an expansion in power series. Up to cubic terms, we get

$$\frac{1}{\alpha}\frac{dm}{dt} = -\varepsilon m - \frac{1}{3}m^3, \tag{12.33}$$

where $\varepsilon = (K_c - K) = (T - T_c)/T$. Multiplying both sides of Eq. (12.33) by m, we arrive at equation

$$\frac{1}{2\alpha}\frac{dm^2}{dt} = -\varepsilon m^2 - \frac{1}{3}m^4, \tag{12.34}$$

whose solution, for $\varepsilon \neq 0$, is

$$m^2 = \frac{3\varepsilon}{ce^{2\varepsilon\alpha t} - 1}, \tag{12.35}$$

where c is a constant which must be determined by the initial condition.

When $t \to \infty$, we get distinct solutions depending on the sign of ε. If $\varepsilon > 0$ $(T > T_c)$, then $m \to 0$, what corresponds to the paramagnetic state. If $\varepsilon < 0$ $(T < T_c)$, then $m \to m^* = \sqrt{3|\varepsilon|}$ so that the order parameter m^* behaves as

$$m^* \sim |\varepsilon|^{1/2}. \tag{12.36}$$

From Eq. (12.35), we see that for temperatures above the critical temperature, $\varepsilon > 0$, and for large times the magnetization m decays to zero exponentially,

$$m = a\,e^{-2\alpha\varepsilon t}, \tag{12.37}$$

with a relaxation time given by $\tau = 1/(2\alpha\varepsilon)$. Similarly, for temperatures below the critical temperature, $\varepsilon < 0$, and for large times the magnetization m decays exponentially to its stationary value

$$m = \pm m^* + b\,e^{-2|\varepsilon|\alpha t}, \tag{12.38}$$

with a relaxation time given by $\tau = 1/(2\alpha|\varepsilon|)$. In both cases, the relaxation time behaves as

$$\tau \sim |\varepsilon|^{-1}. \tag{12.39}$$

At the critical point, $\varepsilon = 0$ and Eq. (12.33) becomes

$$\frac{1}{\alpha}\frac{dm}{dt} = -\frac{1}{3}m^3, \tag{12.40}$$

whose solution is

$$m = \frac{\sqrt{3}}{\sqrt{2\alpha t + c}}, \tag{12.41}$$

and the decay ceases to be exponential. For large times we get the algebraic behavior

$$m \sim t^{-1/2}. \tag{12.42}$$

Susceptibility and variance of magnetization Now, we determine the susceptibility $\chi^* = \partial m/\partial h$ at zero field. Deriving both sides of Eq. (12.30) with respect to h, we get

$$\frac{1}{\alpha}\frac{d\chi^*}{dt} = -\chi^* + (K\chi^* + \beta)\operatorname{sech}^2 Km. \tag{12.43}$$

a result valid for zero field. In the stationary state and at zero field, it is given by

$$\chi^* = \frac{\beta(1 - m^2)}{1 - K(1 - m^2)}, \tag{12.44}$$

where we used the result $\operatorname{sech}^2 Km = 1 - m^2$, obtained from (12.32).

When $K < 1$ ($T > T_c$), $m = 0$ and hence $\chi^* = \beta/(1-K)$. When $K > 1$, we use the result $m^2 = 3(1 - K)$, valid near the critical point, to obtain $\chi^* = \beta/2(K - 1)$. In both cases, the susceptibility diverges according to

$$\chi^* \sim |\varepsilon|^{-1}. \tag{12.45}$$

At the critical point the susceptibility diverges in the limit $t \to \infty$. To determine the behavior of χ at $T = T_c$ for large times, we introduce the solution $m^2 = 3/2\alpha t$ into Eq. (12.43) at zero field and $K = 1$ to get

$$\frac{1}{\alpha}\frac{d\chi^*}{dt} = \beta - \frac{3\chi^*}{2\alpha t}, \tag{12.46}$$

where we have retained in the right-hand side only the dominant terms when t is large. The solution of this equation is $\chi^* = 2\beta\alpha t/5$. Therefore χ^* behaves as

$$\chi^* \sim t, \tag{12.47}$$

for large times.

Next, we determine the variance of the magnetization, given by (12.28). To this end, we first note that

$$s^2 = \frac{1}{N^2}\sum_{ij}\sigma_i\sigma_j = \frac{1}{N^2}\sum_{i\neq j}\sigma_i\sigma_j + \frac{1}{N}. \tag{12.48}$$

Using (12.5) and the rate (12.25), at zero field,

$$\frac{1}{\alpha}\frac{d}{dt}\langle\sigma_i\sigma_j\rangle = -2\langle\sigma_i\sigma_j\rangle + \langle(\sigma_i + \sigma_j)\tanh Ks\rangle, \tag{12.49}$$

valid for $i \neq j$. From this result, we get

$$\frac{1}{\alpha}\frac{d}{dt}\langle s^2\rangle = -2\langle s^2\rangle + \frac{2}{N} + \frac{2(N-1)}{N}\langle s\tanh Ks\rangle. \tag{12.50}$$

On the other hand,

$$\frac{1}{\alpha}\frac{d}{dt}\langle s\rangle^2 = \frac{2m}{\alpha}\frac{dm}{dt} = -2m^2 + 2m\langle\tanh Ks\rangle. \tag{12.51}$$

Therefore,

$$\frac{1}{2\alpha}\frac{d\chi}{dt} = -\chi + 1 - \langle s\tanh Ks\rangle + N\langle(s-m)\tanh Ks\rangle. \tag{12.52}$$

But, in the limit $N \to \infty$, the Gaussian distribution (12.27) leads us to the results $\langle s\tanh Ks\rangle \to m\tanh Km$ and $N\langle(s-m)\tanh Ks\rangle \to \chi K\operatorname{sech}^2 Km$ from which we get the equation which gives the time evolution of the variance

$$\frac{1}{2\alpha}\frac{d\chi}{dt} = -\chi + 1 - m\tanh Km + \chi K\operatorname{sech}^2 Km. \tag{12.53}$$

In the stationary state we see that

$$\chi = \frac{1-m^2}{1-K(1-m^2)}, \tag{12.54}$$

where we used the result $m = \tanh Km$ valid in the stationary state. Comparing with (12.44), we see that $\chi^* = \beta\chi = \chi/kT$ which is the result expected in equilibrium.

Therefore, the critical behavior of χ in equilibrium is the same as that of χ^*, that is, $\chi \sim |\varepsilon|^{-1}$. At the critical point and for large times Eqs. (12.43) and (12.53) become identical so that χ has the same behavior as χ^*, that is, $\chi \sim t$.

The results obtained above within the mean-field theory show that the Ising model presents a phase transition from a paramagnetic phase, at high temperatures, to a ferromagnetic phase, at low temperatures. Since the results refer to a model in which each site interacts with a large number of sites, although weak, they cannot be considered in fact as an evidence of the existence of a phase transition, at a finite temperature, in models in which a site interacts with few neighbors. The one-dimensional model, as we have seen, does not present the ferromagnetic state at finite temperatures. However, it is possible to show, using arguments due to Peierls and Griffiths, that the Ising model with short range interactions indeed exhibits a phase transition at finite temperatures in regular lattices in two or more dimensions. The mean-field theory, on the other hand, gives the correct critical behavior in regular lattices when the number of neighboring sites is large enough, a situation that occurs in high dimensions. For the Ising model it is known that this occurs above four dimensions.

12.5 Critical Exponents and Universality

As we have seen, the various quantities that characterize a system that exhibits a critical point behave around the critical point according to power laws and are characterized by critical exponents. In general, we distinguish the static exponents, which are those related to the behavior of quantities at the stationary state, from those that we call dynamic exponents. Below we list the more important quantities used in characterizing the critical behavior, together with their respective critical exponents.

Specific heat (and variance of energy)

$$c \sim \varepsilon^{-\alpha}, \tag{12.55}$$

where $\varepsilon = |T - T_c|$. Magnetization, which is identified with the order parameter,

$$m \sim \varepsilon^{\beta}, \tag{12.56}$$

valid at temperatures below the critical temperature. Susceptibility (and variance of magnetization)

$$\chi \sim \varepsilon^{-\gamma}. \tag{12.57}$$

Magnetization as a function of time, at the critical temperature,

$$m \sim h^{1/\delta}. \tag{12.58}$$

Correlation length,

$$\xi \sim \varepsilon^{-\nu_\perp}. \tag{12.59}$$

Pair correlation function,

$$\rho(r) \sim \frac{1}{r^{d-2+\eta}}, \tag{12.60}$$

at the critical point. Time correlation length or relaxation time

$$\tau \sim \varepsilon^{-\nu_\parallel}. \tag{12.61}$$

The relation between ξ and τ is characterized by the dynamic critical exponent z,

$$\tau \sim \xi^z, \tag{12.62}$$

from which follows the relation $z = \nu_\parallel / \nu_\perp$. Time decay of the magnetization at the critical point

$$m \sim t^{-\beta/\nu_\parallel}. \tag{12.63}$$

Time increase of the variance at the critical point

$$\chi \sim t^{\gamma/\nu_\parallel}. \tag{12.64}$$

Time increase of the correlation $Q = \langle \mathcal{M}(t)\mathcal{M}(0)\rangle / N$ starting from an uncorrelated configuration,

$$Q \sim t^\theta. \tag{12.65}$$

at the critical point. Time increase of the time correlation $C(t) = \langle \sigma_i(t)\sigma_i(0)\rangle$ starting from an uncorrelated configuration

$$C \sim t^\lambda. \tag{12.66}$$

at the critical point.

Magnetization of a finite system at the stationary state determined at the critical point

$$m \sim L^{-\beta/\nu_\perp}, \tag{12.67}$$

Table 12.2 Critical exponents of the Glauber-Ising model, according to compilations of Pelissetto and Vicari (2002) (static exponents) and Henkel and Pleimling (2010) (dynamic exponents). The errors in the numerical values with decimal point are less or equal to 10^{-n}, where n indicates the decimal position of the last digit

d	α	β	γ	δ	$\nu_\perp$	η	$\nu_\parallel$	z	θ	λ
2	0	1/8	7/4	15	1	1/4	2.17	2.17	0.19	0.73
3	0.110	0.326	1.237	4.79	0.630	0.036	1.28	2.04	0.11	1.4
≥ 4	0	1/2	1	3	1/2	0	1	2	0	2

where L is the linear size of the system. Here we should understand the magnetization in the sense given by Eq. (12.23). Variance of a finite system at the stationary state at the critical point

$$\chi \sim L^{\gamma/\nu_\perp}. \tag{12.68}$$

Here we should understand the variance in the sense given by Eq. (12.24).

Table 12.2 shows the values of several exponents of the Glauber-Ising model determined by various methods. Taking into account that the stationary properties of the Glauber-Ising model are the same as those of the Ising model, the static critical exponents coincide with those of the Ising model. In two dimensions they are determined from the exact solution of the model. In particular, the specific heat diverges logarithmically, that is, $c \sim \ln|T - T_c|$, and is therefore characterized by $\alpha = 0$.

The exponents determined in the previous section, calculated within the mean-field theory, are called classical exponents and become valid, for the case of the Glauber-Ising model, when $d \geq 4$. In $d = 4$, the critical behavior may exhibit logarithmic corrections. Notice that the specific heat according to the mean-field theory presents a discontinuity and is therefore characterized by $\alpha = 0$.

It is worth to note also that the behavior of the quantities for finite systems is particularly useful in the determination of the critical exponents by numerical simulations. The numerical results shown in Figs. 12.1 and 12.2 can therefore be used to determine the critical exponents. For example, a plot of $\ln m$ at the critical point versus $\ln L$ gives the ratio $-\beta/\nu_\perp$ with the slope of the straight lined fitted to the data. Similarly, the plots of $\ln \chi$ and $\ln c$ at the critical point as functions of $\ln L$ give the ratios $\gamma/\nu_\perp$ and $\alpha/\nu_\perp$.

It is worth noticing that systems of different nature present values for the critical exponents that are similar to each other and consistent with those corresponding to the Glauber-Ising model in $d = 3$. These equalities among the exponents of distinct systems is a manifestation of the *principle of the universality of the critical behavior*. According to this principle, the critical behavior depend only on a small number of properties. These include (a) the dimensionality of the system, (b) the dimensionality of the order parameter, and (c) the symmetry. In the present case of the Glauber-Ising model, the system has inversion symmetry, that is, the inversion rate (12.19) is invariant under the transformation $\sigma_i \to -\sigma_i$ and the order parameter is a scalar (one dimensional). All systems that have these properties must have the same critical exponents no matter what is the physical nature of the system. The set of these systems comprises the class of universality of the Glauber-Ising model. We should add also that the universal properties do not depend on the details of the microscopic interaction as long as they are of short range or on the details of the inversion rate as long as they involve sites on a small neighborhood.

12.6   Majority Vote Model

The majority vote model is a stochastic model defined on a lattice, which is irreversible, has the inversion symmetry and presents a transition between a paramagnetic phase and a ferromagnetic phase. Let us imagine a community of individuals each one holding an opinion about a certain issue, being in favor or against it. As the passage of time, the individuals change their opinion according to the opinion of the individuals in their neighborhood. In an easily receptive community, an individual would take the opinion of the majority of the individuals in the neighborhood. However, we face a community of hesitating individuals, that is, the individuals sometimes act contrary to the opinion of the majority. Thus, we introduce a positive parameter q which gives a measure of the hesitation of the individuals, which is understood as the probability of a certain individual to take the opposite opinion of the majority.

To set up a model that describes such a community, we assume that the individuals are located at the sites of a regular lattice. To each site i we associate a stochastic variable σ_i that takes the value $+1$ in case the individual at i is in favor and the value -1 otherwise. The number of individuals with favorable opinion is given by $N_F = \sum_i (1 + \sigma_i)/2$.

The dynamics of the majority vote model, governed by the master equation (12.2), is analogous to the Glauber-Ising with the exception of the rate of inversion which in the present model is

$$w_i(\sigma) = \frac{\alpha}{2}\{1 - \gamma\sigma_i \mathscr{S}(\sum_\delta \sigma_{i+\delta})\}, \tag{12.69}$$

where $\mathscr{S}(x) = -1, 0, +1$ in case $x < 0$, $x = 0$ and $x > 0$, respectively, and the sum extends over the neighbors of site i. The parameter γ is restricted to the interval $0 \leq \gamma \leq 1$. Notice that, if we perform the transformation $\sigma_i \to -\sigma_i$, the rate of inversion (12.69) remains invariant, that is, $w_i(-\sigma) = w_i(\sigma)$.

The simulation of the majority vote model is done as follows. At each time step, we choose a site at random and look to its neighborhood. The chosen site changes sign with probability equal to $q = (1-\gamma)/2$ if the majority of the neighbors has the same site of the chosen site and with probability $p = (1+\gamma)/2$ if the majority of the neighbors has the opposite sign. Equivalently, the chosen site takes the sign of the majority of the neighbors with probability p and the opposite sign with probability

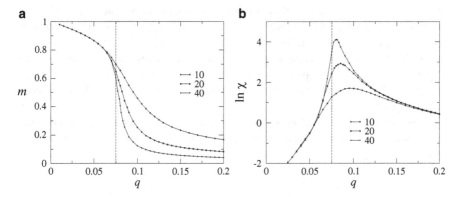

Fig. 12.4 (a) Magnetization m and (b) variance χ versus q for the majority vote model in a square lattice obtained by simulation for various values of L indicated. The *dashed line* indicates the critical value $q_c = 0.075$

q. When the number of plus and minus signs are equal the chosen site takes either sign with equal probability. Figure 12.4 shows the results obtained from simulation for the magnetization m and variance χ as defined by (12.23) and (12.24).

In the stationary regime, in a way similar to what happens to the Glauber-Ising model, the majority vote model exhibits a phase transition in two or more dimensions. For small values of the parameter q, the model presents a ferromagnetic phase characterized by the presence of a majority of individuals with the same opinion. Above a critical value q_c of the parameter q, the model presents a paramagnetic state with equal number, in the average, of individuals with distinct opinions. Results from numerical simulations show that $q_c = 0.075 \pm 0.001$ for the model defined on a square lattice.

In one dimension the majority vote model becomes equivalent to the one-dimensional Glauber model. Indeed, in one dimension

$$\mathscr{S}(\sigma_{i-1} + \sigma_{i+1}) = \frac{1}{2}(\sigma_{i-1} + \sigma_{i+1}), \qquad (12.70)$$

so that the rate of inversion (12.69) becomes equal to the rate of inversion of the one-dimensional Glauber model. Thus in the stationary regime the majority vote model does not have a ferromagnetic phase, except at $\gamma = 0$.

Results of numerical simulations of the majority vote model on a square lattice, indicate that the critical behavior is the same as that of the Glauber-Ising model, what leads us to say that both models are in the same universality class. Notice, however, that the Glauber-Ising is reversible in the stationary state whereas the majority vote model is not. The similarity of the critical behavior between the two models is due to the fact that both models hold the inversion symmetry $\sigma_i \to -\sigma_i$.

The absence of microscopic reversibility in the majority vote model can be seen in a very simple way. But, before, we will make a brief exposition of a procedure used to check if a certain model does or does not hold microscopic reversibility (or detailed balance). Consider a sequence of states and suppose that the sequence is walked in the direct way and in the reversed way. If the probability of occurrence of the direct sequence is equal to the probability of the reverse sequence, for any sequence, then the model presents microscopic reversibility. If these probabilities are different for at least one sequence of states, then the model does not hold microscopic reversibility.

Consider the sequence of four states A, B, C and D, as shown below. Only the central sites change states when one moves from one state to another, while the other remains the same,

$$A = \begin{bmatrix} & + \, + & \\ + & + \, + & - \\ & + \, + & \end{bmatrix} \qquad B = \begin{bmatrix} & + \, + & \\ + & + \, - & - \\ & + \, + & \end{bmatrix}$$

$$C = \begin{bmatrix} & + \, + & \\ + & - \, - & - \\ & + \, + & \end{bmatrix} \qquad D = \begin{bmatrix} & + \, + & \\ + & - \, + & - \\ & + \, + & \end{bmatrix}$$

According to the transition probabilities, the probability of the trajectory $A \to B \to C \to D \to A$ is given by

$$(q)(\tfrac{1}{2})(p)(q) P(A) = \frac{1}{2} pq^2 P(A), \tag{12.71}$$

where $P(A)$ is the probability of state A in the stationary regime, and of its reverse $A \to D \to C \to B \to A$ is given by

$$(p)(q)(\tfrac{1}{2})(p) P(A) = \frac{1}{2} p^2 q P(A). \tag{12.72}$$

Comparing the two expressions, we see that they are different and therefore the microscopic reversibility is not observed in the majority vote model. In other words, the model has irreversible local rules, which do not obey detailed balance.

By way of comparison, we do the same calculation for the case of the Glauber-Ising model. According to the transition probabilities of the Glauber-Ising model, the probability of the trajectory $A \to B \to C \to D \to A$ is given by

$$(e^{-4\beta J})(1)(1)(e^{-4\beta J}) P(A) = e^{-8\beta J} P(A), \tag{12.73}$$

and of its reverse $A \to D \to C \to B \to A$ is given by

$$(1)(e^{-8\beta J})(1)(1) P(A) = e^{-8\beta J} P(A). \tag{12.74}$$

The two expressions are identical. This identity per se does not guarantee the microscopic reversibility. However, we know that the Glauber transition rates obey detailed balance.

12.7 Mean-Field Approximation

We consider here an approximation solution for the majority vote model on a square lattice, where each site has four nearest neighbors. From Eq. (12.4) and using the rate (12.69), we get the following evolution equation for $\langle \sigma_0 \rangle$,

$$\frac{1}{\alpha}\frac{d}{dt}\langle \sigma_0 \rangle = -\langle \sigma_0 \rangle + \gamma \langle \mathscr{S}(\sigma_1 + \sigma_2 + \sigma_3 + \sigma_4) \rangle, \tag{12.75}$$

where the site 0 has as neighboring sites the sites $1, 2, 3, 4$. Using the identity

$$\mathscr{S}(\sigma_1 + \sigma_2 + \sigma_3 + \sigma_4) = \frac{3}{8}(\sigma_1 + \sigma_2 + \sigma_3 + \sigma_4)$$

$$-\frac{1}{8}(\sigma_1\sigma_2\sigma_3 + \sigma_1\sigma_2\sigma_4 + \sigma_1\sigma_3\sigma_4 + \sigma_2\sigma_3\sigma_4). \tag{12.76}$$

we arrive at

$$\frac{1}{\alpha}\frac{d}{dt}\langle \sigma_0 \rangle = -\langle \sigma_0 \rangle + \frac{3\gamma}{8}(\langle \sigma_1 \rangle + \langle \sigma_2 \rangle + \langle \sigma_3 \rangle + \langle \sigma_4 \rangle)+$$

$$-\frac{\gamma}{8}(\langle \sigma_1\sigma_2\sigma_3 \rangle + \langle \sigma_1\sigma_2\sigma_4 \rangle + \langle \sigma_1\sigma_3\sigma_4 \rangle + \langle \sigma_2\sigma_3\sigma_4 \rangle). \tag{12.77}$$

Next, we use the following approximation

$$\langle \sigma_1\sigma_2\sigma_3 \rangle = \langle \sigma_1 \rangle \langle \sigma_2 \rangle \langle \sigma_3 \rangle, \tag{12.78}$$

called simple mean-field approximation. Moreover, we use the translational invariance so that $\langle \sigma_i \rangle = m$, independently of any site, what allows us to write

$$\frac{1}{\alpha}\frac{d}{dt}m = -\varepsilon m - \frac{\gamma}{2}m^3, \tag{12.79}$$

where

$$\varepsilon = 1 - \frac{3}{2}\gamma. \tag{12.80}$$

Multiplying both sides by m, we reach the equation

$$\frac{1}{2\alpha}\frac{d}{dt}m^2 = -\varepsilon m^2 - \frac{\gamma}{2}m^4, \tag{12.81}$$

which can be considered a differential equation for m^2. This differential equation can be solved exactly. For the initial condition $m(0) = m_0$, the solution is

$$m^2 = \frac{2m_0^2\varepsilon}{(2\varepsilon + \gamma m_0^2)e^{2\alpha\varepsilon t} - \gamma m_0^2}. \tag{12.82}$$

When $t \to \infty$ the solutions are: (a) $m = 0$, corresponding to the paramagnetic or disordered state, for $\varepsilon > 0$ and (b) $m = \pm m^*$ corresponding to the ferromagnetic or ordered state for $\varepsilon < 0$, where

$$m^* = \sqrt{\frac{2|\varepsilon|}{\gamma}} = \sqrt{\frac{3\gamma - 2}{\gamma}} = \sqrt{\frac{1 - 6q}{1 - 2q}}, \tag{12.83}$$

that is, the order parameter behaves as

$$m^* \sim (q_c - q)^{-1/2}, \tag{12.84}$$

where $q_c = 1/6 = 0.1666\ldots$.

Notice that the order parameter m is related to the average number of individuals in favor by

$$\langle N_F \rangle = \frac{N}{2}(1 + m), \tag{12.85}$$

where N is the total number of individuals. Thus, for values of q above the critical value q_c, there is, in the average, half of individuals in favor and half against. Below this critical value, there might occur a predominance of individuals in favor, or a predominance of individuals against.

Entropy production The majority vote model is irreversible as we have seen and hence is found in continuous production of entropy. Here we are interested in determining the production of entropy in the stationary state. In the stationary state we have seen in Chap. 8 that he rate of entropy production Π can be written in terms of the transition rates. For models defined by the rates of inversion $w_i(\sigma)$ it is written as

$$\Pi = k \sum_\sigma \sum_i w_i(\sigma) P(\sigma) \ln \frac{w_i(\sigma)}{w_i(\sigma^i)}, \tag{12.86}$$

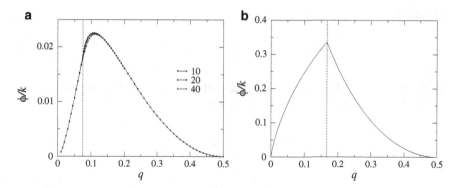

Fig. 12.5 Rate of entropy production per site ϕ versus q for the majority vote model, according to Crochik and Tomé (2005). (**a**) Result obtained by simulation in a square lattice for several values of L indicated. (**b**) Result obtained by mean-field approximation. The *dashed lines* indicate the critical value q_c

which can be written as the average

$$\Pi = k \sum_i \left\langle w_i(\sigma) \ln \frac{w_i(\sigma)}{w_i(\sigma^i)} \right\rangle, \tag{12.87}$$

This last formula is particularly useful because it can be used in simulations. Figure 12.5 shows the entropy production obtained from simulations of the majority vote model for several values of lattice size. A way to calculate it consists in obtaining the average of the expression between brackets. Another equivalent way consists in calculate $s_i(\sigma) = \ln[w_i(\sigma^i)/w_i(\sigma)]$ in any time step in which the change in sign of the variable σ_i is performed with success. The arithmetic average of s_i along the simulation will give an estimate of Π. Notice that s_i takes only two values of opposite signs, which are $\ln(q/p)$ or $\ln(p/q)$. The first occurs when σ_i has the same sign of the majority of its neighbors and the second, when σ_i has the opposite sign.

As seen in Fig. 12.5, the entropy production has an inflexion at the critical point that is analogous to what occurs with the energy, or entropy, of the Glauber-Ising model at the critical point. From this analogy we presume that the slope of Π at the critical point grows without limits when $L \to \infty$. According to this analogy, the critical behavior of Π must be the same as that of u of the Glauber-Ising model.]

To determine Π within the mean-field approximation, we observe first that

$$\ln \frac{w_i(\sigma)}{w_i(\sigma^i)} = \sigma_i \mathscr{S}_i \ln \frac{q}{p}, \tag{12.88}$$

where $\mathscr{S}_i = \mathscr{S}(\sum_\delta \sigma_{i+\delta})$ and we recall that

$$w_i(\sigma) = \frac{\alpha}{2}(1 - \gamma \sigma_i \mathscr{S}_i). \tag{12.89}$$

Therefore

$$w_i(\sigma) \ln \frac{w_i(\sigma)}{w_i(\sigma^i)} = \frac{\alpha}{2}(\sigma_i \mathscr{S}_i - \gamma \mathscr{S}_i^2) \ln \frac{q}{p}. \tag{12.90}$$

Using the identity (12.76), valid for a central site $i = 0$ and its neighbors $j = 1, 2, 3, 4$, and the mean-field approximation, we get

$$\langle \sigma_0 \mathscr{S}_0 \rangle = \frac{1}{2}(3m^2 - m^4), \tag{12.91}$$

$$\langle \mathscr{S}_0^2 \rangle = \frac{1}{8}(5 + 6m^2 - 3m^4). \tag{12.92}$$

Therefore, the rate of entropy production $\phi = \Pi/N$ per site is

$$\phi = k\alpha \left(\frac{1}{4}(3m^2 - m^4) - \frac{\gamma}{16}(5 + 6m^2 - 3m^4) \right) \ln \frac{q}{p}. \tag{12.93}$$

We should distinguish two cases. In the first, $q \geq q_c = 1/6$ and hence $m = 0$ so that

$$\phi = k\alpha \frac{5}{16}(1 - 2q) \ln \frac{1 - q}{q}, \tag{12.94}$$

where we have taken into account that $\gamma = p - q$ and $p + q = 1$. In the second, $q < q_c = 1/6$ and $m^2 = (1 - 6q)/(1 - 2q)$ so that

$$\phi = k\alpha \frac{q(1 - q)}{1 - 2q} \ln \frac{1 - q}{q}. \tag{12.95}$$

Figure 12.5 shows the production of entropy ϕ as a function of q. As we can see, it is continuous in q, but the derivative of ϕ has a jump at the critical point.

Chapter 13
Systems with Absorbing States

13.1 Introduction

In this chapter we continue our study of systems defined on a lattice and governed by master equations. However, we address here systems with absorbing states, which are intrinsically irreversible. The transition rates that define the dynamics forbid them to obey detailed balance. An absorbing state is the one such that the transition from it to any other state is forbidden, although the transition from other state to it might occur. Once in the absorbing state, the system cannot escape from it. All models that exhibit continuous transition to an absorbing state have the same critical behavior, that is, they comprise a universality class. This conjecture, advanced by Janssen (1981) and Grassberger (1982), has been verified by several numerical studies in a great variety of models.

The models studied in this chapter have no analogy with equilibrium models in the sense that they are intrinsically irreversible. No equilibrium model (obeying detailed balance) belongs to this universality class. The simple fact that a model has an absorbing state shows clearly that it cannot obey detailed with respect to the probability distribution of the active state, and therefore they are not described by a Hamiltonian known a priori. The systems are defined in lattices where to each site i we associate a stochastic variable η_i. The total configuration of the system is denoted by

$$\eta = (\eta_1, \eta_2, \ldots, \eta_i, \ldots, \eta_N), \tag{13.1}$$

where N is the total number of sites in the lattice. We will consider dynamics such that, at each time step, only one site is updated. We examine first the models for which the variable η_i takes two values: 1, meaning that the site is occupied by a particle, or 0, meaning that the site is empty.

© Springer International Publishing Switzerland 2015
T. Tomé, M.J. de Oliveira, *Stochastic Dynamics and Irreversibility*, Graduate Texts in Physics,
DOI 10.1007/978-3-319-11770-6_13

The systems are governed by master equations with transition rates $w_i(\eta)$ from $\eta_i \rightarrow 1 - \eta_i$ that comprise the mechanisms that give rise to absorbing states. The time evolution of the probability $P(\eta, t)$ of state η is governed by the master equation

$$\frac{d}{dt} P(\eta, t) = \sum_i \{ w_i(\eta^i) P(\eta^i, t) - w_i(\eta) P(\eta, t) \}, \tag{13.2}$$

where η^i denotes the configuration

$$\eta^i = (\eta_1, \eta_2, \dots, 1 - \eta_i, \dots, \eta_N). \tag{13.3}$$

The evolution equation for the average $\langle \eta_i \rangle$ is obtained from the master equation (13.2) is given by

$$\frac{d}{dt} \langle \eta_i \rangle = \langle (1 - 2\eta_i) w_i(\eta) \rangle. \tag{13.4}$$

The evolution equation for the correlation $\langle \eta_i \eta_j \rangle$ can also be obtained from the master equation (13.2) and is given by the expression

$$\frac{d}{dt} \langle \eta_i \eta_j \rangle = \langle (1 - 2\eta_i) \eta_j w_i(\eta) \rangle + \langle \eta_i (1 - 2\eta_j) w_j(\eta) \rangle. \tag{13.5}$$

13.2 Contact Process

The contact process is defined as a stochastic model with an absorbing state and time evolution governed by a master equation. It is possibly the simplest model exhibiting a phase transition in one dimension. The model was introduced by Harris (1974) who showed that, in the thermodynamic limit, the model presents an active state in addition to the absorbing state. This is a relevant result if we bear in mind that for finite systems, studied in numerical simulations, the absorbing state is always reached if we wait enough time. This time is smaller the closest the system is to the point that determines the transition from the active to the absorbing state.

The contact process can also be viewed as a model for an epidemic spreading. The process is defined on a lattice where at each site resides an individual that might be susceptible (empty site) or infected (occupied site). The susceptible individuals becomes infected, if at least one individual in the neighborhood is infected. The rate of infection is proportional to the number of infected individuals in the neighborhood. In addition to the process of infection, there is a process related to the recovery of the infected. They cease to be infected spontaneously and become susceptible, that is, the recovery does not provide immunity to the individuals. As we see, it is a dynamic process in which the individuals are infected and recover continuously. However, in the time evolution the system may reach a state in which there is no infected individual anymore, all individuals had become susceptible. In this case, the epidemic does not have any means to spread and this state is a stationary absorbing state. The system may also evolve to a stationary state in which the epidemic never ceases completely but also does not contaminate the totality of the individuals because they always can recover spontaneously. Next, we formally present the model.

The contact process consists in a system of interacting particles residing on the sites of a lattice and evolving according to Markovian local rules. Each site of the lattice can be in two states, empty or occupied. At each time step, a site of the lattice is chosen at random.

(a) If the site is empty, it becomes occupied with a transition rate proportional to number n_v of neighboring occupied sites. In a regular lattice of coordination number z, the transition rate is assumed to be $\lambda n_v/z$, where λ is a positive parameter. Thus, in the one-dimensional case, if only one neighboring site is occupied, the rate is $\lambda/2$. If the two neighboring sites are occupied, the chosen site become occupied with rate λ. If there is no neighboring site occupied, then the site remains empty.

(b) If the chosen site is occupied, it becomes empty with rate one.

The first process corresponds to an autocatalytic creation process and the second to a spontaneous annihilation process. For λ large enough, the system presents an active state such that the density of particles is nonzero. Decreasing λ, we reach a critical value of λ_c below which the system will be found in the absorbing state.

The simulation of the contact process defined on a regular lattice of coordination z, with N sites can be preformed as follows. At each time step we choose a site at random, say site i. (a) If i is occupied, than we generate a random number ξ uniformly distributed in the interval $[0, 1]$. If $\xi \leq \alpha = 1/\lambda$, the particle is annihilated and the site becomes empty. Otherwise, the site remains occupied. (b) If i is empty, then one of its neighbors is chose at random. If the neighboring site is occupied then we create a particle at site i. Otherwise, the site i remains empty. At each Monte Carlo step we calculate the total number of particles $n = \sum_i \eta_i$ from which we determine the average density of particles

$$\rho = \frac{1}{N} \langle n \rangle, \tag{13.6}$$

and the variance in the number of particles.

$$\chi = \frac{1}{N} \{ \langle n^2 \rangle - \langle n \rangle^2 \}. \tag{13.7}$$

We used the algorithm presented above to simulate the one-dimensional contact process whose results are shown in Fig. 13.1 for several system sizes. To avoid that the system fall into the absorbing state, we forbid the last particle to be annihilated.

With this requirement it is possible to simulate systems of any size as shown in Fig. 13.1. Notice that with this requirement the density ρ is always finite, but for $\lambda < \lambda_c$ ($\alpha > \alpha_c$) it vanishes in the limit $N \to \infty$, characterizing therefore the absorbing state.

Another way of studying the contact model is done by starting from a single occupied site on an infinite lattice. A better way to perform the simulation is to use a list of active sites, that is, a list of sites that may have their states modified. At each iteration, one site of the list is chose at random and the time is increased by a value equal to the inverse of the numbers of sites in the list. Next, the list is updated, with some sites leaving the list and others entering the list. A simulation performed at the critical point according to this prescription for the one-dimensional model is shown in Fig. 13.2. At the critical point the clusters have a fractal structure with fractal dimension $d_F = 0.748$. If we use this prescription for an ordinary simulation, in

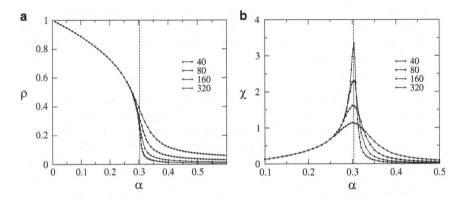

Fig. 13.1 (**a**) Density ρ and (**b**) variance χ versus $\alpha = 1/\lambda$ for the contact model in a one-dimensional lattice obtained by simulations for various values of L indicated. The *dashed line* indicates the critical point, $\alpha_c = 0.303228$

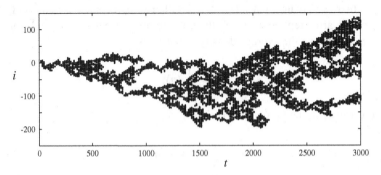

Fig. 13.2 Sites occupied i as functions of time t for the one-dimensional contact process. Simulation performed at the critical point, $\alpha_c = 0.303228$, starting from a single particle located at $i = 0$. The clusters has fractal dimension $d_F = 0.748$

which one of the N sites is chosen at random, we see that the increment in time equals $1/N$ and therefore one unit of time coincides with a Monte Carlo step.

The contact model has a conservative version whose rules are such that the number of particles remain invariant. The rules of this version are as follows. At each time step a particle and an empty site i are chosen at random and independently. The particle is transferred to the empty site with a probability equal to the fraction r_i of particles belonging to the neighborhood of site i. If the empty site has no neighboring particles, the chosen particle remains where it is. In this version, the number of particles works as an external parameter and the rate α is determined as follows. Define n_a as being the weighted number of active sites. An active site is an empty site with at least one neighbor occupied. The weight of an active site i is equal to the fraction r_i so that

$$n_a = \sum_i r_i (1 - \eta_i). \tag{13.8}$$

The rate α is calculate as the ratio between the average weighted number of active sites and the number of particles n,

$$\alpha = \frac{\langle n_a \rangle}{n}. \tag{13.9}$$

The critical rate is determined as the limit of α when $n \to \infty$, in an infinite lattice. Using this procedure we can determine the critical rates of the contact model for various lattices, as shown in Table 13.1.

Evolution equations for the correlations According to the rules above, the transition rate $w_i(\eta)$ of changing the state of site i, that is, $\eta_i \to 1 - \eta_i$, of the contact model is given by

$$w_i(\eta) = \frac{\lambda}{z}(1 - \eta_i) \sum_\delta \eta_{i+\delta} + \eta_i, \tag{13.10}$$

Table 13.1 Critical parameters λ_c and $\alpha_c = 1/\lambda_c$ of the contact model for various lattices of dimension d and coordination number z, according to Jensen and Dickman (1993) and Sabag and Oliveira (2002). The errors are in the last digit

Lattice	d	z	λ_c	α_c
Chain	1	2	3.29785	0.303228
Square	2	4	1.64872	0.60653
Cubic	3	6	1.31683	0.75940
Hipercubic	4	8	1.19511	0.83674
Hipercubic	5	10	1.13847	0.87837

where the summation is over the z nearest neighbors of site i. The first term, in this equation, is related to the autocatalytic creation, which is the process (a) defined above; and the second, to the spontaneous annihilation, which is the process (b) above. Replacing (13.10) into the evolution equation for the average $\langle \eta_i \rangle$, given by (13.4), we get

$$\frac{d}{dt} \langle \eta_i \rangle = \frac{\lambda}{z} \sum_\delta \langle (1 - \eta_i) \eta_{i+\delta} \rangle - \langle \eta_i \rangle. \tag{13.11}$$

and therefore the evolution equation for $\langle \eta_i \rangle$ depends not only on $\langle \eta_i \rangle$ but also on the correlation of two neighboring sites.

The time evolution related to the correlation of two neighboring sites, $\langle \eta_i \eta_j \rangle$, is obtained from (13.5). Replacing the rate (13.10) into this equation, we get

$$\frac{d}{dt} \langle \eta_i \eta_j \rangle = -2 \langle \eta_i \eta_j \rangle + \frac{\lambda}{z} \sum_\delta \langle (1 - \eta_i) \eta_j \eta_{i+\delta} \rangle + \frac{\lambda}{z} \sum_\delta \langle (1 - \eta_j) \eta_i \eta_{j+\delta} \rangle. \tag{13.12}$$

We should bear in mind that the first sum in δ contains a term such that $i + \delta = j$. In this case, $\eta_j \eta_{i+\delta} = \eta_j$ so that

$$\sum_\delta \langle (1 - \eta_i) \eta_j \eta_{i+\delta} \rangle = \sum_{\delta (i+\delta \neq j)} \langle (1 - \eta_i) \eta_j \eta_{i+\delta} \rangle + \langle (1 - \eta_i) \eta_j \rangle \tag{13.13}$$

Similarly

$$\sum_\delta \langle (1 - \eta_j) \eta_i \eta_{j+\delta} \rangle = \sum_{\delta (j+\delta \neq i)} \langle (1 - \eta_j) \eta_i \eta_{j+\delta} \rangle + \langle (1 - \eta_j) \eta_i \rangle. \tag{13.14}$$

We see that the evolution equation for the correlation of two sites depend on the correlation of three sites. The evolution equation for the correlation of three sites will depend on the correlation of four sites and so on. The equations for the various correlations form therefore a hierarchy such that the equation for a certain correlation depend on correlation involving a larger number of sites. An approximate form of solving this hierarchy of equations consists in a truncation scheme, as we will see next.

13.3 Mean-Filed Approximation

Assuming a homogeneous and isotropic solution, then $\langle \eta_i \rangle = \rho$ and $\langle \eta_i \eta_{i+\delta} \rangle = \phi$, independently of i and δ, and Eq. (13.11) becomes

$$\frac{d}{dt} \rho = \lambda(\rho - \phi) - \rho. \tag{13.15}$$

Employing the simple mean-field approximation, which amounts to use the approximation $P(\eta_i, \eta_j) = P(\eta_i)P(\eta_j)$, and therefore, $\phi = \rho^2$, Eq. (13.15) can be written in the following form

$$\frac{d}{dt}\rho = (\lambda - 1)\rho - \lambda\rho^2. \tag{13.16}$$

The stationary solutions of these equations, which characterize the stationary states of the model, are $\rho = 0$, the trivial solution, which refers to the absorbing state, and

$$\rho = \frac{\lambda - 1}{\lambda}, \tag{13.17}$$

valid for $\lambda > 1$, which is related with the active state, characterized by a nonzero density of particles. Figure 13.3 show ρ as a function of the parameter $\alpha = 1/\lambda$. As seen, in the stationary state, there is a phase transition between an active state, characterized by $\rho \neq 0$, which is identified with the order parameter, and an absorbing state, characterized by $\rho = 0$. Close to the critical point, $\lambda_c = 1$, ρ behaves as

$$\rho \sim (\lambda - \lambda_c). \tag{13.18}$$

Equation (13.16) can be exactly solved with the following solution for the case $\lambda \neq 1$,

$$\rho = \frac{\lambda - 1}{\lambda - ce^{-(\lambda - 1)t}}, \tag{13.19}$$

where c is a constant which must be determined by the initial conditions. When $t \to \infty$, the stationary solution $\rho = 0$, corresponding to the absorbing state, is reached if $\lambda < 1$, while the solution $\rho = (\lambda - 1)/\lambda$, corresponding to the active state, if $\lambda > 1$. In the first case, $\lambda < 1$, the density ρ decays to zero exponentially,

$$\rho = ae^{-(1-\lambda)t}. \tag{13.20}$$

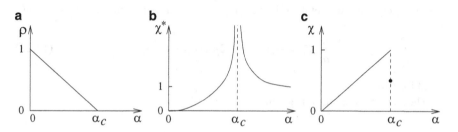

Fig. 13.3 (a) Density ρ, (b) susceptibility χ^*, and (c) variance χ versus $\alpha = 1/\lambda$ for the contact model according to the mean-field theory, $\alpha_c = 1/\lambda_c$

Similarly, in the second case, the density relaxes to the nonzero solution exponentially,

$$\rho = \frac{\lambda - 1}{\lambda} - be^{-(\lambda-1)t}. \tag{13.21}$$

In both cases, the relaxation time τ is given by

$$\tau \sim |\lambda - \lambda_c|^{-1}, \tag{13.22}$$

where $\lambda_c = 1$.

At the critical point, $\lambda_c = 1$, the relaxation time diverges and the relaxation ceases to be exponential. The time evolution of the density becomes

$$\frac{d\rho}{dt} = -\rho^2, \tag{13.23}$$

whose solution is $\rho = 1/(t+c)$, so that for large times the decay becomes algebraic,

$$\rho \sim t^{-1}. \tag{13.24}$$

Now, we modify the transition rate so that the absorbing state is absent. This is carried out by introducing a spontaneous creation of particles in empty sites. The transition rate becomes

$$w_i(\eta) = \frac{\lambda}{z}(1 - \eta_i) \sum_\delta \eta_{i+\delta} + \eta_i + h(1 - \eta_i), \tag{13.25}$$

where h is a parameter associated to the spontaneous creation. Using this rate and within the simple mean-field approximation, the equation for the evolution of the density of particles becomes

$$\frac{d}{dt}\rho = (\lambda - 1)\rho - \lambda\rho^2 + h(1 - \rho). \tag{13.26}$$

From this equation we get the following equation for the susceptibility $\chi^* = \partial\rho/\partial h$,

$$\frac{d}{dt}\chi^* = -(1 - \lambda + 2\rho\lambda)\chi^* + (1 - \rho), \tag{13.27}$$

valid for $h = 0$.

The stationary solution is given by

$$\chi^* = \frac{1 - \rho}{1 - \lambda + 2\rho\lambda}, \tag{13.28}$$

and is shown in Fig. 13.3. When $\lambda < \lambda_c = 1$, $\rho = 0$ and $\chi^* = 1/(1 - \lambda)$. When $\lambda > \lambda_c$, $\rho = (\lambda - 1)/\lambda$ and $\chi^* = 1/\lambda(\lambda - 1)$. In both cases the susceptibility diverges as

$$\chi^* = |\lambda - \lambda_c|^{-1}. \tag{13.29}$$

At the critical point, $\lambda = 1$, and for large times, the equation for the susceptibility is reduced to

$$\frac{d}{dt}\chi^* = 1 - \frac{2}{t}\chi^*, \tag{13.30}$$

whose solution is $\chi^* = t/3$. Therefore, for large times the susceptibility has the following behavior at the critical point

$$\chi^* \sim t. \tag{13.31}$$

To obtain an approximation of the second order, we use the time evolution equation related to the correlation of two neighboring sites, $\langle \eta_i \eta_j \rangle$, given by (13.5). The mean-field approximation of second order, called pair mean-field approximation, is that in which the correlation of three sites is determined by using the approximation

$$P(\eta_j, \eta_i, \eta_k) = \frac{P(\eta_j, \eta_i) P(\eta_i, \eta_k)}{P(\eta_i)}, \tag{13.32}$$

where j and k are distinct neighbors of site i. Through this procedure, Eqs. (13.4) and (13.5) become closed equations for $\langle \eta_i \rangle$ and $\langle \eta_i \eta_j \rangle$. Assuming isotropy, Eq. (13.12) is transformed into the following equation

$$\frac{d}{dt}\phi = \frac{2\lambda(z - 1)}{z}\frac{(\rho - \phi)^2}{(1 - \rho)} + \frac{2\lambda}{z}(\rho - \phi) - 2\phi, \tag{13.33}$$

which together with Eq. (13.15) constitute a close set of equations for ρ and ϕ. A stationary solution is given by $\rho = \phi = 0$, which corresponds to the absorbing state. The other stationary solution, corresponding to the active state, is given by

$$\rho = \frac{\lambda(z - 1) - z}{\lambda(z - 1) - 1}, \tag{13.34}$$

and $\phi = (\lambda - 1)\rho/\lambda$, valid for $\lambda > \lambda_c$, where

$$\lambda_c = \frac{z}{z - 1} \tag{13.35}$$

is the critical rate in the pair approximation. The critical behavior of the order parameter, $\rho \sim (\lambda - \lambda_c)$, remains the same as that of the simple mean-field approximation, although the critical rate has a different value.

13.4 Mean-Field Theory

An exact solution of the contact model can be obtained when the number of neighbors is very large. According to this idea we modify the rate (13.10) so that the sum extends over all sites,

$$w_i(\eta) = \lambda(1 - \eta_i)x + \eta_i, \tag{13.36}$$

where the stochastic variable x is defined by

$$x = \frac{1}{N} \sum_j \eta_j. \tag{13.37}$$

For N large enough, we assume that x is distributed according to a Gaussian of mean ρ and variance χ/N,

$$P(x) = \frac{\sqrt{N}}{\sqrt{2\pi\chi}} e^{-N(x-\rho)^2/2\chi}, \tag{13.38}$$

where ρ and χ depend on t. Notice that

$$\chi = N\{\langle x^2 \rangle - \langle x \rangle^2\}, \tag{13.39}$$

and hence χ coincides with the definition (13.7) since $n = Nx$.

Next, we replace (13.36) into Eq. (13.4), which gives the evolution of the density of particles, and use the definition of x to get

$$\frac{d}{dt}\langle x \rangle = \lambda(\langle x \rangle - \langle x^2 \rangle) - \langle x \rangle, \tag{13.40}$$

In the limit $N \to \infty$, the distribution $P(x)$ becomes a Dirac delta function and $\langle x^2 \rangle \to \rho^2$ and we arrive at the following equation for the density of particles,

$$\frac{d}{dt}\rho = \lambda(\rho - \rho^2) - \rho, \tag{13.41}$$

which is Eq. (13.16) obtained previously.

Next, we determined the time evolution of χ. To this end, we observe first that

$$x^2 = \frac{1}{N^2} \sum_{ij} \eta_i \eta_j = \frac{1}{N^2} \sum_{i \neq j} \eta_i \eta_j + \frac{x}{N}. \tag{13.42}$$

Using the rate (13.36) and Eq. (13.5),

$$\frac{d}{dt} \langle \eta_i \eta_j \rangle = -2 \langle \eta_i \eta_j \rangle - 2\lambda \langle \eta_i \eta_j x \rangle + \lambda \langle (\eta_i + \eta_j) x \rangle, \tag{13.43}$$

valid for $i \neq j$. From this result, we get

$$\frac{d}{dt} \langle x^2 \rangle = 2(\lambda - 1) \langle x^2 \rangle - 2\lambda \langle x^3 \rangle + \frac{1}{N} [\lambda \rho (1 - \rho) + \rho]. \tag{13.44}$$

On the other hand

$$\frac{d}{dt} \langle x \rangle^2 = 2 \langle x \rangle \frac{d}{dt} \langle x \rangle = 2(\lambda - 1) \rho^2 - 2\lambda \rho \langle x^2 \rangle, \tag{13.45}$$

where we used the result (13.40). From these results we get the time evolution of the variance

$$\frac{d\chi}{dt} = 2(\lambda - 1)\chi + \lambda \rho (1 - \rho) + \rho - 2\lambda N \langle (x - \rho) x^2 \rangle. \tag{13.46}$$

But in the limit $N \to \infty$, the Gaussian distribution (13.38) leads us to the following result $N \langle (x - \rho) x^2 \rangle \to 2\rho \chi$ from which we get

$$\frac{d\chi}{dt} = 2(\lambda - 1 - 2\lambda \rho)\chi + \lambda \rho (1 - \rho) + \rho. \tag{13.47}$$

In the stationary state

$$\chi = \frac{\lambda \rho (1 - \rho) + \rho}{2(-\lambda + 1 + 2\lambda \rho)}, \tag{13.48}$$

which is shown in Fig. 13.3. We distinguish two cases. In the first, $\lambda < \lambda_c = 1$ and $\rho = 0$ from which we conclude that $\chi = 0$. In the second case, $\lambda > \lambda_c$, we get $\rho = (\lambda - 1)/\lambda$ so that

$$\chi = \frac{1}{\lambda}, \tag{13.49}$$

and therefore χ remains finite when $\lambda \to \lambda_c$. It is worth to note that, at $\lambda = \lambda_c = 1$, we get $\chi = 1/2$. Therefore, the variance is finite but has a jump at the critical point equal to 1.

The variance χ is distinct from the susceptibility χ^*, defined as follows. The transition rate (13.36) is modified by the addition of a spontaneous creation of particle in empty sites, given by $h(1 - \eta_i)$ similarly to what we have done in (13.25). From this new rate we see that the time evolution of ρ, given by (13.41), is modified and become identical to Eq. (13.26). The susceptibility is defined by $\chi^* = \partial\rho/\partial h$ and therefore in the stationary state it is given by formula (13.28), which is distinct from χ, given by formula (13.48). Notice that the susceptibility χ^* diverges at the critical point whereas the variance χ remains finite, as seen in Fig. 13.3.

13.5   Critical Exponents and Universality

It is worth to note that the contact process undergoes a phase transition in any dimension, including in one dimension as revealed by the simulations whose results are shown in Fig. 13.1. In this aspect, it has a behavior distinct from the Ising model and other equilibrium models with short range interactions, which do not present a phase transition, at finite temperature, in one dimension. Table 13.1 shows the values of the critical parameter λ_c of the contact model for various regular lattices.

The behavior around the critical point of the various quantities related to the contact model is characterized by critical exponents, which are the same as those of the directed percolation, to be studied later on. These two models, and others having the same exponents, comprise the universality class of the directed percolation, whose exponents are shown in Table 13.2.

The most important quantities that characterize the critical behavior and their respective exponents are presented next. Density of particles, which is identified as the order parameter,

$$\rho \sim \varepsilon^\beta, \tag{13.50}$$

where $\lambda - \lambda_c$. Susceptibility,

$$\chi^* \sim \varepsilon^{-\gamma}. \tag{13.51}$$

Variance,

$$\chi \sim \varepsilon^{-\gamma'}. \tag{13.52}$$

Unlike what happens to the Glauber-Ising model, the exponents γ' and γ are distinct. Spatial correlation length,

$$\xi \sim \varepsilon^{-\nu_\perp}. \tag{13.53}$$

Table 13.2 Critical exponents of models belonging to the universality class of directed perco-lation, which includes the contact process, according to compilations by Muñoz et al. (1999), Hinrichsen (2000) and Henkel et al. (2008). The errors in the numerical values with decimal point are less or equal to 10^{-n}, where n indicates the decimal position of the last digit

d	β	γ	γ'	$\nu_\perp$	$\nu_\parallel$	z	δ	θ
1	0.27649	2.27773	0.54388	1.09685	1.73385	1.58074	0.15946	0.31369
2	0.583	1.59	0.30	0.73	1.30	1.766	0.451	0.230
3	0.81	1.24	0.13	0.58	1.11	1.90	0.73	0.11
≥ 4	1	1	0	1/2	1	2	1	0

Relaxation time or time correlation length,

$$\tau \sim \varepsilon^{-\nu_\parallel}. \tag{13.54}$$

Some exponents are defined from the time behavior of certain quantities at the critical point. These quantities are calculated by assuming that at the initial time there is a single particle, placed at the origin $\mathbf{r} = 0$ of a coordination system. From this seed, a cluster of particles grows. We define the average number of particles by

$$n_p = \sum_{\mathbf{r}} \langle \eta_{\mathbf{r}} \rangle, \tag{13.55}$$

where the sum extends over the sites of the lattice. At the critical point, the average number of particles grows according to

$$n_p \sim t^\theta. \tag{13.56}$$

The spreading of particles is defined by

$$R^2 = \frac{1}{n_p} \sum_{\mathbf{r}} \langle r^2 \eta_{\mathbf{r}} \rangle. \tag{13.57}$$

At the critical point, it behaves as

$$R \sim t^{1/z}. \tag{13.58}$$

At the critical point the surviving probability $\mathscr{P}$ behaves as

$$\mathscr{P} \sim t^{-\delta}. \tag{13.59}$$

In the limit $t \to \infty$ and out of the critical point the surviving probability $\mathscr{P}^*$ is finite and behaves around the critical point according to

$$\mathscr{P}^* \sim \varepsilon^{\beta'}. \tag{13.60}$$

For the contact model, the surviving probability coincides with the density of particles so that $\beta' = \beta$. Later on, we will analyze a model for which this relation is not valid. The fractal dimension d_F is defined as the number of particles n_p that are found inside a region of linear size L, determined at the critical point, that is,

$$n_p \sim L^{d_F}. \tag{13.61}$$

13.6   Evolution Operator

The master equation (13.2) related to the contact process can be written in terms of operators. To this end, we start by building a vector space whose basis vectors are given by $|\eta\rangle = |\eta_1\eta_2\ldots\eta_N\rangle$. Next, we define the probability vector $|\Psi(t)\rangle$ by

$$|\Psi(t)\rangle = \sum_{\eta} P(\eta, t)|\eta\rangle, \tag{13.62}$$

where the sum extends over all basis vectors. Using the master equation (13.2), we determine the time evolution of the probability vector. Deriving this equation with respect to time and using the master equation, we get

$$\frac{d}{dt}|\Psi(t)\rangle = \sum_{\eta}\sum_{i} w_i(\eta)P(\eta, t)\{|\eta^i\rangle - |\eta\rangle\}, \tag{13.63}$$

where $|\eta^i\rangle = |\eta_1\eta_2\ldots 1 - \eta_i \ldots \eta_N\rangle$. Defining the operator $\mathscr{F}_i$ by

$$\mathscr{F}_i|\eta\rangle = |\eta^i\rangle, \tag{13.64}$$

and the operator $\mathscr{Z}_i$ by

$$\mathscr{Z}_i|\eta\rangle = w_i(\eta)|\eta\rangle, \tag{13.65}$$

then, from (13.63), we get

$$\frac{d}{dt}|\Psi(t)\rangle = \sum_{i}(\mathscr{F}_i - 1)\mathscr{Z}_i|\Psi(t)\rangle. \tag{13.66}$$

Therefore, we may write

$$\frac{d}{dt}|\Psi(t)\rangle = \mathscr{S}|\Psi(t)\rangle, \tag{13.67}$$

where the evolution operator $\mathscr{S}$ is

$$\mathscr{S} = \sum_{i=1}^{N}(\mathscr{F}_i - 1)\mathscr{Z}_i. \tag{13.68}$$

Using the rate $w_i(\eta)$, given by (13.10), the operator $\mathscr{Z}_i$ is defined by

$$\mathscr{Z}_i|\eta\rangle = \left(\frac{\lambda}{z}(1 - \eta_i)\sum_{\delta}\eta_{i+\delta} + \eta_i\right)|\eta\rangle. \tag{13.69}$$

The operators $\mathscr{F}_i$ and $\mathscr{Z}_i$ can be written in terms of creation and annihilation operators $\mathscr{A}_i$ and $\mathscr{A}_i^{+}$, defined by

$$\mathscr{A}_i|\eta\rangle = \eta_i|\eta^i\rangle, \qquad \mathscr{A}_i^{+}|\eta\rangle = (1 - \eta_i)|\eta^i\rangle. \tag{13.70}$$

We see that

$$\mathscr{F}_i = \mathscr{A}_i + \mathscr{A}_i^{+}, \tag{13.71}$$

and hence

$$\mathscr{Z}_i = \frac{\lambda}{z}(1 - \mathscr{N}_i)\sum_{\delta}\mathscr{N}_{i+\delta} + \mathscr{N}_i, \tag{13.72}$$

where

$$\mathscr{N}_i = \mathscr{A}_i^{+}\mathscr{A}_i \tag{13.73}$$

is the number operator. Therefore, the evolution operator reads

$$\mathscr{S} = \frac{\lambda}{z}\sum_{i}(\mathscr{A}_i + \mathscr{A}_i^{+} - 1)(1 - \mathscr{N}_i)\sum_{\delta}\mathscr{N}_{i+\delta} + \sum_{i}\mathscr{N}_i. \tag{13.74}$$

Using the property $\mathscr{A}_i^{+}\mathscr{A}_i + \mathscr{A}_i\mathscr{A}_i^{+} = 1$ we get

$$\mathscr{S} = \frac{\lambda}{z}\sum_{i,\delta}(1 - \mathscr{A}_i)\mathscr{A}_i^{+}\mathscr{A}_{i+\delta}^{+}\mathscr{A}_{i+\delta} + \sum_{i}(1 - \mathscr{A}_i^{+})\mathscr{A}_i. \tag{13.75}$$

Notice that $\mathscr{S}$ is an operator which is not Hermitian. This expression for the evolution operator is the starting point for obtaining time series and perturbation series for various quantities that characterize the critical behavior of the contact process.

With the aim of comparing the evolution operator $\mathscr{S}$ of the contact process with the evolution operator associated to the Regge field theory, and which is not Hermitian, we do the following unitary transformation

$$\mathscr{H} = -\mathscr{V}\,\mathscr{S}\,\mathscr{V}^{-1}, \tag{13.76}$$

with

$$\mathscr{V} = \prod_i (1 + \mathscr{A}_i), \qquad \mathscr{V}^{-1} = \prod_i (1 - \mathscr{A}_i). \tag{13.77}$$

Hence the operator $\mathscr{H}$ defined in (13.76) is given by

$$\mathscr{H} = \frac{\lambda}{2} \sum_{i,\delta} \mathscr{A}_i^+ (\mathscr{A}_i - 1)(\mathscr{A}_{i+\delta}^+ + 1)\mathscr{A}_{i+\delta} + \sum_i \mathscr{A}_i^+ \mathscr{A}_i. \tag{13.78}$$

This operator is related with the evolution operators that describe processes in the Regge field theory. The correspondence of the operator $\mathscr{H}$ and the spin models of the Regge field theory was shown by Grassberger and de la Torre within a formalism similar to the one presented here.

13.7   Creation by Two or More Particles

The interaction mechanisms contained in the contact process, which are those of the first Schlögl model, studied in Sect. 10.5, provide the basis for the formulation of a variety of lattice models used to describe epidemic spreading, population dynamics, and chemical reactions. Moreover, several models based on the contact process have been used to study the microscopic ingredients that are relevant in characterizing the kinetic phase transitions in models with absorbing states. The very development of a lattice model, comprising the reactions of the first Schlögl model, led to the conjecture mentioned in Sect. 13.1.

The second Schlögl model has the same mechanisms of the first, but involves an autocatalytic creation by pairs. In the analysis by Schlögl, based on the equations of chemical kinetics and seen in Sect. 10.5, a discontinuous transition is predicted. In the construction of a lattice model that evolves according to local Markovian rules that comprises the reactions of the second Schögl model, fluctuations are implicit introduced. This formulation allows to observe that in one-dimension the model indeed exhibit a phase transition, but the transition is continuous, with a critical behavior in the same universality class of the ordinary contact model. Therefore, the introduction of local rules with stochastic evolution changes in a sensible way, in the one-dimensional case, the predictions of the chemical kinetics.

Possible generalizations of the contact process, inspired in the second Schlögl model, consist therefore in considering that the autocatalysis can only occur when there is a certain minimum number of particles taking part in the reaction. In the contact process, the minimum number of particles is equal to one. Models with pair creation correspond to a minimum of two, triplet creation, a minimum of three, and so on. In addition to the autocatalysis and to the spontaneous annihilation, it is also possible to include the diffusive process. In general one does not add a diffusive process in the ordinary contact process because it has already a proper diffusion. This intrinsic diffusion occurs when an autocatalytic process is followed by a spontaneous annihilation as in the example: $01000 \to 01100 \to 00100$. All these generalizations are models of one component, which have an absorbing state corresponding to the complete absence of particles.

The generalization of the contact model which we study here consists in the pair-creation model. The creation of a particle requires the presence of at least two particles in the neighborhood. When there is just one particle or none the creation is forbidden. The annihilation of particles is spontaneous. Various models can be defined with these properties. Here we consider only those such that the creation of particles is proportional to the number of pairs of occupied sites of a certain neighborhood. To each site of a regular lattice, we denote by V_i the set of pairs of sites (j, k) belonging to a certain neighborhood of site i. The transition rate $w_i(\eta)$ from η_i to $1 - \eta_i$ is defined by

$$w_i(\eta) = \frac{\lambda}{z'}(1 - \eta_i) \sum_{(j,k) \in V_j} \eta_j \eta_k + \eta_i, \tag{13.79}$$

where z' is equal to the number of pairs of sites in the neighborhood, that is, of elements of the set V_i, which we consider to be the same for all sites.

The model so defined is quite generic since we did not specify the neighborhood V_i. For the one-dimensional case, we consider the neighboring pairs of site i as the pairs $(i - 2, i - 1)$ and $(i + 1, i + 2)$. In this case, the rate is

$$w_i(\eta) = \frac{\lambda}{2}(1 - \eta_i)(\eta_{i-2}\eta_{i-1} + \eta_{i+1}\eta_{i+2}) + \eta_i, \tag{13.80}$$

Numerical simulations show that this model shows a phase transition that occurs at $\lambda_c = 7.45$. When $\lambda < \lambda_c$, the model shows an absorbing state. When $\lambda > \lambda_c$, the state is active. As we said above, the critical behavior is the same as that of the ordinary contact model.

In a regular lattice of dimension larger or equal to two, we may consider a neighborhood formed by sites that are nearest neighbors of a certain site. In a regular lattice of coordination number z, we consider the model defined by the transition rate

$$w_i(\eta) = \frac{\lambda}{2z'}(1 - \eta_i)n_i(n_i - 1) + \eta_i, \tag{13.81}$$

where $z' = z(z-1)/2$ is the number of pairs of neighboring sites and n_i is the number of neighboring sites occupied, given by

$$n_i = \sum_\delta \eta_{i+\delta}, \tag{13.82}$$

where the summation is over the nearest neighbors.

Mean-field theory The solution of the pair-creation model defined by the rate (13.81) can be obtained when the number of neighbors is very large. According to this idea, the transition rate that we will regard, in the place of the rate (13.81) and in analogy with the rate (13.36), is the following

$$w_i(\eta) = \lambda(1 - \eta_i)x^2 + \eta_i, \tag{13.83}$$

where the stochastic variable x is defined by

$$x = \frac{1}{N}\sum_j \eta_j. \tag{13.84}$$

Proceeding in a way analogous to that followed in Sect. 13.4, we get

$$\frac{d}{dt}\langle x \rangle = \lambda(\langle x^2 \rangle - \langle x^3 \rangle) - \langle x \rangle. \tag{13.85}$$

In the thermodynamic limit, we obtain the following equation for $\rho = \langle x \rangle$,

$$\frac{d\rho}{dt} = \lambda(\rho^2 - \rho^3) - \rho. \tag{13.86}$$

In the stationary state $\rho = 0$ or $\rho^2 - \rho + \alpha = 0$, that is,

$$\rho = \frac{1}{2}\{1 + \sqrt{1 - 4\alpha}\} = \rho_0, \tag{13.87}$$

where $\alpha = 1/\lambda$. Using the initial condition $\rho = 1$, the stationary solution is shown in Fig. 13.4. When $\alpha > \alpha_0 = 1/4$, or $\lambda < \lambda_0 = 4$, then $\rho = 0$. Otherwise, the stationary solution is that given by (13.87), $\rho = \rho_0$.

To determine the behavior for long times, we expand the right-hand site of (13.86) around the stationary solution. For $\lambda < \lambda_0$, up to linear terms, $d\rho/dt = -\rho$ and hence the decay is exponential, $\rho \sim e^{-t}$ with a relaxation time $\tau = 1$. For $\lambda > \lambda_0$,

$$\frac{d}{dt}\rho = -2(\lambda - \lambda_0)^{1/2}\rho_0(\rho - \rho_0), \tag{13.88}$$

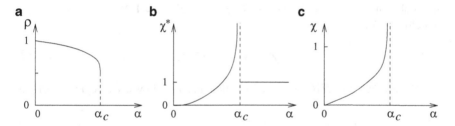

Fig. 13.4 (a) Density ρ, (b) susceptibility χ^*, and (c) variance χ versus $\alpha = 1/\lambda$ for the pair-creation model, according to the mean-field theory, $\alpha_c = 1/\lambda_c$

whose solution is exponential

$$\rho = \rho_0 - Ke^{-t/\tau}, \tag{13.89}$$

with a relaxation time equal to

$$\tau \sim (\lambda - \lambda_0)^{-1/2}. \tag{13.90}$$

When $\lambda = \lambda_0$ and close to the stationary solution, for which $\rho = \rho_0 = 1/2$,

$$\frac{d}{dt}\rho = -2(\rho - \frac{1}{2})^2, \tag{13.91}$$

whose solution for large times is algebraic,

$$\rho - \rho_0 \sim t^{-1}. \tag{13.92}$$

To determine the susceptibility we proceed as previously, that is, we add a rate of spontaneous creation $h(1 - \eta_i)$ in (13.83), which results in the following equation for ρ

$$\frac{d}{dt}\rho = \lambda(\rho^2 - \rho^3) - \rho + h(1 - \rho). \tag{13.93}$$

Deriving with respect to the parameter h, we get an equation for the susceptibility $\chi^* = \partial\rho/\partial h$,

$$\frac{d}{dt}\chi^* = \lambda(2\rho - 3\rho^2)\chi^* - \chi^* + (1 - \rho), \tag{13.94}$$

valid for $h = 0$. The stationary solution is given by

$$\chi^* = \frac{1 - \rho}{1 - \lambda(2\rho - 3\rho^2)}, \tag{13.95}$$

and is shown in Fig. 13.4. When $\lambda < \lambda_0$, $\rho = 0$ and $\chi^* = 1$. When $\lambda > \lambda_0$, $\rho = \rho_0$ and close to λ_0,

$$\chi^* \sim (\lambda - \lambda_0)^{-1/2}. \tag{13.96}$$

The variance $\chi = N[\langle x^2 \rangle - \langle x \rangle^2]$ is calculated as follows. We begin by writing

$$\frac{d}{dt}\langle x^2 \rangle = -2\lambda\langle x^4 \rangle + 2\lambda\langle x^3 \rangle - 2\langle x^2 \rangle + \frac{1}{N}\left(-\lambda\langle x^3 \rangle + \lambda\langle x^2 \rangle + \langle x \rangle\right), \tag{13.97}$$

$$\frac{d}{dt}\langle x \rangle^2 = 2\langle x \rangle\frac{d}{dt}\langle x \rangle = 2\langle x \rangle\left(\lambda\langle x^2 \rangle - \lambda\langle x^3 \rangle - \langle x \rangle\right). \tag{13.98}$$

Subtracting these equations and multiplying by N,

$$\frac{d}{dt}\chi = 2N\lambda\langle (x - \rho)(x^2 - x^3) \rangle - 2\chi - \lambda\langle x^3 \rangle + \lambda\langle x^2 \rangle + \rho. \tag{13.99}$$

In the limit $N \to \infty$, $N\langle (x - \rho)(x^2 - x^3) \rangle \to \chi(2\rho - 3\rho^2)$ and

$$\frac{d}{dt}\chi = 2[\lambda\rho(2 - 3\rho) - 1]\chi + \lambda\rho^2(1 - \rho) + \rho. \tag{13.100}$$

In the stationary state

$$\chi = \frac{\lambda\rho^2(1 - \rho) + \rho}{2[1 - \lambda\rho(2 - 3\rho)]}. \tag{13.101}$$

When $\lambda > 4$, $\rho = 0$ and hence $\chi = 0$. When $\lambda < 4$, $\rho = \rho_0$ and therefore

$$\chi = \frac{1}{\lambda(2\rho_0 - 1)} = \frac{1}{\sqrt{\lambda(\lambda - 4)}}. \tag{13.102}$$

Therefore near $\lambda = \lambda_0 = 4$,

$$\chi = (\lambda - \lambda_0)^{-1/2}. \tag{13.103}$$

13.8 Models with Two Absorbing States

Here we analyze models with two absorbing states. We imagine a community of individuals, each one holding a opinion about a certain issue, being in favor or against it. The opinion of an individual changes with time according to certain rules that involve the opinion of the neighbors. At each time step an individual chooses at random one of his neighbors. If the neighbor's opinion is different from his, the opinion is accepted with a certain probability to be specified later. If the opinions are the same, the individual's opinion does not change. According to this rule, if all individuals of the community are favorable, there will be no change of opinion anymore and this state is absorbing. Similarly, if all individuals are against the issue, there will be no change of opinion. Therefore, the model so defined has two absorbing states.

Next we specify the probability of opinion acceptance when the opinions are different. (a) If the individual is against and the neighbor is favorable, then the individual becomes favorable with probability p and therefore remains against with probability $q = 1 - p$. (b) If the individual is in favor and the neighbor is against, then the individual becomes against with probability q and therefore remains favorable with probability p. These rules indicate that it will be a bias to the favorable opinion if $p > 1/2$ and a bias against the opinion if $p < 1/2$.

With the purpose of describing the model in an analytical way, we consider the individuals to be located on the sites of a regular lattice. Each site has a stochastic variable σ_i that takes the value $+1$ or -1 according to whether the individual at site i is favorable or against, respectively. According to the rules above, the probability of the transition $\sigma_i = -1 \rightarrow \sigma_i = +1$ is $p n_i / z$, where z is the number of neighbors and n_i is the number of neighbors in state $+1$. Similarly, the probability of the transition $\sigma_i = +1 \rightarrow \sigma_i = -1$ is $q m_i / z$, where $m_i = z - n_i$ is the number of neighbors in state -1. Introducing the parameter μ such that $p = (1 + \mu)/2$, and therefore $q = (1 - \mu)/2$, and taking into account that

$$n_i = \sum_\delta \frac{1}{2}(1 + \sigma_{i+\delta}), \qquad m_i = \sum_\delta \frac{1}{2}(1 - \sigma_{i+\delta}), \qquad (13.104)$$

where the sum extends over the neighbors of i, then the transition rate $w_i(\sigma)$ from σ_i to $-\sigma_i$ can be written as

$$w_i(\sigma) = \alpha(1 - \mu\sigma_i)\frac{1}{2z}\sum_\delta (1 - \sigma_i\sigma_{i+\delta}), \qquad (13.105)$$

where α is a parameter that defines the time scale. Indeed, when $\sigma_i = -1$, the rate is $2\alpha p n_i / z$ and when $\sigma_i = +1$, the rate is $2\alpha q m_i / z$. The parameter μ is restricted to the interval $-1 \leq \mu \leq 1$. When $\mu = 0$, the model is reduced to the voter model seen in Sect. 11.4.

If $\mu \neq 0$, and for long times, the system approaches one of the two absorbing states. If $\mu > 0$, the stationary state is the absorbing state in which all sites are in state $+1$. If $\mu < 0$, the stationary state is the absorbing state in which all sites are in state -1. Thus the model presents, in the stationary state, a phase transition between two absorbing states. At $\mu = 0$, the system is found at the critical state, which we had the opportunity to study in Sect. 11.4 when we studied the voter model. Here we focus only on the situation in which $\mu \neq 0$.

We start by considering the one-dimensional case for which the transition rate is given by

$$w_i(\sigma) = \alpha(1 - \mu\sigma_i)[1 - \frac{1}{2}\sigma_i(\sigma_{i-1} + \sigma_{i+1})]. \tag{13.106}$$

Consider an infinite lattice and suppose that initially only the site $i = 0$ is in state -1 while all the others are in state $+1$. After a while the system will have several sites in state $+1$, but, according to the transition rate, theses sites form a single cluster of contiguous sites in state -1. Denoting by $P_n(t)$ the probability of occurrence, at time t, of a cluster with n sites in state -1, then according to the transition rates, we get the following equation

$$\frac{dP_0}{dt} = aP_1, \tag{13.107}$$

$$\frac{dP_1}{dt} = -(a + b)P_1 + aP_2, \tag{13.108}$$

$$\frac{dP_n}{dt} = bP_{n-1} - (a + b)P_n + aP_{n+1}, \tag{13.109}$$

for $n \geq 2$, where $a = 2\alpha p$ and $b = 2\alpha q$. Notice that P_0 is the probability of the absorbing state in which all sites are in state $+1$.

These equations are identical to the model of the random walk with an absorbing state, seen in Sect. 8.5. One of the results obtained in that section is that the probability of entrance in the absorbing state at time t, per unit time, starting from the state n_0, is given by

$$\mathscr{R}(t) = At^{-3/2}e^{-(\sqrt{a}-\sqrt{b})^2 t}, \tag{13.110}$$

where $A = bn_0r^{-n_0}/\sqrt{\pi}(ab)^{3/4}$. Therefore, the decay to the absorbing state is exponential with a relaxation time

$$\tau = (\sqrt{a} - \sqrt{b})^{-2}. \tag{13.111}$$

We see thus that for this model the exponent $\nu_\parallel = 2$ in $d = 1$. Table 13.3 shows this exponent and other critical exponent for this model. The universality class of this model is called compact direct percolation.

Table 13.3 Critical exponents of compact direct percolation, according to Henkel et al. (2008)

d	β	β'	$\nu_\perp$	$\nu_\parallel$	z	δ	θ	d_F
1	0	1	1	2	2	1/2	0	1
≥ 2	0	1	1/2	1	2	1	0	2

A cluster of sites in state -1 can grow forever or it can grow up to a certain point and then shrink. The surviving probability is identified as the probability of permanence of the particle that performs the random walk with absorbing state. As we have seen in Sect. 8.5, the probability of permanence, starting from the state n_0, is

$$\mathscr{P}^* = 1 - \left(\frac{b}{a}\right)^{n_0}, \tag{13.112}$$

for $a > b$. Around the critical point, which occurs at $a = b$, the probability of surviving behaves as

$$\mathscr{P}^* \sim (a - b), \tag{13.113}$$

which gives an exponent $\beta' = 1$.

At the critical point, $a = b$, the absorbing rate $\mathscr{R}(t)$ behaves as $\mathscr{R}(t) \sim t^{-3/2}$. Recalling that the surviving probability $\mathscr{P}(t)$ is related to the absorbing rate through $d\mathscr{P}(t)/dt = -\mathscr{R}(t)$, then

$$\mathscr{P}(t) \sim t^{-1/2}, \tag{13.114}$$

which leads us to the exponent $\delta = 1/2$.

Chapter 14
Population Dynamics

14.1 Predator-Prey Model

Within the context of population dynamics, a variety of models has been proposed with the purpose of describing the mechanisms of competition between biological populations. Among these models the most known is the Lotka-Volterra, used in describing the time behavior of populations of two biological species that coexist in a certain region. One of them is prey and the other is predator. The prey feed on plants, considered abundant. The predators live at the expense of prey. Suppose that at certain moment the number of predators is large. This implies that the prey are annihilated quickly, not even having the chance to reproduce. The decline in prey population, in turn, causes the decrease in the number of predators. Having no food, many do not reproduce and disappear. When this occurs, conditions are created for the prey reproduction, increasing, thus, the prey population. As a consequence, the predator population increases again. And so on. With the passage of time, these situations repeat periodically in time. Under these conditions, the system predator-prey presents auto-organization. The predator and prey populations oscillate in time with a period determined by parameters inherent to the predator-prey interactions. In other terms, we face an auto-organization in the sense of Prigogine, which is expressed here by means of time oscillations in the prey and predator populations.

The Lotka-Volterra model is defined through the equations

$$\frac{dx}{dt} = k_1 ax - k_2 xy, \tag{14.1}$$

$$\frac{dy}{dt} = k_2 xy - k_3 y, \tag{14.2}$$

where x and y represent the densities of prey and predator respectively, and a is the density of prey food, considered to be constant because it is abundant. Preys

© Springer International Publishing Switzerland 2015
T. Tomé, M.J. de Oliveira, *Stochastic Dynamics*
and Irreversibility, Graduate Texts in Physics,
DOI 10.1007/978-3-319-11770-6__14

reproduce with rate k_1. Predators reproduce with rate k_2, as long as there are preys, and disappear spontaneously with rate k_3. One solution of these equations predicts time oscillations of preys and predators.

The description of population dynamics can also be done by the introduction of stochastic models defined on a lattice, as we have been studying so far. We describe below a model of this type for the predator-prey system, that was introduced by Satulovsky and Tomé (1994). In the construction of this model, it is aimed the microscopic description of the process inherent to the dynamics of the Lotka-Volterra. To do so, one proposes a stochastic model defined on a lattice, that describes the predator-prey dynamics by means of local Markovian rules, which simulate interactions similar to those of the contact process. Therefore, the model comprehends mechanisms of local interactions that take into account the spatial structure.

We consider a square lattice with N sites, each one being occupied by a prey, occupied by a predator or empty. To each site we associate a stochastic variable η_i that takes the values 1, 2, or 0, according to whether the site is occupied by a prey, occupied by a predator or empty, respectively.

The master equation, that governs the time evolution of the probability distribution $P(\eta, t)$ of state

$$\eta = (\eta_1, \eta_2, \ldots, \eta_i, \ldots, \eta_N), \tag{14.3}$$

at time t, is given by

$$\frac{d}{dt} P(\eta, t) = \sum_{i=1}^{N} \{w_i(R_i\eta) P(R_i\eta, t) - w_i(\eta) P(\eta, t)\}, \tag{14.4}$$

where $w_i(\eta)$ is the transition rate related to the transformation $0 \to 1$, $1 \to 2$, $2 \to 0$ according to whether $\eta_i = 0, 1, 2$, respectively, and R_i makes the inverse transformation, that is,

$$R_i\eta = (\eta_1, \ldots, \eta_i', \ldots, \eta_N), \tag{14.5}$$

where $\eta_i' = 2, 0, 1$ according to whether $\eta_i = 0, 1, 2$, respectively.

The transition rate $w_i(\eta)$ comprehends the following processes of interaction between species:

(a) Creation of prey. If a site i is in empty state ($\eta_i = 0$), then a prey can be created in this site with a rate equal to $an_i/4$, where a is a parameter and n_i is the number of prey in the neighborhood of site i. That is, a prey is born if there is at least one prey in the neighborhood of the chosen empty site. Moreover, the probability of birth is proportional to the number of neighboring preys.

(b) Creation of predator and annihilation of prey. If a site i is occupied by a prey ($\eta_i = 1$), then a predator will occupy site i, with an instantaneous annihilation of prey, with a rate bm_i, where b is a parameter and m_i is the number of predators in the neighborhood of site i. Thus a predator is created only if there is at least one predator in the neighborhood of the prey. Since the prey "becomes" a predator for being next to the predator, the model may be viewed, pictorially, as a model for "vampire predators". Moreover, the rate in which the preys disappear and the predators are created is proportional to the number of neighboring sites occupied by predators.

(c) Annihilation of predator. If a site i is occupied by a predator ($\eta_i = 2$), then the site will become empty with rate c independently of the neighborhood. The disappearance of predator is spontaneous.

According to the rules above, the rate w_i related to site i of a regular lattice of coordination ζ is written in an explicit form as

$$w_i(\eta) = a\,\delta(\eta_i, 0)\frac{n_i}{\zeta} + b\,\delta(\eta_i, 1)\frac{m_i}{\zeta} + c\,\delta(\eta_i, 2), \qquad (14.6)$$

where n_i is the number of prey and m_i is the number of predators present in the neighborhood of site i, which are given by

$$n_i = \sum_e \delta(\eta_{i+e}, 1), \qquad m_i = \sum_e \delta(\eta_{i+e}, 2), \qquad (14.7)$$

where the sum in e extends over the neighbors of site i.

These rules of evolution are similar to the rules of the contact process. The first two are in full analogy with the autocatalytic creation of particles in the contact model and the last one is analogous to the annihilation process in the contact model. The difference between the models resides in the fact that the contact process is a model with two states per site whereas the predator-prey model has three states per site. In addition, only one species is annihilated spontaneously.

This system can also be interpreted in terms of a generic epidemic as follows: preys are interpreted as susceptible individuals and predators as infected. The empty sites are interpreted as recovered individuals.

14.2 Numerical Simulations

The model has two absorbing states. One in which all sites of the lattice are occupied by prey and the other where all lattices are empty. The system also exhibits stationary active states with nonzero densities of predators and preys.

From the point of view of population dynamics, one of the most relevant results predicted by the model is a stationary active state with oscillations in time and space. The stationary active states are those in which the predators and prey are continuously being created and annihilated. In a certain region of the phase diagram defined by the parameters a, b and c, the active states are characterized

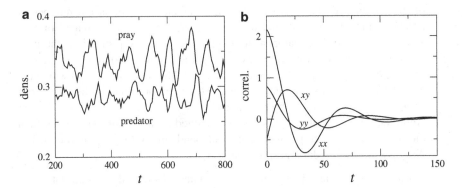

Fig. 14.1 Results of numerical simulations of the predator-prey model on a square lattice for the parameters: $a = b = 0,45$ and $c = 0,1$, according to Tomé et al. (2009). (**a**) Density of preys and predators as functions of time, for a lattice of size 40×40. (**b**) Time correlations between prey and predator (xy), prey and prey (xx) and predator and predator (yy) for a lattice of size 100×100

by a constant densities of prey and predator. In another region of this diagram, the densities of prey and predator oscillate in time and in space. The oscillations obtained by numerical simulations, as those shown in Fig. 14.1, are very similar to the actual time oscillations in the number of animals belonging to the two competing populations reported in statistical data registered along several years.

As seen in Fig. 14.1 the oscillations are not deterministic, but are characterized by the presence of a stochastic noise which induce the phase loss of these oscillations. The phase loss can be distinguished by the time correlation function as shown in Fig. 14.1. The time correlation function between two stochastic variables A and B is defined by

$$C(t) = \langle (A(t+s)B(s)) - \langle A(s) \rangle \langle B(s) \rangle, \qquad (14.8)$$

where s is the waiting time, which we consider to be large enough, and t, the observation time. When the observation time is large enough, we expect an exponential decay of the correlation, except when the system displays a critical behavior. The exponential decay can be pure or oscillatory, as those shown in Fig. 14.1, called phase-forgetting oscillations.

The oscillations observed in this model, when defined on a square lattice, seems to be local. That is, for systems large enough, these oscillations are not global but cover small regions in space. Thus, the system appears to be divided into uncorrelated subsystems. This supposition is based in the observation of the behavior of the amplitude of the oscillations in the number of prey and predator, which decays as $1/\sqrt{N}$, where N is the number of sites of the lattice.

14.3 Epidemic Spreading

Other models defined on a lattice, similar to the predator-prey model, can be set up. The model called susceptible-infected-removed-susceptible (SIRS) describes the epidemic spreading in a community of individuals that lose the immunity and can become susceptible again. Each site of a lattice can be occupied by a susceptible (S), by an infected (I) or by a removed (R). This last state can also be understood as immune. The rules of the model are as follows. At each time interval an individual is chosen at random. (a) If the individual is susceptible, he becomes infected if there is at least one infected in the neighborhood. The transition rate is equal to b times the fraction of infected individuals present in the neighborhood. (b) If the individual is infected, he becomes removed spontaneously with rate c. (c) If the individual is removed he becomes susceptible, that is, loses the immunity, spontaneously with rate a. Denoting by $\eta_i = 1, 2, 0$ the occupation of site i by a susceptible, infected and removed, respectively, and using the previous notation, then the transition rate $w_i(\eta)$ related to site i is given by

$$w_i(\eta) = a\,\delta(\eta_i, 0) + b\,\delta(\eta_i, 1)\frac{m_i}{\zeta} + c\,\delta(\eta_i, 2), \qquad (14.9)$$

where n_i is the number of susceptible and m_i is the number or infected present in the neighborhood of site i.

We see that the SIRS is constituted by three states and three reactions and in this sense it is similar to the predator-prey model. However, the SIRS model has one autocatalytic and two spontaneous reactions whereas the predator-prey model has two autocatalytic and one spontaneous reactions. When the parameter a vanishes, the SIRS model reduces to a model called susceptible-infected-removed (SIR) model, that corresponds thus to the case in which an individual remains immune forever. The transition rate for this model is thus

$$w_i(\eta) = b\,\delta(\eta_i, 1)\frac{m_i}{\zeta} + c\,\delta(\eta_i, 2), \qquad (14.10)$$

where m_i is the number of infected individuals present in the neighborhood of site i. It is worth to note that the same model is obtained from the predator-prey model when $a = 0$.

The contact process seen previously can also be interpreted as a model for the epidemic spreading called susceptible-infected-susceptible (SIS). In this model, each site can be occupied by a susceptible (S) individual or by an infected (I) individual. At each time interval a site of the lattice is chosen at random. (a) If the individual at the site is susceptible, he becomes infected if there is at least one infected individual in the neighborhood. The transition rate is equal to b times the fraction of infected present in the neighborhood. (b) If the individual is infected he becomes susceptible spontaneously with rate c. Using the notation $\eta_i = 0$ for a site S and and $\eta_i = 1$ for a site I, then the transition rate is

$$w_i(\eta) = b\,(1 - \eta_i)\frac{m_i}{\zeta} + c\,\eta_i, \tag{14.11}$$

where $m_i = \sum_e \eta_{i+e}$ is the number of infected present in the neighborhood of site i.

14.4 Mean-Field Theory

The models defined above can be analyzed within the mean-field theory. We start by the predator-prey model whose transition rate is modified so that

$$w_i(\eta) = a\,\delta(\eta_i, 0)\frac{n}{N} + b\,\delta(\eta_i, 1)\frac{m}{N} + c\,\delta(\eta_i, 2), \tag{14.12}$$

where n is the number of preys, m is the number of predators and N is the number of sites. With this modification, we see that the system can be described by means of the stochastic variables n and m only. For convenience, we define the auxiliary variable $k = N - n - m$, which is the number of empty sites. The three subprocesses are described by means of these variables as follows. (a) Creation of a prey in an empty site, $n \to n + 1$, $k \to k - 1$; (b) annihilation of a prey and simultaneous creation of a predator, $n \to n - 1$, $m \to m + 1$; (c) spontaneous annihilation of a predator leaving a site empty, $m \to m - 1$, $k \to k + 1$. The rates of these subprocesses are, respectively,

$$\alpha_{nm} = ak\frac{n}{N}, \qquad \beta_{nm} = bn\frac{m}{N}, \qquad \gamma_{nm} = cm. \tag{14.13}$$

Numerical simulations performed according to these rates are shown in Fig. 14.2.

Denoting by $P_{n,m}(t)$ the probability of the occurrence of n preys and m predators at time t, then the master equation that governs the evolution of this probability distribution is given by

$$\frac{d}{dt}P_{nm} = \alpha_{n-1,m}P_{n-1,m} - \alpha_n P_{nm}$$
$$+\beta_{n+1,m-1}P_{n+1,m-1} - \beta_{nm}P_{nm} + \gamma_{n,m+1}P_{n,m+1} - \gamma_{nm}P_{nm}. \tag{14.14}$$

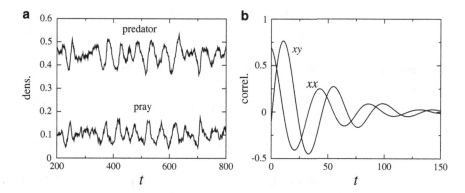

Fig. 14.2 Results obtained for the predator-prey from the transition rates (14.13), according to Tomé and de Oliveira (2009). (**a**) Densities of prey and predator as functions of time for $N = 100$, $a = b = 0.475$ and $c = 0.05$. (**b**) Corresponding time correlations, prey-predator (xy) and prey-prey (xx)

Next, we consider an expansion of this equation for large values of N. To this end, it is convenient to look at the probability density $\rho(x, y, t)$ of the variables $x = n/N$ and $y = m/N$. The expansion up to order $\varepsilon = 1/N$ results in the following Fokker-Planck equation

$$\frac{\partial \rho}{\partial t} = -\frac{\partial}{\partial x}(f_1 \rho) - \frac{\partial}{\partial y}(f_2 \rho)$$

$$+\frac{\varepsilon}{2}\frac{\partial^2}{\partial x^2}(g_{11}\rho) + \frac{\varepsilon}{2}\frac{\partial^2}{\partial y^2}(g_{22}\rho) + \varepsilon\frac{\partial^2}{\partial x \partial y}(g_{12}\rho), \qquad (14.15)$$

where

$$f_1 = ax(1 - x - y) - bxy, \qquad\qquad f_2 = bxy - cy, \qquad (14.16)$$

$$g_{11} = axy + bxy, \qquad\qquad g_{22} = bxy + cy, \qquad\qquad g_{12} = bxy. \qquad (14.17)$$

The Fokker-Planck equation is associated to the following Langevin equations

$$\frac{dx}{dt} = f_1 + \zeta(t), \qquad (14.18)$$

$$\frac{dy}{dt} = f_2 + \xi(t), \qquad (14.19)$$

where ζ and ξ are stochastic noises with the properties

$$\langle \zeta(t) \rangle = 0, \qquad\qquad \langle \xi(t) \rangle = 0, \qquad (14.20)$$

$$\langle \zeta(t)\zeta(t') \rangle = g_{11}\delta(t - t'), \qquad (14.21)$$

$$\langle \xi(t)\xi(t')\rangle = g_{22}\delta(t - t'),$$ (14.22)

$$\langle \zeta(t)\xi(t')\rangle = g_{12}\delta(t - t').$$ (14.23)

Averages The evolution of the averages $\langle x \rangle$ and $\langle y \rangle$ is given by

$$\frac{d}{dt}\langle x \rangle = \langle f_1(x, y)\rangle,$$ (14.24)

$$\frac{d}{dt}\langle y \rangle = \langle f_2(x, y)\rangle.$$ (14.25)

When N is very large we may replace $\langle f_i(x, y)\rangle$ by $f_i(x_1, x_2)$, where $x_1 = \langle x \rangle$ and $x_2 = \langle y \rangle$, to get the equations

$$\frac{dx_1}{dt} = ax_1(1 - x_1 - x_2) - bx_1x_2,$$ (14.26)

$$\frac{dx_2}{dt} = bx_1x_2 - cx_2.$$ (14.27)

According to the theory of reactive systems, seen in Chap. 10, these equation describe the reactions

$$C + A \rightarrow 2A, \qquad A + B \rightarrow 2B, \qquad B \rightarrow C,$$ (14.28)

where A, B and C represent, respectively, a prey, a predator and an empty site.

For large times we assume that x_1 and x_2 reach stationary values. The stability of the stationary solution is obtained by means of the Hessian matrix H, constituted by the elements $H_{ij} = \partial f_i/\partial x_j$, that is,

$$H = \begin{pmatrix} a(1 - 2x_1) - (a + b)x_2 & -(a + b)x_1 \\ bx_2 & bx_1 - c \end{pmatrix}.$$ (14.29)

If the real part of each eigenvalue is negative, then the solution is stable. A trivial stationary solution of Eqs. (14.26) and (14.27) is $x_1 = 0$ and $x_2 = 0$. The corresponding Hessian matrix has an eigenvalue equal to a and therefore this solution is unstable and does not occur. The other trivial solution is $x_1 = 1$ and $x_2 = 0$ and corresponds to a lattice full of prey and complete absence of predator. In this case the Hessian matrix has an eigenvalue equal to $-a$ and the other equal to $b - c$. Therefore, this solution is stable when $b < c$.

The nontrivial stationary solution is given by

$$x_1 = \frac{c}{b}, \qquad x_2 = \frac{a(b - c)}{b(a + b)},$$ (14.30)

and corresponds to the coexistence of the two species. Since $x_2 > 0$, then this solution exists as long as $b > c$. The Hessian matrix is

$$H = \begin{pmatrix} -ac/b & -(a+b)c/b \\ a(b-c)/(a+b) & 0 \end{pmatrix}, \qquad (14.31)$$

and the eigenvalues are the roots of the equation

$$b\lambda^2 + ac\lambda + ac(b-c) = 0. \qquad (14.32)$$

Since $b > c$, then the product of the roots is positive. The sum of the roots being negative, then (a) if the roots are real they are both negative, (b) if the roots are complex the real parts are negative. In both case, the solution is stable.

According to the results above we see that there is a phase transition between the absorbing and the species coexisting phases occurring when $b = c$. In Fig. 14.3 we show the phase diagram for the case where the parameter hold the relation $a + b + c = 1$, in which case it is convenient to use another parameter p such that $a = (1 - p - c)/2$ and $b = (1 + p - c)/2$. The transition line in space (c, p) is described by $c = (1 + p)/3$.

Fluctuation and correlation The densities x and y of the two species as functions of time are determined by the Langevin equations and hence they present fluctuations. They are not characterized only by their averages but also by their covariances. To characterize the behavior of x and y as functions of time we should also consider

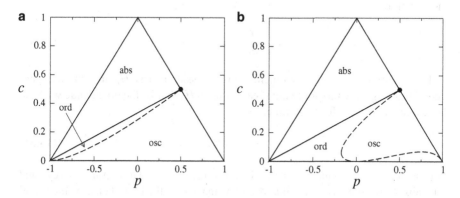

Fig. 14.3 Phase diagram of the predator-prey (**a**) and SIRS (**b**) models according to mean-field theory, in the plane c versus p, obtained by Satulovsky and Tomé (1994) and Souza and Tomé (2010), respectively. The *solid line* separates the absorbing (*abs*) phase and the species coexistence phase. In this phase, the time behavior can be ordinary (*ord*), with pure exponential decay of the correlations, or can be oscillatory (*osc*), with oscillatory decay of the correlations. Along the right side of the triangle, both models reduce to the SIR model, with a phase transition at the point represented by a *full circle*

their time correlations. The covariances predator-prey is defined by

$$\chi_{12} = N\left(\langle xy \rangle - \langle x \rangle \langle y \rangle\right). \tag{14.33}$$

Similarly, we define the covariances χ_{11}, χ_{22} and χ_{21}. Notice that $\chi_{21} = \chi_{12}$. The time correlation predator-prey is defined by

$$\gamma_{12} = N\left(\langle x(t)y(0) \rangle - \langle x(t) \rangle \langle y(0) \rangle\right). \tag{14.34}$$

Similarly, we define γ_{11}, γ_{22} and γ_{21}.

Defining the matrix X whose elements are the covariances χ_{ij} and the matrix G whose elements are g_{ij}, then the following equation determines the covariances

$$\frac{d}{dt}X = HX + XH^\dagger + G, \tag{14.35}$$

where $H^\dagger$ is the transpose of H.

Defining the matrix Γ whose elements are the correlations γ_{ij}, then the following equation determines the correlations

$$\Gamma = CX, \tag{14.36}$$

where C is the matrix whose elements C_{ij}, called correlation function, are defined as the derivatives of x_i with respect to the initial conditions. Denoting by x_i^0 the initial conditions, then $C_{ij} = \partial x_i / \partial x_i^0$. From $dx_i/dt = f_i$ we see that the matrix C obeys the equation

$$\frac{dC}{dt} = HC. \tag{14.37}$$

In the stationary state, the elements of the Hessian matrix H_{ij} are time independent. Denoting by λ the dominant eigenvalue of H, that is, the eigenvalue with the largest real part, then for large times

$$C_{ij} \sim e^{\lambda t}. \tag{14.38}$$

For the state $x_1 = 1$ and $x_2 = 0$, which correspond to a lattice full of preys and total absence of predators, and that occurs when $b < c$, the largest eigenvalue is real and negative so that the correlation function has exponential decay. For the state in which the species coexist, and that occurs when $b > c$, λ is the root of (14.32). The discriminant of Eq. (14.32) is

$$\Delta = a^2 c^2 - 4bac(b - c). \tag{14.39}$$

Therefore, if $\Delta > 0$, then λ is real and negative and the decay of the correlation function is pure exponential. This result says that x and y, the densities of prey and predators, fluctuate stochastically around their mean value and become uncorrelated with a characteristic time equal to $1/|\lambda|$.

If $\Delta < 0$, then λ has an imaginary part. Writing $\lambda = -\alpha \pm i\omega$, then

$$C_{ij} \sim e^{-\alpha t} \cos(\omega t + \phi_{ij}). \tag{14.40}$$

This result means that the densities of prey and predators, x and y as functions of time, present an oscillation with angular frequency $\omega = \sqrt{|\Delta|}$ superimposed to the noise that causes the loss of phase. The time it takes for the phase loss is equal to $1/\alpha$. The line that separates the two behaviors, ordinary and oscillating, is described by $\Delta = 0$, that is, by $ac = 4b(b - c)$ and shown in Fig. 14.3.

SIRS model In this case, the equations similar to Eqs. (14.26) and (14.27) are

$$\frac{dx}{dt} = a(1 - x - y) - bxy, \tag{14.41}$$

$$\frac{dy}{dt} = bxy - cy. \tag{14.42}$$

Here, we denote by x and y the mean densities of susceptible and infected, in the place of x_1 and x_2. According to the theory of reactive systems seen in Chap. 10, these equations describe the reactions

$$R \to S, \qquad\qquad S + I \to 2I, \qquad\qquad I \to R. \tag{14.43}$$

The first reaction occurs with rate a, the second with rate b and the third with rate c.

The Hessian matrix H is

$$H = \begin{pmatrix} -a - by & -a - bx \\ by & bx - c \end{pmatrix}. \tag{14.44}$$

A stationary solution of Eqs. (14.41) and (14.42) is $x = 1$ and $y = 0$ and correspond to a lattice full of susceptible and complete absence of infected. The corresponding Hessian matrix has an eigenvalue equal to $-a$ and the other equal to $b-c$. Therefore, this solution is stable as long as $b < c$.

The non-trivial stationary solution is given by

$$x = \frac{c}{b}, \qquad\qquad y = \frac{a(b - c)}{b(a + c)}, \tag{14.45}$$

and correspond to the coexistence of susceptible and infected individuals. Since $y > 0$, then this solution exists when $b > c$. The Hessian matrix is

$$H = \begin{pmatrix} -a(a+b)/(a+c) & -(a+c) \\ a(b-c)/(a+c) & 0 \end{pmatrix}, \qquad (14.46)$$

and the eigenvalues are the roots of the equation

$$(a+c)\lambda^2 + a(a+b)\lambda + a(b-c)(a+c) = 0. \qquad (14.47)$$

Since $b > c$, then the product of the roots is positive. Being the sum of the roots negative, then (a) if the roots are real then they are negative, (b) if the roots are complex the real parts are negative. In both cases, the solution is stable. We see, therefore, that there is a phase transition between the absorbing state and a state where the susceptible and infected individuals coexist, which occur when $b = c$ as shown in the phase diagram of Fig. 14.3 in the variables p and c. Again, we consider $a + b + c = 1$ and the parametrization $a = (1 - p - c)/2$ and $b = (1 + p - c)/2$.

The discriminant of Eq. (14.47) is

$$\Delta = a^2(a+b)^2 - 4a(b-c)(a+c)^2. \qquad (14.48)$$

If $\Delta > 0$, then λ is real and negative and the decay of the correlation function is pure exponential. If $\Delta < 0$, then λ has a imaginary part and the decay is of the type (14.40). The separation between the two behaviors occurs when $\Delta = 0$ or $a(a+b)^2 = 4(b-c)(a+c)^2$ and is shown in Fig. 14.3.

SIR model The predator-prey and SIRS models are reduced to the SIR model when the parameter $a = 0$. Thus, within the mean-field theory the evolution equations for x, y and z, which are the density of susceptible, infected and removed, respectively, are given by

$$\frac{dx}{dt} = -bxy, \qquad (14.49)$$

$$\frac{dy}{dt} = bxy - cy, \qquad (14.50)$$

$$\frac{dz}{dt} = cy, \qquad (14.51)$$

which are the Kermack and MacKendrick equations that describe the epidemic spreading in which the individuals acquire perennial immunity. It is clear that only two equations are independent. According to the theory of reactive systems seen in Chap. 10, these equations describe the reactions

$$S + I \rightarrow 2I, \qquad I \rightarrow R. \qquad (14.52)$$

The first reaction occur with rate b and the second with rate c.

When $t \rightarrow \infty$, the density of infected y vanishes. To determine the density of susceptible and removed in this limit, we proceed as follows. Dividing the second equation by the first, we obtain the equation

$$\frac{dy}{dx} = -1 + \frac{c}{bx}, \qquad (14.53)$$

which can be solved, with the solution

$$y = 1 - x + \frac{c}{b} \ln \frac{x}{1-h}, \qquad (14.54)$$

where the constant of integration was obtained according to the initial condition, which we consider to be $y = h$ and $z = 0$ so that $x = 1 - h$. Since in the stationary state $y = 0$, then x, in the stationary state, is given by

$$x = (1 - h)e^{-b(1-x)/c}. \qquad (14.55)$$

As $z = 1 - x$, since $y = 0$, then the equation for z is

$$1 - z = (1 - h)e^{-bz/c}. \qquad (14.56)$$

Figure 14.4 shows the final concentration z as a function of b/c for several values of h. The concentration z grows monotonically with b/c. When $h \rightarrow 0$,

$$1 - z = e^{-bz/c}, \qquad (14.57)$$

and we see that $z \rightarrow 0$ for $b/c < 1$. For $b/c > 1$, the final concentration is nonzero. Therefore, at $b/c = 1$, there is a change in behavior, which can be understood as a phase transition.

SIR model with spontaneous recovery We study now a modification of the model that we have just studied in such a way that an infected may become susceptible instead of becoming recovered. In this case, the equations are the following

$$\frac{dx}{dt} = ay - bxy, \qquad (14.58)$$

$$\frac{dy}{dt} = bxy - cy - ay. \qquad (14.59)$$

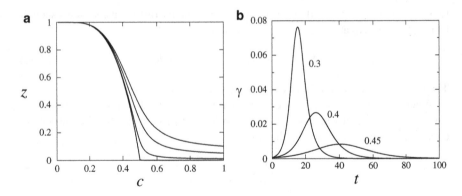

Fig. 14.4 (a) Final density z of removed as a function of c for various values of h, obtained from Eq. (14.56). The curves, from bottom to top, correspond to the values $h = 0$, $h = 0.01$, $h = 0.05$, $h = 0.1$. (b) Infection curve or density of susceptible individuals that is converting into infected per unit time, $\gamma = -dx/dt$, versus t, obtained from the solution of (14.49) and (14.50). The values of c are indicated and $h = 0.001$

These equations describe the reactions

$$I \rightarrow S, \qquad\qquad S + I \rightarrow 2I, \qquad\qquad I \rightarrow R. \qquad\qquad (14.60)$$

The first equation occurs with rate a and correspond to the spontaneous recovery of an infected. The second occurs with rate b and the third with rate c.

Dividing the second by the first, we get the equation

$$\frac{dy}{dx} = -1 + \frac{c}{bx - a}, \qquad\qquad (14.61)$$

whose solution is

$$y = 1 - x + \frac{c}{b} \ln \frac{bx - a}{b - a}, \qquad\qquad (14.62)$$

where we used as initial condition $y = 0$ and $x = 1$. In the stationary state, $y = 0$ and x is given by

$$\frac{bx - a}{b - a} = e^{-b(1-x)/c}, \qquad\qquad (14.63)$$

or, in terms of $z = 1 - x$,

$$1 - \frac{bz}{b - a} = e^{-bz/c}. \qquad\qquad (14.64)$$

Thus we see that, if $b < c + a$, the only solution is $z = 0$. When $b > c + a$, there appears a nontrivial solution $z \neq 0$. Therefore, there is a phase transition that occurs when $b = c + a$. We should notice that the transition is similar to the case of the strict SIR model ($a = 0$) except when $c = 0$. In this case, the model reduces to the SIS model.

SIR model with reinfection Next, we consider another modification of the SIR model in such a way that the removed individuals may lose their immunity becoming infected by a catalytic reaction. The equations are as follows

$$\frac{dx}{dt} = -bxy,$$

(14.65)

$$\frac{dy}{dt} = bxy - cy + a(1 - x - y)y,$$

(14.66)

and they describe the reactions

$$R + I \to 2I, \qquad S + I \to 2I, \qquad I \to R. \qquad (14.67)$$

The first reaction occurs with rate a and correspond to a catalytic reinfection. The second occurs with rate b and the third with rate c. When $b = 0$, the model becomes a contact process with rate of infection a/c.

Dividing the second by the first, we obtain the equation

$$\frac{dy}{dx} = \frac{a-b}{b} - \frac{a-c}{bx} + \frac{ay}{bx}.$$

(14.68)

To solve it, we use an auxiliary variable Y defined by $y = -x + (a - c)/a + Y$. Replacing in (14.68), we get the equation $dY/dx = aY/bx$ whose solution is $Y = Kx^{a/b}$. Using the initial condition $y = 0$ and $x = 1$, we determine the constant of integration K and we find the following relation between y and x,

$$y = -x + \frac{a-c}{a} + \frac{c}{a}x^{a/b}.$$

(14.69)

In the stationary state, $y = 0$ and x is given by

$$x - \frac{a-c}{a} = \frac{c}{a}x^{a/b}.$$

(14.70)

or, in terms of $z = 1 - x$,

$$1 - \frac{az}{c} = (1 - z)^{a/b}.$$

(14.71)

In the interval $a < c$, we conclude from this equation that there is a phase transition that occurs when $b = c$. When $b > c$, the state is characterized by $z \neq 0$ and when $b < c$, by $z = 0$.

Chapter 15
Probabilistic Cellular Automata

15.1 Introduction

Probabilistic cellular automata are Markovian processes in discrete time described by a set of stochastic variables that reside on the sites of a lattice. At regular time interval, all variables are updated simultaneously according to probabilistic rules. We may say that a cellular automaton has a synchronous update, which should be distinguished from the asynchronous update of the continuous time Markovian processes, described by a master equation, as, for example, the Glauber-Ising model.

To each site i of a lattice of N sites one associates a stochastic variable η_i. The microscopic configurations of a probabilistic cellular automaton are described by the set of stochastic variables $\eta = (\eta_1, \eta_2, \ldots, \eta_i, \ldots, \eta_N)$ and the system evolves in time through discrete time steps. The evolution of the probability $P_\ell(\eta)$ of state η, at time step ℓ, is governed by the equation

$$P_{\ell+1}(\eta) = \sum_{\eta'} W(\eta|\eta') P_\ell(\eta'), \tag{15.1}$$

where $W(\eta|\eta')$ is the (conditional) probability of transition from state η' to state η and therefore must obey the following properties

$$W(\eta|\eta') \geq 0, \tag{15.2}$$

$$\sum_{\eta} W(\eta|\eta') = 1. \tag{15.3}$$

In a cellular automaton, all sites are updated in an independent and simultaneous way, so that the transition probability $W(\eta|\eta')$ must be expressed in the form of a

© Springer International Publishing Switzerland 2015
T. Tomé, M.J. de Oliveira, *Stochastic Dynamics*
and Irreversibility, Graduate Texts in Physics,
DOI 10.1007/978-3-319-11770-6_15

product, that is,

$$W(\eta|\eta') = \prod_{i=1}^{N} w_i(\eta_i|\eta'), \tag{15.4}$$

where $w_i(\eta_i|\eta') \geq 0$ is the (conditional) transition probability that the state of site i at time $\ell + 1$, is η_i, given that, at time ℓ, the state of the system is η', and has the following property

$$\sum_{\eta_i} w_i(\eta_i|\eta') = 1, \tag{15.5}$$

what implies that the conditions (15.2) and (15.3) are fulfilled.

A probabilistic cellular automaton is simulated by starting from any initial configuration. From this configuration a sequence of configurations is generated, each one obtained from the previous one through a synchronous update of all sites. The i-th site is updated according to the transition probability w_i. That is, the site i takes the value η_i with a probability $w_i(\eta_i|\eta')$, where η' is the previous configuration. The explicit form of $w_i(\eta_i|\eta')$ and the values taken by the variables depend on the specific model that we wish to study. A model very well known is the probabilistic cellular automaton of Domany and Kinzel which we will study in the next section. In this automaton, the discrete variables take only two values and reside on the sites of a one-dimensional chain.

15.2 Domany-Kinzel Cellular Automaton

The cellular automaton introduced by Domany and Kinzel (1984) is defined by irreversible local transition probabilities. It has two states per site, is defined in one-dimensional lattice and has an absorbing state. In the stationary state, it displays a phase transition from an active state to an absorbing state which belongs to the universality class of the models with an absorbing state, as the ones seen in Chap. 13.

From the viewpoint of statistical mechanics of nonequilibrium phase transitions, this automaton and the contact process (studied in Chap. 13) are as fundamental as the Ising model to statistical mechanics of equilibrium phase transitions. They are models with local rules, that involve the sites in a small neighborhood and contains the basic elements of irreversibility. In addition, they exhibit phase transitions even in one dimension. Phase transitions in one dimension are not observed in the Ising model nor in any other equilibrium model with short range interactions.

The Domany-Kinzel automaton is defined in a one-dimensional lattice of N sites, each one being associated to a stochastic variable η_i that assumes two values: $\eta_i = 0$ or $\eta_i = 1$, according to whether the site i is empty or occupied by a particle, respectively. The evolution of the probability $P_\ell(\eta)$ of state $\eta = (\eta_1, \eta_2, \ldots, \eta_i, \ldots, \eta_N)$ at time step ℓ is given by the evolution equation (15.1). Since the sites are updated in an independent an simultaneous way, the transition probability $W(\eta|\eta')$ is given as a product of the transition probability per site as in (15.4).

The transition probabilities per site of the Domany-Kinzel cellular automaton are the same for any site and have the form

$$w_j(\eta_j|\eta') = w_{DK}(\eta_j|\eta'_{j-1}, \eta'_{j+1}). \tag{15.6}$$

The state of site j at time step $\ell + 1$ depends only of the states of the sites $j - 1$ and $j + 1$, which are the nearest neighbors of site j, at time step ℓ. The transition probabilities w_{DK} are summarized in Table 15.1, where p_1 and p_2 are the two parameters of the automaton such that $0 \le p_1 \le 1$ and $0 \le p_2 \le 1$. From Table 15.1 we see, for instance, that $w_{DK}(1|1,0) = p_1$. The last column of Table 15.1 implies the existence of an absorbing state where all sites of the lattice are empty, $\eta_i = 0$ for any site i, and called frozen state. If all sites are empty, it is not possible for a site to be occupied by a particle because $w_{DK}(1|0,0) = 0$.

To analyze the time evolution of the automaton (for N even) it is convenient to do a partition of the space-time lattice in two sublattices, as shown in Fig. 15.1. We observe that the rules of the Domany-Kinzel cellular automaton allows the separation in two systems residing in distinct sublattices and evolving in an independent way. This property permits to write the transition probability w_{DK} for one of the sublattices as $w_{DK}(\eta_i|\eta'_i, \eta'_{i+1})$, where the sites are reenumerate so that $i = (j - \ell)/2$, mod $N/2$. From now on, we will refer always to one sublattice.

Table 15.1 Transition probability of the Domany-Kinzel cellular automaton

	1,1	1,0	0,1	0,0
1	p_2	p_1	p_1	0
0	$1 - p_2$	$1 - p_1$	$1 - p_1$	1

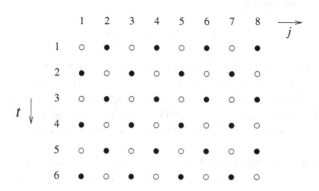

Fig. 15.1 Space-time lattice for the Domany-Kinzel probabilistic cellular automaton

Denoting by $P_\ell(\eta_i)$ the probability distribution related to a site and using Eqs. (15.1) and (15.4), we get

$$P_{\ell+1}(\eta_i) = \sum_{\eta_i'} \sum_{\eta_{i+1}'} w_{DK}(\eta_i | \eta_i', \eta_{i+1}') P_\ell(\eta_i', \eta_{i+1}'), \qquad (15.7)$$

where $P_\ell(\eta_i', \eta_{i+1}')$ is the probability distribution related to two consecutive sites. Using the transition probability given on Table 15.1, we get

$$P_{\ell+1}(1) = p_2 P_\ell(11) + 2p_1 P_\ell(10). \qquad (15.8)$$

The equation for $P_\ell(0)$ is not necessary since

$$P_\ell(0) + P_\ell(1) = 1. \qquad (15.9)$$

The evolution equation for the probability distribution $P_\ell(\eta_i, \eta_{i+1})$ related to two neighboring sites can be obtained by a similar procedure and gives

$$P_{\ell+1}(\eta_i, \eta_{i+1})$$

$$= \sum_{\eta_i'} \sum_{\eta_{i+1}'} \sum_{\eta_{i+2}'} w_{DK}(\eta_i | \eta_i', \eta_{i+1}') w_{DK}(\eta_{i+1} | \eta_{i+1}', \eta_{i+2}') P_\ell(\eta_i', \eta_{i+1}', \eta_{i+2}'), \quad (15.10)$$

where $P_\ell(\eta_i', \eta_{i+1}', \eta_{i+2}')$ is the probability distribution related to three consecutive sites. Replacing the rules in Table 15.1 in the equation above, one obtains the evolution for $P_\ell(11)$

$$P_{\ell+1}(11) = p_1^2 P_\ell(010) + 2p_1 p_2 P_\ell(110) + p_1^2 P_\ell(101) + p_2^2 P_\ell(111), \quad (15.11)$$

The evolution equation for $P_\ell(10)$ is not necessary since this quantity can be obtained from the equality

$$P_\ell(10) + P_\ell(11) = P_\ell(1). \qquad (15.12)$$

The evolution equation for the probabilities related to three sites $P_\ell(111)$, which appear in Eq. (15.11), involve probabilities related to four sites. The evolution equation for these probabilities involve the probability of five sites and so on, so that an infinite set of equations is generated. This hierarchic set can be analyzed by means of a truncation scheme that results in approximations called dynamic mean field.

The treatment of these equations, by mean-field approximations, will be given here at the level of one and two sites. Within the one-site or simple mean-field approach, we use the approximation

$$P_\ell(\eta_i, \eta_{i+1}) = P_\ell(\eta_i) P_\ell(\eta_{i+1}), \qquad (15.13)$$

so that Eq. (15.8) becomes

$$P_{\ell+1}(1) = p_2 P_\ell(1) P_\ell(1) + 2p_1 P_\ell(1) P_\ell(0), \tag{15.14}$$

Using the notation $x_\ell = P_\ell(1)$, then $P_\ell(0) = 1 - x_\ell$ so that

$$x_{\ell+1} = p_2 x_\ell^2 + 2p_1 x_\ell(1 - x_\ell). \tag{15.15}$$

In the stationary regime

$$x = p_2 x^2 + 2p_1 x(1 - x), \tag{15.16}$$

whose solutions are $x = 0$ and

$$x = \frac{2p_1 - 1}{2p_1 - p_2}. \tag{15.17}$$

The first of these equations correspond to the frozen phase and the second to the active phase. The transition from the frozen phase to the active phase occurs at $p_1 = 1/2$, in this approximation.

Within the mean-field approach at the level of two sites, we use the approximation

$$P_\ell(\eta_1, \eta_2, \eta_3) = \frac{P_\ell(\eta_1, \eta_2) P_\ell(\eta_2, \eta_3)}{P_\ell(\eta_2)}, \tag{15.18}$$

that inserted into (15.11) gives

$$P_{\ell+1}(11) = p_1^2 \frac{P_\ell(01) P_\ell(10)}{P_\ell(1)} + 2p_1 p_2 \frac{P_\ell(01) P_\ell(11)}{P_\ell(1)}$$
$$+ p_2^2 \frac{P_\ell(11) P_\ell(11)}{P_\ell(1)} + p_1^2 \frac{P_\ell(10) P_\ell(01)}{P_\ell(0)}. \tag{15.19}$$

Using the notation $z_\ell = P_\ell(11)$, we write Eq. (15.8) as

$$x_{\ell+1} = p_2 z_\ell + 2p_1(x_\ell - z_\ell) \tag{15.20}$$

since $P_\ell(10) = x_\ell - z_\ell$, and Eq. (15.19) as

$$z_{\ell+1} = p_1^2 \frac{(x_\ell - z_\ell)^2}{x_\ell} + 2p_1 p_2 \frac{(x_\ell - z_\ell) z_\ell}{x_\ell} + p_2^2 \frac{z_\ell^2}{x_\ell} + p_1^2 \frac{(x_\ell - z_\ell)^2}{1 - x_\ell}. \tag{15.21}$$

In the stationary state,

$$x = p_2 z + 2p_1(x - z), \tag{15.22}$$

$$z = p_1^2 \frac{(x-z)^2}{x} + 2p_1 p_2 \frac{(x-z)z}{x} + p_2^2 \frac{z^2}{x} + p_1^2 \frac{(x-z)^2}{1-x}. \tag{15.23}$$

From Eq. (15.22), we get the result

$$z = \frac{1 - 2p_1}{p_2 - 2p_1} x, \tag{15.24}$$

which, replaced into Eq. (15.23), gives

$$x = \frac{(3p_1 - 2)p_1 + (p_1 - 1)^2 p_2}{(2p_1 - 1)(2p_1 - p_2)}. \tag{15.25}$$

The transition line is obtained when $x \to 0$, that is, for

$$p_2 = \frac{(2 - 3p_1)p_1}{(1 - p_1)^2}. \tag{15.26}$$

We see that when $p_2 = 0$, then $p_1 = 2/3$, when $p_2 = 1$, then $p_1 = 1/2$ and, along $p_2 = p_1$, we get $p_2 = p_1 = (\sqrt{5} - 1)/2 = 0.61803$.

Close to the transition line, the order parameter, which is identified as the particle density x, has the same behavior along the whole line, except around the region around the point $p_1 = 1/2$ and $p_2 = 1$. For example, for $p_2 = 0$ we get;

$$x = \frac{(3p_1 - 2)}{2(2p_1 - 1)}, \tag{15.27}$$

or

$$x \sim (p_1 - p_c), \tag{15.28}$$

where $p_c = 2/3$, implying the critical exponent $\beta = 1$.

15.3   Direct Percolation

The model for direct percolation is constructed by considering a lattice composed by layers with a certain number of sites in each of them. The layers are labeled by an index ℓ from top to bottom by the numbers $0, 1, 2, 3, \ldots$. Each site of a certain layer is connected to sites belonging to the upper and lower layers, but not to sites of the same layer. Each site can be active with probability p or inert with probability

$1 - p$. In same manner, each bond between two neighboring sites can be intact with probability q or broken with probability $1 - q$.

Given a configuration with active sites and intact bonds, we wish to know the probability that at a site of a certain layer is connected to a site of layer 0, the top layer. Before that, we should define what we mean by connection between any two sites and in particular between two neighboring sites.

(a) Two neighboring sites are connected if they are active and if the bond between then is intact.
(b) Any two sites are connected if there is at least one path between them (i) formed by pairs of neighboring connected sites and (ii) that the path is always upward (or downward).

Notice that two sites belonging to the same layer are never connected because it is impossible to exist an upward (or downward) path between them.

To each site i of a layer ℓ, we associate a variable $\eta_{i\ell}$ that indicates whether the site is connected or not to the top layer. We set $\eta_{i\ell} = 1$ in case it is connected and $\eta_{i\ell} = 0$ otherwise. We denote by $P_\ell(\eta) = P_\ell(\eta_1, \eta_2, \ldots, \eta_N)$ the probability that site 1 of layer ℓ is in state η_1, the site 2 of the layer ℓ is in state η_2, the site 3 of layer ℓ is in state η_3 etc.

From $P_\ell(\eta)$ we can determine the marginal probabilities $P_\ell(\eta_i)$ related to one site, and $P_\ell(\eta_i, \eta_{i+1})$ related to two consecutive sites. The relation between these two probabilities is

$$P_{\ell+1}(\eta_i) = \sum_{\eta_i'} \sum_{\eta_{i+1}'} w(\eta_i | \eta_i', \eta_{i+1}') P_\ell(\eta_i', \eta_{i+1}'), \tag{15.29}$$

where $w(\eta_i | \eta_i', \eta_{i+1}')$ is the probability that site i of layer $\ell + 1$ is in state η_i, given that site i and $i + 1$ of layer ℓ are in sates η_i' and η_{i+1}', respectively.

We notice that $w(1|11)$ is understood as the conditional probability that a site is connected to the top layer, given that the two neighbors are connected to the top layer. We see then that

$$w(1|11) = 2pq(1 - q) + pq^2 = pq(2 - q). \tag{15.30}$$

In same manner $w(1|10)$ or $w(1|01)$ are understood as the conditional probabilities that a site is connected to the top layer, given that one of the two neighbors at the upper layer is connected to the top layer and the other is not. Therefore,

$$w(1|10) = w(1|01) = pq. \tag{15.31}$$

Taking into account that a site can never be connected to the top layer if the two neighbors at the upper layer are not connected, then

$$w(1|00) = 0. \tag{15.32}$$

The quantity $P_\ell(1) = P_\ell(\eta_i = 1)$ is the probability that the site i of layer ℓ is connected to the top layer. We wish to determine the behavior of $P_\ell(1)$ in the limit $\ell \to \infty$. To this end, we define $P = \lim_{\ell \to \infty} P_\ell(1)$. Thus, if $P \neq 0$, then a site infinitely far from the top layer is connected to the top layer and we face a percolating state. If $P = 0$, there is no percolation. Therefore, the probability of percolation P can be considered as the order parameter of the direct percolation model.

Generically we may write

$$P_{\ell+1}(\eta) = \sum_{\eta'} W(\eta|\eta') P_\ell(\eta'), \tag{15.33}$$

where

$$W(\eta|\eta') = \prod_i w(\eta_i|\eta'_i, \eta'_{i+1}), \tag{15.34}$$

so that the problem of direct percolation is mapped into the Domany-Kinzel probabilistic cellular automaton. The parameters p_1 and p_2 of the automaton are related to p and q by

$$p_1 = pq, \qquad\qquad p_2 = pq(2 - q). \tag{15.35}$$

The inverse transformation is given by

$$p = \frac{p_1^2}{2p_1 - p_2}, \qquad\qquad q = \frac{2p_1 - p_2}{p_1}. \tag{15.36}$$

It is worth to note that the mapping is not biunivocal since there are values of p_1 and p_2 that correspond to values of p and q that are outside the interval $[0, 1]$.

The case of site percolation corresponds to $q = 1$, so that we get the relation $p_2 = p_1$, whereas the bond percolation corresponds to $p = 1$, from which we get $p_2 = 2p_1 - p_1^2$.

With this correspondence, we see that the percolation probability P is identified with the variable x of the previous section. Using the results of the simple mean-field approximation, we see that one solution is $P = 0$ and the other is obtained from (15.17) and is given by

$$P = \frac{2pq - 1}{pq^2}, \tag{15.37}$$

valid for $2pq \geq 1$. For site percolation ($q = 1$), we get

$$P = \frac{2p - 1}{p}, \tag{15.38}$$

from which we get $p_c = 1/2$, and for bond percolation ($p = 1$), we get

$$P = \frac{2q - 1}{q^2},$$

(15.39)

from which we obtain $q_c = 1/2$.

Using the pair mean-field approximation, we get a solution $P = 0$ and the other is obtained from (15.25) and is given by

$$P = p - \frac{(pq - 1)^2}{2pq - 1}.$$

(15.40)

For the site percolation ($q = 1$), we get

$$P = \frac{p^2 + p - 1}{2p - 1},$$

(15.41)

which gives $p_c = (\sqrt{5}-1)/2 = 0.61803$. For the case of bond percolation ($p = 1$), we get

$$P = \frac{-q^2 + 4q - 2}{2q - 1},$$

(15.42)

implying $q_c = 2 - \sqrt{2} = 0.58579$.

From this procedure, the correspondence between the Domany-Kinzel cellular automaton and direct percolation becomes well established. Due to this correspondence, it becomes clear that the two models are in the same universality class which is the same as the models with an absorbing state studied in Chap. 13, which includes the contact model.

15.4 Damage Spreading

Suppose that the Domany-Kinzel cellular automaton is simulated for certain values of the parameters p_1 and p_2 starting from a given initial configuration. Suppose that, at the same time, the automaton is simulated using the same sequence of random numbers, that is, with the same noise but distinct initial condition. The second automaton is a replica of the original one. The initial configuration of the replica is similar to the initial configuration of the original with a certain number of sites in distinct states, which we call a damage. The original and the replica evolve in time simultaneously according to the same noise. We may then ask if they will remain forever in the distinct states or will eventually reach the same state. In other words, if the damage will spread of not.

The measure of the damage spreading can be done through the Hamming distance, which will be defined below. If it is distinct from zero, there is a damage spreading. If the Hamming distance vanishes, there is no damage spreading. The results from simulations shows that the phase diagram of the Domany-Kinzel with its replica exhibits three phases: frozen, active without damage and active with damage.

In the description of the stationary states with damage, it is necessary to consider configurations for both the system and the replica. These two must evolve under the same noise, but must keep their identities. Thus, if $\sigma = (\sigma_1, \sigma_2, \ldots, \sigma_N)$ and $\tau = (\tau_1, \tau_2, \ldots, \tau_N)$ denote the configurations of the system and replica, respectively, then the site i of the system must be updated according to $w_{DK}(\sigma_i | \sigma'_i, \sigma'_{i+1})$ and the site i of the replica according to $w_{DK}(\tau_i | \tau'_i, \tau'_{i+1})$, which are the replicas of the Domany-Kinzel cellular automaton. The fact that the evolution of the system and replica are ruled by the same noise or by the same sequence of random numbers, requires an analytical formulation, to be introduced below. The Hamming distance, which characterize the phase with damage, is an example of a quantity that needs to be evaluated by the establishment of a prescription of the joint evolution of the system and replica.

The Hamming distance ψ_ℓ is defined by the expression

$$\psi_\ell = \frac{1}{2} \langle (\sigma_i - \tau_i)^2 \rangle_\ell, \tag{15.43}$$

where the average considered here is performed using the joint probability distribution $P_\ell(\sigma; \tau)$, defined as the probability of the system and replica being in state σ and τ, respectively, at time t. It is clear that if, at time ℓ, the configurations σ and τ are identical, then $\psi_\ell = 0$.

To obtain the time evolution of ψ_ℓ, we need to known the evolution equation for $P_\ell(\sigma; \tau)$. The evolution equation for the joint probability distribution $P_\ell(\sigma; \tau)$ is written through the expression

$$P_{\ell+1}(\sigma; \tau) = \sum_{\sigma'} \sum_{\tau'} W(\sigma; \tau | \sigma'; \tau') P_\ell(\sigma'; \tau'), \tag{15.44}$$

where

$$W(\sigma; \tau | \sigma'; \tau') = \prod_i w(\sigma_i; \tau_i | \sigma'_i, \sigma'_{i+1}; \tau'_i, \tau'_{i+1}) \tag{15.45}$$

is the joint transition probability, that is, the probability that the system and the replica are in states σ and τ at time $\ell + 1$, respectively, given that they were in states σ' and τ', respectively, at time ℓ. It is worth to observe that the joint dynamic of the system and replica must be understood as a dynamics of an enlarged system with twice the number of sites of the original system.

The evolution equation for ψ_ℓ, obtained from the definition (15.43) and using (15.44) and (15.45), is

$$\psi_{\ell+1} = P_{\ell+1}(1;0) = \langle w(1;0|\sigma_i', \sigma_{i+1}'; \tau_i', \tau_{i+1}')\rangle_\ell. \qquad (15.46)$$

This equation can be evaluated once we have the expressions for the joint transition probabilities.

The problem we face now concerns the establishment of the prescription that will lead us to the joint transition probability or, in other terms, to the construction of a prescription for $w(\sigma_i; \tau_i | \sigma_i', \sigma_{i+1}'; \tau_i', \tau_{i+1}')$ which is capable to take into account the evolution of the system and replica under the same noise. The following properties should be observed:

(a) The probability $w(\sigma_i; \tau_i | \sigma_i', \sigma_{i+1}'; \tau_i', \tau_{i+1}')$ must be normalized, that is, the sum over σ_i and τ_i must be equal to one;
(b) The marginal probability obtained from the sum of $w(\sigma_i; \tau_i | \sigma_i', \sigma_{i+1}'; \tau_i', \tau_{i+1}')$ over the replica variables τ_i should reproduce the probabilities w_{DK} $(\sigma_i | \sigma_i', \sigma_{i+1}')$ and vice-versa;
(c) The rules must reflect the symmetry of the lattice and of the original model; and
(d) The system and replica must be updated with the same noise.

Taking into account these properties, the rules are given in Table 15.2, where $a + b + c + d = 1$. The choice of the parameters a, b, and c, which takes into account the fact that the joint evolution is under the same noise, is given by the prescription of Tomé (1994),

$$a = p_1, \qquad b = p_2 - p_1, \qquad c = 0, \qquad d = 1 - p_2, \qquad p_1 < p_2, \qquad (15.47)$$

$$a = p_2, \qquad b = 0, \qquad c = p_1 - p_2, \qquad d = 1 - p_1, \qquad p_1 > p_2. \qquad (15.48)$$

The evolution equation for the Hamming distance $\psi_\ell = P_\ell(1;0)$ can then be evaluated by using the prescription above in Eq. (15.46)

$$P_{\ell+1}(1;0) = p_2 P_\ell(11;00) + 2p_1 P_\ell(10;00) + 2(b+c)P_\ell(11;10). \qquad (15.49)$$

Using the simple mean-field approximation, that is, the approximation such that $P_\ell(\sigma_1, \sigma_2; \tau_1, \tau_2) = P_\ell(\sigma_1; \tau_1)P_\ell(\sigma_2; \tau_2)$, we get

$$P_{\ell+1}(1;0) = p_2 P_\ell(1;0)P_\ell(1;0)$$
$$+2p_1 P_\ell(1;0)P_\ell(0;0) + 2(b+c)P_\ell(1;1)P_\ell(1;0). \qquad (15.50)$$

Table 15.2 Transition probabilities of the Domany-Kinzel probabilistic cellular automaton with damage spreading

	11;11	10;10	00;00	11;00	10;00	11;10
1;1	p_2	p_1	0	0	0	a
1;0	0	0	0	p_2	p_1	b
0;1	0	0	0	0	0	c
0;0	$1 - p_2$	$1 - p_1$	1	$1 - p_2$	$1 - p_1$	d

In the stationary state we arrive at the equation

$$P(1;0) = p_2 P(1;0) P(1;0)$$
$$+2p_1 P(1;0) P(0;0) + 2(b+c) P(1;1) P(1;0). \tag{15.51}$$

To solve this equation, we need the equations for $P(1;1)$ and for $P(0;0)$. In view of

$$P(1;1) + P(1;0) = P(1), \tag{15.52}$$

and that

$$P(1;0) + P(0;0) = P(0) = 1 - P(1), \tag{15.53}$$

then $P(1;1) = x - \psi$ and $P(0;0) = 1 - x - \psi$, where $x = P(1)$, so that the Hamming distance at the stationary state, ψ, is given by

$$\psi = p_2 \psi^2 + 2p_1 \psi (1 - x - \psi) + 2|p_1 - p_2|(x - \psi)\psi. \tag{15.54}$$

Using the simple mean-field solution for x, we can draw the following conclusions.

(a) For $p_1 < 1/2$, we have $x = 0$ and $\psi = 0$, which describes the frozen phase.
(b) For $p_1 > 1/2$, and $p_2 > 2p_1/3$, we have $x \neq 0$ given by (15.17) and $\psi = 0$, which describes the active phase without damage.
(c) For $p_1 > 1/2$, and $p_2 < 2p_1/3$, we have $x \neq 0$ given by (15.17) and ψ given by

$$\psi = \frac{(2p_1 - 1)(2p_1 - 3p_2)}{(2p_1 - p_2)(4p_1 - 3p_2)}, \tag{15.55}$$

which describes the active phase with damage.

The critical behavior around the transition line between the active phase without damage to the active phase with damage of the Domany-Kinzel cellular automaton belongs to the universality class of direct percolation. When the system and replica reach the same configuration, the evolution of both system and replica become identical, that is, the configurations σ and τ will be forever identical. This is a consequence of the fact that they are subject to the same noise. In other terms, from the moment in which σ and τ become identical, the trajectory is confined into the subspace $\sigma = \tau$, which is thus an absorbing subspace.

15.5   Model for the Immune System

Models defined on a lattice has been used in the analysis of biological systems for some years. The first model proposed in this context has been introduced by Kauffman (1969) and is known as the Kauffman automaton. From this, there followed several lattice models which aimed to describe distinct biological systems. Particularly, the several parts of the immune system have been studied, with success, by lattice models.

Next we present a probabilistic cellular automaton to study the part of the immune system comprised by the T-helper cells, also known as T_H. The simulation of this system by an automaton leads us to consider a simple model that has up-down symmetry. The model, defined on a square lattice, displays kinetic phase transitions which occur via symmetry breaking. From the viewpoint of nonequilibrium statistical mechanics, the model is able to contribute to the understanding of the critical behavior of irreversible models. From the viewpoint of immunology, the proposed mechanisms for the study of the T_H through cellular automata seems to give a good description of this part of the immune system since the T_H are found packed in the lymph nodes and communicate by short-range interactions. In this case the spatial structure seems to be really important in predicting the type of immune response given by the system. Lattice models, like the one to be explained below, are able to show that, when the parasitic infection is high enough, the immune system is able to self-organize as to provide complete immunity or susceptibility with respect to certain parasites. This process, called "polarization" of the T_H cells, may be considered in the analysis of the immune status for various parasitic infections. Next we move on to the analysis of the specific model.

We consider a probabilistic cellular automaton in which three types of cells, T_H0, T_H1 and T_H2, occupy the sites of a lattice. Each site is associated to a stochastic variable σ_i which takes the values 0, 1, or -1, according to whether the site i is occupied by a cell of type T_H0, T_H1, or T_H2, respectively. The dynamic rules that describe the interactions among the T_H cells are stochastic and local rules defined through the transition probabilities $w_i(\sigma_i|\sigma_i', s_i')$, such that

$$w_i(\pm 1|\sigma_i', s_i') = p\delta(\sigma_i', 0)\{\delta(\sigma_i', \pm 1) + \frac{1}{2}\delta(\sigma_i', 0)\} + (1-r)\delta(\sigma_i', \pm 1), \quad (15.56)$$

and

$$w_i(0|\sigma_i, s_i') = (1-p)\delta(\sigma_i', 0) + r\{\delta(\sigma_i', 1) + \delta(\sigma_i', -1)\}. \quad (15.57)$$

with s_i' given by

$$s_i' = S(\sum_\delta \sigma_{i+\delta}'), \quad (15.58)$$

where $S(\xi) = 1$ if $\xi > 0$, $S(\xi) = 0$ if $\xi = 0$, and $S(\xi) = -1$ if $\xi < 0$. The summation is over the nearest neighbors of a site i of the lattice.

The evolution of the probability $P_\ell(\sigma)$ is given as in (15.1) and the transition probability $W(\sigma|\sigma')$ is given through the product of the probabilities $w_i(\sigma_i|\sigma_i', S_i')$ as in (15.4) because the sites are updated simultaneously.

The densities of the T_H1, T_H2 and T_H0 cells, at step ℓ, are defined through the expressions

$$x_\ell = \langle \delta(\sigma_i, 1) \rangle_\ell, \qquad y_\ell = \langle \delta(\sigma_i, -1) \rangle_\ell, \qquad z_\ell = \langle \delta(\sigma_i, 0) \rangle_\ell. \tag{15.59}$$

The evolution equations for the densities of each type of cell are obtained by the definition of these quantities and by the use of the evolution equation for $P_\ell(\sigma)$ given by (15.1). These equations are given by

$$x_{\ell+1} = P_{\ell+1}(1) = \langle w_i(1|\sigma') \rangle_\ell, \tag{15.60}$$

$$y_{\ell+1} = P_{\ell+1}(-1) = \langle w_i(-1|\sigma') \rangle_\ell, \tag{15.61}$$

$$z_{\ell+1} = (1-p)z_\ell + r(1-z_\ell). \tag{15.62}$$

The equation for the evolution of the correlation of two nearest neighbor sites i and j, that are occupied by T_H1 and T_H0 cells, respectively, is given by

$$P_{\ell+1}(1, 0) = \langle w_i(-1|\sigma_i, S_i) w_j(0|\sigma'_j, S_j) \rangle_\ell. \tag{15.63}$$

Similar equations can be derived for the other correlations of two or more neighboring sites.

On observing the above equations, we see that the rules of the model has invariance by the change of signs of all variables used in the definition of the transition probability $w_i(\sigma_i|\sigma'_i, S'_i)$. However, the interaction mechanisms among the cells contained in the model, through the transition probabilities are such that the states -1 and $+1$ are symmetrical. Thus, the three-state model here examined has the up-down symmetry, that is, the transition probability is invariant by the transformation $\sigma_i \rightarrow -\sigma_i$ applied to all sites of the lattice. This symmetry also occurs in the Blume-Emery-Griffiths equilibrium model, which is also a model with three state per site.

The up-down symmetry allows us to obtain a subjacent dynamics from which we may get some exact results. The subjacent dynamics is defined over a reduced space of configurations, described by the dynamic variables $\{\eta_i\}$ related to the variables σ_i by $\eta_i = \sigma_i^2$. The marginal probability distribution of the reduced variables $\{\eta_i\}$ obeys also an evolution equation of the type (15.1), which we call subjacent dynamics.

The subjacent dynamics has simple rules such that the reduced state of a site is modified independently of the others. The transition probabilities $w_S(\eta_i|\eta'_i)$ of the subjacent dynamics are $w_S(1|0) = p$, $w_S(0|0) = 1 - p$, $w_S(0|1) = r$ and $w_S(1|1) = 1 - r$. We may say that $w_S(\eta_i|\eta'_i)$ are the elements of the stochastic matrix

$$w_s = \begin{pmatrix} 1-p & r \\ p & 1-r \end{pmatrix}, \tag{15.64}$$

whose stationary probability is

$$P(0) = \frac{r}{p+r}, \qquad P(1) = \frac{p}{p+r}. \tag{15.65}$$

From the definition of z_ℓ, we see that $z_\ell = \langle \delta(\eta_i, 0) \rangle_\ell = P_\ell(1)$ so that

$$z_{\ell+1} = w_S(0|0)z_\ell + w_S(0|1)(1 - z_\ell), \tag{15.66}$$

that is,

$$z_{\ell+1} = (1 - p)z_\ell + r(1 - z_\ell). \tag{15.67}$$

The stationary density of T_H0 cells is obtained from this last equation and is given by

$$z = P(0) = \frac{r}{p+r}, \tag{15.68}$$

which is an exact result. Since $x + y + z = 1$, then the sum of the stationary concentrations of the cells of type T_H1 and T_H2 is given by

$$x + y = P(1) + P(-1) = \frac{p}{p+r}. \tag{15.69}$$

It is also possible to determine exactly the probability of finding clusters having K cells of type T_H0, given by

$$P(0, 0, \ldots, 0) = \left(\frac{r}{p+r}\right)^K. \tag{15.70}$$

The above mentioned results, as well as the subjacent dynamics itself, are relevant in the description of the stationary states. They are important, for instance, in the study of the behavior of the cell concentrations as functions of the parameter of the model, inasmuch as the Monte Carlo simulations cannot provide definite results.

The model has a trivial stationary solution where the concentration of T_H1 cell is equal to the concentration of T_H2 cells, that is, $x = y$. Using (15.69), we conclude that in this state $x = y = (1/2)p/(p + r)$. Thus, we have an exact description of this phase which we call disordered.

By means of dynamic mean-field approximation and Monte Carlo simulations in a square lattice, we see that the disordered phase ($x = y$) ceases to be stable for values of r less that a critical value r_c, for a fixed values of p. For $r < r_c$, the stable phase is the ordered phase for which $x \neq y$. Defining the order parameter α as the difference between the densities of the T_H1 and T_H2 cells, that is, $\alpha = x - y$, then in the ordered phase ($\alpha \neq 0$) there is a polarization of the T_H cells, that is, there is a majority of one type of cells: either T_H1 or T_H2. Since the transition rules has up-down symmetry, the transition from the ordered phase to the disordered phase belongs to the universality class of the Glauber-Ising model. This result is confirmed through the estimates of the critical exponents obtained by numerical simulations.

From the viewpoint of immunology, these results can be interpreted in terms of distinct immune responses. The difference between the concentration x of the T_H1 cells and the concentration y of the T_H2 cells determine the type of immune response that the organism may exhibit. The parameter r is related with the inverse of the infection rate (the smaller r the greater the infection rate). For a fixed p and for $r > r_c$, we have $x = y$, so that the immune response is undifferentiated. An organism can have immune responses of the type T_H1 or type T_H2. From the point where there occurs the spontaneous symmetry breaking ($r = r_c$ for fixed p), states arise in which a type of cell predominates over the other ($\alpha \neq 0$), that is, in which the immune response of the system is differentiated. Therefore, from a critical threshold of infection, the organism becomes immune (or susceptible) with respect to certain parasites depending of the type of T_H cells that, in this regime ($r < r_c$), predominates over the other. The immunity and the susceptibility, granted by the predominance of one type of T_H cells, depend of the type of parasite to which the individual is exposed.

Chapter 16
Reaction-Diffusion Processes

16.1 Introduction

The reaction-diffusion processes occur in systems comprising particles, atoms or molecules of various types which diffuse and can react with each other. However, we will study only systems constituted by particles of a single type. Thus the possible reaction involve the annihilation or creation of particles of the same type. The particles are located on the sites of a regular lattice, moving at random from one site to another, and capable of react when next to each other. We will restrict to the case in which each site can be occupied by at most one particle. Thus, we associate to each site i an occupation variable η_i which takes the value 0 when the site is empty and the value 1 when the site is occupied. The state of the system is defined by the vector $\eta = (\eta_1, \eta_2, \ldots, \eta_N)$.

The process of reaction-diffusion is defined as follows. At each time step two neighboring sites, say i and j, are chosen at random. The states η_i and η_j of these two sites are modified to η_i' and η_j' according to certain transition probabilities proportional to the transition rate $\tilde{w}(\eta_i', \eta_j'; \eta_i, \eta_j)$. The time evolution of the probability $P(\eta, t)$ of occurrence of state η at time t is governed by the master equation

$$\frac{d}{dt} P(\eta, t) = \sum_{(ij)} \sum_{\eta_i', \eta_j'} \{ \tilde{w}(\eta_i, \eta_j; \eta_i', \eta_j') P(\eta', t) - \tilde{w}(\eta_i', \eta_j'; \eta_i, \eta_j) P(\eta, t) \},$$

(16.1)

where the first summation extends over the pair of neighboring sites. The state η' is defined by $\eta' = (\eta_1, \eta_2, \ldots, \eta_i', \ldots, \eta_j', \ldots, \eta_N)$.

The possible processes are:

1. Creation of two particles, with rate $\tilde{w}(11, 00)$,

$$0 + 0 \rightarrow A + A,$$

(16.2)

© Springer International Publishing Switzerland 2015
T. Tomé, M.J. de Oliveira, *Stochastic Dynamics and Irreversibility*, Graduate Texts in Physics,
DOI 10.1007/978-3-319-11770-6_16

2. Diffusion, with rates $\tilde{w}(10,01)$ and $\tilde{w}(01,10)$,

$$A + O \rightarrow O + A \quad \text{and} \quad O + A \rightarrow A + O, \tag{16.3}$$

3. Annihilation of two particles, with rate $\tilde{w}(00,11)$,

$$A + A \rightarrow O + O, \tag{16.4}$$

4. Creation of one particle, with rates $\tilde{w}(10,00)$ and $\tilde{w}(01,00)$,

$$O + O \rightarrow A + O \quad \text{and} \quad O + O \rightarrow O + A, \tag{16.5}$$

5. Annihilation of one particle, with rates $\tilde{w}(00,10)$ and $\tilde{w}(00,01)$,

$$A + O \rightarrow O + O \quad \text{and} \quad O + A \rightarrow O + O, \tag{16.6}$$

6. Autocatalysis, with rates $\tilde{w}(11,01)$ and $\tilde{w}(11,10)$,

$$O + A \rightarrow A + A \quad \text{and} \quad A + O \rightarrow A + A, \tag{16.7}$$

7. Coagulation, with rates $\tilde{w}(01,11)$ and $\tilde{w}(10,11)$,

$$A + A \rightarrow O + A \quad \text{and} \quad A + A \rightarrow A + O. \tag{16.8}$$

An empty site is denoted by O and a site occupied by a particle by A.

16.2 Models with Parity Conservation

We study models that include only the processes 1–4 above. These processes are such that the two sites i and j are modified simultaneously, that is, such that $\eta_i \rightarrow \eta_i' = 1 - \eta_i$ and $\eta_j \rightarrow \eta_j' = 1 - \eta_j$. Defining $w(\eta_i, \eta_j) = \tilde{w}(\overline{\eta}_i, \overline{\eta}_j; \eta_i, \eta_j)$, where we use the notation $\overline{\eta}_i = 1 - \eta_i$, the master equation becomes

$$\frac{d}{dt} P(\eta, t) = \sum_{(ij)} \{ w(\overline{\eta}_i, \overline{\eta}_j) P(\eta^{ij}, t) - w(\eta_i, \eta_j) P(\eta, t) \}, \tag{16.9}$$

where $w(\eta_i, \eta_j)$ is the transition rate $(\eta_i, \eta_j) \rightarrow (\overline{\eta}_i, \overline{\eta}_j)$, and the state η^{ij} is defined by $\eta^{ij} = (\eta_1, \eta_2, \ldots, \overline{\eta}_i, \ldots, \overline{\eta}_j, \ldots, \eta_N)$.

The most generic form of symmetrical transition rate is given by

$$w(\eta_i, \eta_j) = A \eta_i \eta_j + C \overline{\eta}_i \overline{\eta}_j + D \overline{\eta}_i \eta_j + D \eta_i \overline{\eta}_j, \tag{16.10}$$

where the parameters A, C, and D are respectively the annihilation, creation and diffusion rates, and therefore must be such that $A \geq 0$, $C \geq 0$ and $D \geq 0$. However, we analyze here only the case in which D is strictly positive, case in which the detailed balance is satisfied. The cases in which $D = 0$ include the models called random sequential adsorption to be studied in the next chapter.

The model defined by the rates (16.10) conserves parity, which means that at any time the number of particle is always even or always odd. If, at $t = 0$, the configuration has an even (odd) number of particles, then at any time $t > 0$ the number of particle will be even (odd). We may say that the evolution matrix splits into two sectors. One involving only configurations with an even number of particles and the other configurations with an odd number of particles.

To determine the stationary probability $P_e(\eta)$, we use the detailed balance condition

$$w(\overline{\eta}_i, \overline{\eta}_j) P_e(\eta^{ij}) = w(\eta_i, \eta_j) P_e(\eta). \tag{16.11}$$

Assuming solutions of the type

$$P_e(\eta) = \prod_i \phi(\eta_i), \tag{16.12}$$

then

$$w(\overline{\eta}_i, \overline{\eta}_j)\phi(\overline{\eta}_i)\phi(\overline{\eta}_j) = w(\eta_i, \eta_j)\phi(\eta_i)\phi(\eta_j). \tag{16.13}$$

Setting $\eta_i = 1$ and $\eta_j = 1$, we get

$$A[\phi(1)]^2 = C[\phi(0)]^2, \tag{16.14}$$

which has two solutions

$$\phi(0) = q, \qquad \phi(1) = p, \tag{16.15}$$

$$\phi(0) = q, \qquad \phi(1) = -p, \tag{16.16}$$

where

$$q = \frac{\sqrt{A}}{\sqrt{A} + \sqrt{C}}, \qquad p = \frac{\sqrt{C}}{\sqrt{A} + \sqrt{C}}. \tag{16.17}$$

We can thus set up two stationary solutions that are

$$P_e^{(1)}(\eta) = q^{N-n} p^n, \tag{16.18}$$

$$P_e^{(2)}(\eta) = q^{N-n}(-p)^n, \tag{16.19}$$

where N is the total number of sites and $n = \sum_i \eta_i$. Summing and subtracting these two solutions, we find the stationary probabilities $P_e^{(+)}(\eta)$ and $P_e^{(-)}(\eta)$ with even and odd parities, respectively, given by

$$P_e^{(\pm)}(\eta) = \frac{1}{N_\pm}[1 \pm (-1)^n]q^{N-n}p^n, \tag{16.20}$$

where N_+ and N_- are normalization factors given by

$$N_\pm = 1 \pm (q - p)^N. \tag{16.21}$$

From them we determine the particle densities $\rho = \langle \eta_i \rangle$, which is given by

$$\rho = \frac{1}{N_\pm}\{1 \mp (q - p)^{N-1}\}p, \tag{16.22}$$

where the upper sign is valid for even parity and the lower sign for odd parity. In both cases we obtain, in the limit $N \to \infty$, the result

$$\rho = p = \frac{\sqrt{C}}{\sqrt{A} + \sqrt{C}}, \tag{16.23}$$

which is independent of the diffusion rate D, as long as it is nonzero. The extremal cases occur when $A = 0$, which gives $\rho = 1$ and when $C = 0$, which gives $\rho = 0$. Notice moreover that the formula above ceases to be valid when A and C are both zero, corresponding to pure diffusion. In this last case, the particle density is a constant because the number of particles is a conserved quantity in pure diffusion.

16.3 Evolution Operator

We start by a change of variables from η_i to σ_i, where $\sigma_i = 2\eta_i - 1$. The new variables σ_i are such that, if site i is occupied, then $\sigma_i = 1$; if the site is empty, $\sigma_i = -1$. Replacing the inverse transformation $\eta_i = (1 + \sigma_i)/2$ in (16.10), we get the transition rate in the form

$$w(\sigma_i, \sigma_j) = a + b(\sigma_i + \sigma_j) + c\sigma_i\sigma_j, \tag{16.24}$$

where

$$a = \frac{1}{4}(A + C + 2D), \tag{16.25}$$

$$b = \frac{1}{4}(A - C), \tag{16.26}$$

$$c = \frac{1}{4}(A + C - 2D). \tag{16.27}$$

The master equation becomes then

$$\frac{d}{dt}P(\sigma,t) = \sum_{(ij)}\{w(-\sigma_i,-\sigma_j)P(\sigma^{ij},t) - w(\sigma_i,\sigma_j)P(\sigma,t)\}, \tag{16.28}$$

where $\sigma = (\sigma_1,\sigma_2,\ldots,\sigma_N)$ and $\sigma^{ij} = (\sigma_1,\sigma_2,\ldots,-\sigma_i,\ldots,-\sigma_j,\ldots,\sigma_N)$.

Next, we set up a vector space whose basis vectors $|\sigma\rangle$ are given by $|\sigma\rangle = |\sigma_1\sigma_2\ldots\sigma_N\rangle$. On this space we define the Pauli operators σ_j^x, σ_j^y and σ_j^z by

$$\sigma_j^x|\sigma\rangle = |\sigma^j\rangle, \qquad \sigma_j^y|\sigma\rangle = i\sigma_j|\sigma^j\rangle, \qquad \sigma_j^z|\sigma\rangle = \sigma_j|\sigma\rangle, \tag{16.29}$$

where in these equation i is the pure imaginary number and we use the notation $|\sigma^j\rangle = |\sigma_1\sigma_2\ldots(-\sigma_j)\ldots\sigma_N\rangle$.

Consider now the vector

$$|P(t)\rangle = \sum_{\sigma} P(\sigma,t)|\sigma\rangle. \tag{16.30}$$

Deriving both sides with respect to time and using the master equation (16.28), we get the evolution equation

$$\frac{d}{dt}|P(t)\rangle = W|P(t)\rangle, \tag{16.31}$$

where W, the evolution operator, is given by

$$W = \sum_{(ij)}(\sigma_i^x\sigma_j^x - 1)Z_{ij}, \tag{16.32}$$

and the operator Z_{ij} is defined by

$$Z_{ij}|\sigma\rangle = w(\sigma_i,\sigma_j)|\sigma\rangle, \tag{16.33}$$

that is, it is given by

$$Z_{ij} = a + b(\sigma_i^z + \sigma_j^z) + c\sigma_i^z\sigma_j^z. \tag{16.34}$$

In explicit form, the evolution operator is given by

$$W = \sum_{(ij)}(\sigma_i^x\sigma_j^x - 1)(a + b\sigma_i^z + b\sigma_j^z + c\sigma_i^z\sigma_j^z). \tag{16.35}$$

Using the property $\sigma_j^x \sigma_j^z = -i\sigma_j^y$ between the Pauli operators, we can write the evolution operator in the form

$$W = \sum_{(ij)} W_{ij}, \tag{16.36}$$

where

$$W_{ij} = a\sigma_i^x \sigma_j^x - c(\sigma_i^y \sigma_j^y + \sigma_i^z \sigma_j^z) - ib(\sigma_i^y \sigma_j^x + \sigma_i^x \sigma_j^y) - b(\sigma_i^z + \sigma_j^z) - a. \tag{16.37}$$

In the case of pure diffusion $A = C = 0$, the transition rate is given by

$$w(\sigma_i, \sigma_j) = \frac{1}{2} D(1 - \sigma_i \sigma_j) \tag{16.38}$$

since $a = D/2, b = 0$ and $c = -D/2$, so that the evolution operator is

$$W = \frac{D}{2} \sum_{(ij)} (\sigma_i^x \sigma_j^x + \sigma_i^y \sigma_j^y + \sigma_i^z \sigma_j^z - 1), \tag{16.39}$$

which can be written in the form

$$W = \frac{D}{2} \sum_{(ij)} (\vec{\sigma}_i \cdot \vec{\sigma}_j - 1), \tag{16.40}$$

where $\vec{\sigma}_i$ is the vector operator $\vec{\sigma}_i = (\sigma_i^x, \sigma_i^y, \sigma_i^z)$.

It is worth to note that the operator W is identical, except by an additive constant, to the antiferromagnetic Heisenberg Hamiltonian

$$H = J \sum_{(ij)} \vec{S}_i \cdot \vec{S}_j, \tag{16.41}$$

where $\vec{S}_i = \vec{\sigma}_i/2$ and $J > 0$.

16.4 Models with Diffusion and Reaction

We consider particles moving in a regular lattice. The time evolution of $\langle \eta_i \rangle$ is obtained from the master equation (16.9) and is given by

$$\frac{d}{dt} \langle \eta_i \rangle = \sum_{\delta} \langle (\bar{\eta}_i - \eta_i) w(\eta_i, \eta_{i+\delta}) \rangle, \tag{16.42}$$

where the sum extends over the neighbors of site i or

$$\frac{d}{dt}\langle \eta_i \rangle = \sum_\delta \{C - (C + D)\langle \eta_i \rangle - (C - D)\langle \eta_{i+\delta} \rangle + (C - A)\langle \eta_i \eta_{i+\delta} \rangle\}, \quad (16.43)$$

where we used the rate (16.10). This equation per se cannot be solved because it involves the pair correlations. The equation for the pair correlations involve the three-site correlations. The equation for three-site correlations involve the four-site correlations and so on. We have a hierarchy set of coupled equations to be solved. An approximative form of solving the set of equations is by truncation at a certain order. Here we use the simplest truncation which consists in using only Eq. (16.43) and approximate the pair correlations $\langle \eta_i \eta_j \rangle$ by $\langle \eta_i \rangle \langle \eta_j \rangle$. Using the notation $\rho_i = \langle \eta_i \rangle$, we get

$$\frac{d}{dt}\rho_i = \sum_\delta \{C - (C + D)\rho_i - (C - D)\rho_{i+\delta} + (C - A)\rho_i \rho_{i+\delta}\}. \quad (16.44)$$

Searching for homogeneous solutions $\rho_i = \rho$, independent of the site, we get

$$\frac{d}{dt}\rho = z\{C - 2C\rho + (C - A)\rho^2\}, \quad (16.45)$$

where z is the coordination number of the regular lattice, that is, the number of neighbors of a site. The stationary solution is given by

$$C - 2C\rho + (C - A)\rho^2 = 0, \quad (16.46)$$

whose solution is

$$\rho_e = \frac{\sqrt{C}}{\sqrt{A} + \sqrt{C}}. \quad (16.47)$$

We now perform an expansion around the stationary solution. To this end, we define $x = \rho - \rho_e$ which satisfies the equation

$$\frac{d}{dt}x = -2z\sqrt{AC}x + z(C - A)x^2, \quad (16.48)$$

whose solution for $AC \neq 0$ is

$$x = \frac{2z\sqrt{AC}}{ke^{2z\sqrt{AC}t} + z(C - A)}, \quad (16.49)$$

where the constant k is determined from the initial conditions. For large times, we get

$$x = x_0 \, e^{-2z\sqrt{AC}t}, \tag{16.50}$$

so that the decay of ρ to the stationary value is exponential with a relaxation time τ given by

$$\tau = \frac{1}{2z}(AC)^{-1/2}. \tag{16.51}$$

Notice that τ diverges when either A or C vanish.

Next, we consider the case in which $C = 0$. In this case the stationary density vanishes and we get the equation

$$\frac{d}{dt}\rho = -zA\rho^2, \tag{16.52}$$

whose solution is

$$\rho = \frac{1}{zAt + \alpha}, \tag{16.53}$$

where α is a constant. The decay ceases to be exponential and for large times it becomes algebraic, $\rho \sim t^{-1}$.

16.5   One-Dimensional Model

We examine here the one-dimensional model for which it is possible to determine in an explicit form the density of particles when the restriction $A+C = 2D$ is fulfilled. To this end, we use a formulation in terms of the variable σ_i which takes the value $+1$ if site i is occupied and the value -1 if site i is empty. Using the restriction $A + C = 2D$, we see that the transition rate (16.24) is written in the form

$$w(\sigma_i, \sigma_j) = a + b(\sigma_i + \sigma_j) \tag{16.54}$$

because the constant c, given by (16.27), vanishes. The other two constants are

$$a = D, \qquad b = \frac{1}{4}(A - C). \tag{16.55}$$

The time evolution for $\langle \sigma_1 \rangle$ is given by

$$\frac{d}{dt}\langle \sigma_1 \rangle = -2\langle \sigma_1 w(\sigma_0, \sigma_1) \rangle - 2\langle \sigma_1 w(\sigma_1, \sigma_2) \rangle, \tag{16.56}$$

or

$$\frac{d}{dt}\langle\sigma_1\rangle = -4b - 4a\langle\sigma_1\rangle - 4b\langle\sigma_1\sigma_2\rangle, \tag{16.57}$$

where we used the translational invariance to make the substitutions $\langle\sigma_0\sigma_1\rangle = \langle\sigma_1\sigma_2\rangle$.

Next, we write the time evolution for the correlation $\langle\sigma_1\sigma_2\ldots\sigma_\ell\rangle \equiv \phi_\ell$. For $\ell \geq 2$, we get

$$\frac{d}{dt}\phi_\ell = -2\langle\sigma_1\sigma_2\ldots\sigma_\ell w(\sigma_0,\sigma_1)\rangle - 2\langle\sigma_1\sigma_2\ldots\sigma_\ell w(\sigma_\ell,\sigma_{\ell+1})\rangle. \tag{16.58}$$

Using the rate (16.54) and the translational invariance, we obtain the equation

$$\frac{d}{dt}\phi_\ell = -4a\phi_\ell - 4b\phi_{\ell-1} - 4b\phi_{\ell+1}, \tag{16.59}$$

which is valid for $\ell \geq 2$. Defining $\phi_0 = 1$, Eq. (16.57) is written as

$$\frac{d}{dt}\phi_1 = -4a\phi_1 - 4b\phi_0 - 4b\phi_2, \tag{16.60}$$

so that we may consider Eq. (16.59) also valid for $\ell = 1$.

The equation is similar to that seen in Chap. 11 in the study of the one-dimensional Glauber model. Using the same technique, we arrive at the solution

$$\frac{d}{dt}\phi_\ell = \frac{2D\gamma}{\pi}\int_0^{2\pi}\sin k \sin k\ell e^{-4D(1-\gamma\cos k)t}dk, \tag{16.61}$$

where

$$\gamma = \frac{C-A}{2D}. \tag{16.62}$$

Now, the density $\rho = \langle\eta_1\rangle = (1 + \langle\sigma_1\rangle) = (1 + \phi_1)/2$ so that

$$\frac{d}{dt}\rho = \frac{D\gamma}{\pi}\int_0^{2\pi}(\sin k)^2 e^{-4D(1-\gamma\cos k)t}dk. \tag{16.63}$$

Performing the time integral, we get an expression for the density.

Now, we study the asymptotic behavior for large times. In this regime the integrand is dominated by small values of k so that

$$\frac{d}{dt}\rho = \frac{D\gamma}{\pi}e^{-4D(1-\gamma)t}\int_{-\infty}^{\infty}k^2 e^{-2D\gamma k^2 t}dk = \frac{D\gamma}{2\sqrt{\pi}}e^{-4D(1-\gamma)t}(2D\gamma t)^{-3/2}. \tag{16.64}$$

Therefore, as long as $\gamma \neq 1$, that is, $C - A \neq 2D$, the derivative $d\rho/dt$ decays exponentially to zero. The density in turn also decays exponentially to its stationary value. When $\gamma = 1$, or $C - A = 2D$, we get

$$\frac{d}{dt}\rho = \frac{2D}{4\sqrt{\pi}}(2Dt)^{-3/2}, \qquad (16.65)$$

so that

$$\rho = \rho_e - \frac{1}{2\sqrt{\pi}}(2Dt)^{-1/2}, \qquad (16.66)$$

and the density decays algebraically to its stationary value with an exponent $1/2$.

Chapter 17
Random Sequential Adsorption

17.1 Introduction

In this chapter we study the random sequential adsorption of atoms or molecules on a lattice. Whenever an atom is adsorbed on a site of the lattice, it prevents the neighboring sites to absorb atoms, which become blocked forever. At each time step a site is chosen at random. If the site is empty and all its neighbor sites are empty, then an atom is adsorbed at the chosen site. This process is repeated until there is no sites where an atom can be adsorbed, as can be seen in Fig. 17.1 for the one-dimensional case.

We may also imagine other types of adsorption. In the adsorption of dimers (a molecule of two atoms), two neighboring sites are chosen at random. If they are empty, then a dimer is adsorbed in these two sites. The process is repeated until when there is no more pairs of neighboring empty sites. We may also consider the case of trimers (three atoms), quadrimers (four atoms), etc. Here, however, we consider only the adsorption of single atoms, defined above. Notice that in one dimension, the adsorption on an atom that blocks the two neighboring sites is equivalent to the adsorption of dimers, as can be seen in Fig. 17.1.

Starting from an empty lattice, we wish to determine the density $\rho(t)$ of atoms as a function of time and in particular to find the final density ρ^*. It is clear that the final density will not be equal to the maximum possible density because there will be many empty sites that cannot be filled with an atom, as seen in Fig. 17.1.

Consider a regular lattice of N sites, where to each site i we associate a stochastic variable η_i that takes the values 1 or 0 according to whether the site is occupied or empty, respectively. The time evolution of the probability $P(\eta, t)$ of occurrence of configuration $\eta = (\eta_1, \eta_2, \ldots, \eta_N)$ at time t, is governed by the master equation

$$\frac{d}{dt} P(\eta, t) = \sum_i \{ w_i(\eta^i) P(\eta^i, t) - w_i(\eta) P(\eta, t) \}, \tag{17.1}$$

© Springer International Publishing Switzerland 2015
T. Tomé, M.J. de Oliveira, *Stochastic Dynamics
and Irreversibility*, Graduate Texts in Physics,
DOI 10.1007/978-3-319-11770-6_17

Fig. 17.1 Final configuration of a random sequential adsorption of dimers on the one-dimensional lattice

where $\eta^i = (\eta_1, \eta_2, \ldots, \overline{\eta}_i, \ldots, \eta_N)$ and $\overline{\eta}_i = 1 - \eta_i$. The transition rate $w_i(\eta)$ from state η_i to state $\overline{\eta}_i$ is given by

$$w_i(\eta) = \overline{\eta}_i \prod_\delta \overline{\eta}_{i+\delta}, \qquad (17.2)$$

where the product extends over all neighbors of site i. Notice that the rate is nonzero only when the site i is empty and all neighbor sites are also empty

To determine the time evolution equation of the correlation

$$C_A = \langle \prod_{j \in A} \overline{\eta}_j \rangle, \qquad (17.3)$$

where the product extends over the sites of a cluster A, we proceed as follows. From Eq. (17.1), we get the equation

$$\frac{d}{dt} \langle \prod_{j \in A} \overline{\eta}_j \rangle = - \langle \prod_{j \in A} \overline{\eta}_j \sum_{i \in A} w_i(\eta) \rangle. \qquad (17.4)$$

Defining A_i as the cluster A with the addition of all neighbors of site i (which already belongs to A), we get

$$\frac{d}{dt} \langle \prod_{j \in A} \overline{\eta}_j \rangle = - \sum_{i \in A} \langle \prod_{j \in A_i} \overline{\eta}_j \rangle. \qquad (17.5)$$

Notice that, if all neighbors of site i belong to the cluster A, then $A_i = A$.

17.2   One-Dimensional Case

We analyze here the adsorption of atoms on a one-dimensional chain. The adsorption of an atom in a site blocks the two nearest neighbor sites. The transition rate is then given by

$$w_i(\eta) = \overline{\eta}_i \overline{\eta}_{i+1} \overline{\eta}_{i-1}. \qquad (17.6)$$

The evolution equation of $\langle \overline{\eta}_\ell \rangle$ is given by

$$-\frac{d}{dt}\langle \overline{\eta}_\ell \rangle = \langle \overline{\eta}_{\ell-1} \overline{\eta}_\ell \overline{\eta}_{\ell+1} \rangle, \tag{17.7}$$

and that of $\langle \overline{\eta}_\ell \overline{\eta}_{\ell+1} \rangle$ is given by

$$-\frac{d}{dt}\langle \overline{\eta}_\ell \overline{\eta}_{\ell+1} \rangle = \langle \overline{\eta}_{\ell-1} \overline{\eta}_\ell \overline{\eta}_{\ell+1} \rangle + \langle \overline{\eta}_\ell \overline{\eta}_{\ell+1} \overline{\eta}_{\ell+2} \rangle, \tag{17.8}$$

and that of $\langle \overline{\eta}_\ell \overline{\eta}_{\ell+1} \overline{\eta}_{\ell+2} \rangle$ is given by

$$-\frac{d}{dt}\langle \overline{\eta}_\ell \overline{\eta}_{\ell+1} \overline{\eta}_{\ell+2} \rangle = \langle \overline{\eta}_\ell \overline{\eta}_{\ell+1} \overline{\eta}_{\ell+2} \rangle$$
$$+\langle \overline{\eta}_{\ell-1} \overline{\eta}_\ell \overline{\eta}_{\ell+1} \overline{\eta}_{\ell+2} \rangle + \langle \overline{\eta}_\ell \overline{\eta}_{\ell+1} \overline{\eta}_{\ell+2} \overline{\eta}_{\ell+3} \rangle. \tag{17.9}$$

The evolution of the other correlations can be obtained similarly.

We introduce next the following notation

$$c_1 = \langle \overline{\eta}_1 \rangle, \tag{17.10}$$

$$c_2 = \langle \overline{\eta}_1 \overline{\eta}_2 \rangle, \tag{17.11}$$

$$c_k = \langle \overline{\eta}_1 \overline{\eta}_2 \overline{\eta}_3 \ldots \overline{\eta}_k \rangle. \tag{17.12}$$

Using the translation invariance, the evolution equations for the correlation reads

$$-\frac{d}{dt}c_1 = c_3, \tag{17.13}$$

$$-\frac{d}{dt}c_2 = 2c_3, \tag{17.14}$$

$$-\frac{d}{dt}c_3 = c_3 + 2c_4. \tag{17.15}$$

For $k \geq 2$, we can show that

$$-\frac{d}{dt}c_k = (k-2)c_k + 2c_{k+1}, \tag{17.16}$$

These equations must be solved with the initial condition such that all sites of the lattice are empty. This means that at $t = 0$ we should have $c_k(0) = 1$ for any $k \geq 1$.

Equations (17.13) and (17.14) imply $c_2(t) = 2c_1(t) + C$. Using the initial condition, we see that the constant $C = -1$ so that $c_2(t) = 2c_1(t) - 1$. The density of atoms ρ is given by $\rho = \langle \eta_i \rangle = 1 - \langle \overline{\eta}_i \rangle = 1 - c_1 = (1 - c_2)/2$.

Next, we solve the set of equations given by (17.16). To this end, we assume a solution of the form

$$c_k(t) = u_k(t)e^{-(k-2)t}. \tag{17.17}$$

The equation for u_k is thus

$$\frac{d}{dt}u_k = -2u_{k+1}e^{-t}. \tag{17.18}$$

Defining the auxiliary variable y by

$$y = e^{-t}, \tag{17.19}$$

we get the following equation for $u_k(y)$,

$$\frac{d}{dy}u_k = 2u_{k+1}, \tag{17.20}$$

valid for $k \geq 2$. Assuming that $u_k(y)$ is independent of k, we conclude that

$$u_k(y) = Ce^{2y}, \tag{17.21}$$

where the constant C must be determined by the initial conditions. From (17.17), (17.19) and (17.21), we arrive at the solution

$$c_k(t) = Ce^{-(k-2)t} \exp\{2e^{-t}\}. \tag{17.22}$$

Since $c_k(0) = 1$, then $C = e^{-2}$ so that

$$c_k(t) = \exp\{2e^{-t} - 2 - (k-2)t\}. \tag{17.23}$$

In particular

$$c_2(t) = \exp\{2e^{-t} - 2\}, \tag{17.24}$$

from which we get the density $\rho = (1 - c_2)/2$, given by

$$\rho(t) = \frac{1}{2}\{1 - \exp(2e^{-t} - 2)\}, \tag{17.25}$$

so that $\rho^* = \rho(\infty)$ is given by

$$\rho^* = \frac{1}{2}(1 - e^{-2}) = 0.43233\ldots \tag{17.26}$$

a result obtained by Flory. This density is smaller then $1/2$, which is the result that we would obtain if the atoms were placed in an alternate way.

17.3 Solution on a Cayley Tree

A Cayley tree, can be understood as a lattice of sites formed by concentric layers as follows. To a central site we connect z sites, that constitute the first layer. To each site of the first layer, we connect $z - 1$ new sites that, constitute the second layer. To each site of the second layer, we connect $z - 1$ new sites, that constitute the third layer. Using this procedure we construct other layers. Each site is connected to $z - 1$ sites of the outside layer and to one site of the inside layer, having thus z neighbors.

A Cayley tree with coordination number z has a fundamental property, which we describe next. Consider a connected cluster of k sites and consider all sites that are connected to any site of the cluster but do not belong to the cluster. The number n_k of these peripheral sites is

$$n_k = (z - 2)k + 2, \tag{17.27}$$

and does not depend on the form of the cluster.

This property can be shown as follows. Consider a cluster A_k with k sites. To form a cluster A_{k+1} with $k + 1$ sites from the cluster A_k, it suffices to include all peripheral sites of A_k into the cluster A_{k+1}. In this process one loses one peripheral site (which becomes part of the cluster) and we gain $z - 1$ peripheral sites (those that are connected to the new site of the cluster, except the one that already belongs to the cluster). Therefore, $n_{k+1} = n_k + (z - 2)$, from which we get $n_k = (z - 2)k + c$. Since $n_1 = z$, the constant $c = 2$.

Consider the correlation

$$c_{k-1} = \langle \prod_i \overline{\eta}_i \rangle, \tag{17.28}$$

where the product extends over all sites of a certain cluster of k sites. Here we assume that the correlation above does not depend of the form of the cluster but only on its size, that is, on k.

The evolution equation for $\langle \overline{\eta}_i \rangle = c_0$ is given by

$$-\frac{d}{dt}c_0 = c_1, \tag{17.29}$$

and the equation for c_1, by

$$-\frac{d}{dt}c_1 = c_1 + zc_2. \tag{17.30}$$

For $k \geq 1$, we can show that

$$-\frac{d}{dt}c_k = kc_k + n_k c_{k+1}.$$ (17.31)

Using the same scheme of the previous section, we suppose that the solution is of the form

$$c_k(t) = u_k(t)e^{-kt}.$$ (17.32)

The equation for u_k is, thus

$$-\frac{d}{dt}u_k = n_k u_{k+1} e^{-t}.$$ (17.33)

Again, we define a variable y by

$$y = e^{-t},$$ (17.34)

from which we get

$$\frac{d}{dy}u_k = n_k u_{k+1},$$ (17.35)

or

$$\frac{d}{dy}u_k = [(z-2)k + 2]u_{k+1}.$$ (17.36)

Assuming a solution of the type

$$u_k = Y^{-k-a},$$ (17.37)

where $a = 2/(z-2)$ we see that

$$\frac{dY}{dy} = -(z-2),$$ (17.38)

whose solution is

$$Y = A - (z-2)y,$$ (17.39)

where A is a constant. To determine the constant A, we should use the initial condition corresponding to an empty lattice. This condition implies the result $c_k(0) = 1$, from which we get $u_k = 1$ when $y = 1$. We conclude thus that $Y = 1$ when $y = 1$, from which we get $A = z - 1$ so that $Y = z - 1 - (z-2)y$.

Defining

$$\gamma = z - 2, \tag{17.40}$$

then $Y = 1 + \gamma - \gamma y$ and

$$u_k(y) = [1 + \gamma - \gamma y]^{-k-2/\gamma}. \tag{17.41}$$

Thus

$$c_k(t) = (1 + \gamma - \gamma e^{-t})^{-k-2/\gamma} e^{-kt}, \tag{17.42}$$

which is valid for $k \geq 1$. In particular

$$c_1(t) = (1 + \gamma - \gamma e^{-t})^{-1-2/\gamma} e^{-t}, \tag{17.43}$$

from which we obtain, by integration of equation $-dc_0/dt = c_1$,

$$c_0(t) = \frac{1}{2}(1 + \gamma - \gamma e^{-t})^{-2/\gamma} + \frac{1}{2}, \tag{17.44}$$

where the constant was chosen so that $c_0(0) = 1$. The density $\rho = 1 - c_0$ is given by

$$\rho(t) = \frac{1}{2} - \frac{1}{2}(1 + \gamma - \gamma e^{-t})^{-2/\gamma}. \tag{17.45}$$

The final density $\rho^* = \rho(\infty)$ is thus

$$\rho^* = \frac{1}{2}[1 - (1 + \gamma)^{-2/\gamma}] = \frac{1}{2}[1 - (z - 1)^{-2/(z-2)}]. \tag{17.46}$$

For large values of z, we get the result $\rho^* = (\ln z)/z$. When $z \to \infty$ the final density $\rho^* \to 0$.

17.4   Adsorption on Regular Lattices

In regular lattices we have to resort to numerical simulations to determine the density as a function of time. Starting from an empty lattice, the sites are being occupied successively. At each time interval we choose a site at random. If it is empty and the neighboring sites are also empty, then the site becomes occupied. This procedure is repeated until there is no more sites to be occupied.

Table 17.1 Final density for
the random sequential
adsorption in which the
adsorption of an atom blocks
the nearest neighboring sites.
The errors are in the last digit

Lattice	d	z	ρ^*
Linear chain	1	2	0.432332[a]
Honeycomb	2	3	0.379
Square	2	4	0.36413
Triangular	2	6	0.23136
Cubic	3	6	0.304
Hypercubic	4	8	0.264

[a]$(1 - e^{-2})/2$

The algorithm can be improved as follows. In a regular lattice the sites occupied by an atom are labeled by the number 1. The sites that are neighbors of an occupied site, which are empty, are labeled by the number 2 and are called inactive. The other empty sites are labeled by the number 0 and are called active. The difference between the active and inactive empty sites is that the former can be occupied whereas the latter remain forever empty. At each time step t, we update the number N_1 of occupied sites, the number N_2 of inactive sites and the number N_0 of active sites. The sum of these numbers remains constant and equal to the number N of lattice sites, $N_1 + N_2 + N_0 = N$.

At each time step of the numerical procedure, we choose at random one of the N_0 active sites and place an atom on it. Next, we determine how many active sites becomes inactive and call this number N_{02}. The new values of the number of occupied sites, active and inactive sites will be

$$N_1' = N_1 + 1, \tag{17.47}$$

$$N_2' = N_2 + N_{02}, \tag{17.48}$$

$$N_0' = N_0 - N_{02} - 1. \tag{17.49}$$

The time is updated according to $t' = t + 1/N_0$. This procedure is repeated until $N_0 = 0$. The density at a given instant of time t is $\rho = N_1/N$.

In Table 17.1 we show the results obtained from numerical simulation and other methods, as time series expansion, for the final density of occupied sites ρ^* for the random sequential adsorption on a regular lattice, where each atom adsorbed blocks the z nearest neighbors.

Chapter 18
Percolation

18.1 Introduction

Percolation means the passage of a liquid through a porous medium. It can be understood as a random process, as much as diffusion is a random process. However, there is a fundamental difference between the two processes. In diffusion, particles with random motion spread in a deterministic medium, in percolation, particles with deterministic motion spread on a random medium. Since the motion is deterministic, the percolation models focus on the description of the medium through which the particles move, understood as a porous medium. We imagine the porous medium as a solid material inside which there is a certain number of porous localized randomly. If the number of porous is large enough, they are connected and the medium is permeable to the passage of the liquid. However, if the number is very small, they are isolated and the passage of the liquid is blocked. Thus we identify two regimes depending on the concentration of porous. In one of them there is percolation, in the other there is not.

A simple model for percolation is constructed by assuming that the porous are located in specific sites inside a solid block. We imagine the sites forming a regular lattice inside the block, for example, forming a cubic lattice. In addition, the porous are of a certain size so that they create a passage between them when they are neighbors, but do not when they are not neighbors. The basic problem of percolation is to find the minimum concentration above which there is a percolating state. Alternatively, assuming that the probability of occurrence of a porous in a site is p, we wish to know the minimum value of p for the occurrence of percolation.

To describe appropriately the location of the porous, we associate to each site i a variable η_i which takes the value 1 or 0 according to whether the site i has or has not a porous. Using a more convenient language, we say that the site is occupied or empty according to whether the variable takes the value 1 or 0. The global configuration is defined by the vector $\eta = (\eta_1, \eta_2, \ldots, \eta_N)$, where N is the

© Springer International Publishing Switzerland 2015
T. Tomé, M.J. de Oliveira, *Stochastic Dynamics
and Irreversibility*, Graduate Texts in Physics,
DOI 10.1007/978-3-319-11770-6__18

total number or sites of the lattice. The probability that η_i takes the value 1 or 0 is p
or $q = 1-p$, respectively, that is, it equals $p^{\eta_i} q^{1-\eta_i}$. The variables η_i are considered
to be independent so that the probability $\mathscr{P}(\eta)$ of configuration η is

$$\mathscr{P}(\eta) = \prod_i p^{\eta_i} q^{1-\eta_i}. \tag{18.1}$$

Notice that the average number n of occupied sites is pN what allows us to interpret
the probability p as the fraction of occupied sites.

18.2 Clusters

Given a configuration η, we may partition it in clusters of occupied sites, as seen
in Fig. 18.1. Two sites belong to the same cluster if there is between them at least
one path made up by occupied sites. A path is defined as a sequence of occupied
sites such that two successive sites are nearest neighbors. There are several types
of neighborhood. Here we consider the case in which the neighborhood of a site is
formed by the nearest neighbor sites. For convenience, we assume that an isolated
occupied site constitutes a cluster of one site. Two occupied nearest neighbor sites
form a cluster of two sites.

The percolation model we considered above is called site percolation. We may
also define bond percolation. In this case each bond can be active with probability
p and inactive with probability $1 - p$. A bond is a connection between two
nearest neighbor sites. The clusters are defined similarly and examples are shown in
Fig. 18.1. Two sites belong to the same cluster if there is between them a path made
up of sites such that two successive sites are connected by an active bond. Here,
however, we will consider only site percolation.

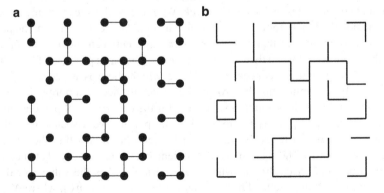

Fig. 18.1 (a) Site percolation and (b) bond percolation on a square lattice

The characterization of a percolating cluster is determined by the size distribution of clusters. This distribution is described by the average number n_s of clusters of size s, that is, clusters having s sites. From n_s we may obtain the following quantities. The average number of clusters

$$\zeta = \sum_s n_s, \tag{18.2}$$

the average number of occupied sites

$$n = \sum_s s\, n_s, \tag{18.3}$$

and the quantity

$$\phi = \sum_s s^2 n_s. \tag{18.4}$$

Since $n = pN$, then

$$\sum_s s\, n_s = pN. \tag{18.5}$$

It is convenient to define density type quantities to be used in the thermodynamic limit, $N \to \infty$. Thus, we introduce the density $\rho_s = n_s/N$ of clusters of size s from which we define the quantities

$$f = \sum_s \rho_s, \tag{18.6}$$

which is the density of clusters and

$$S = \sum_s s^2 \rho_s, \tag{18.7}$$

interpreted as the average cluster size. Taking into account the relation (18.5), then

$$\sum_s s\rho_s = p. \tag{18.8}$$

Up to this point we have considered lattices with a finite number of sites so that we may have only finite clusters. In an infinite lattice we should consider the possibility of the occurrence of a cluster of infinite size. Thus, a certain site may belong to a finite size cluster, to the infinite cluster or may be empty. The probability of a certain site to belong to a cluster of size s, finite, is $s\rho_s$, the probability to belong to the infinite cluster is denoted by P and the probability of being empty is q. Therefore,

$$\sum_{s=1}^{\infty} s\rho_s + q + P = 1, \tag{18.9}$$

or

$$\sum_{s=1}^{\infty} s\rho_s + P = p, \tag{18.10}$$

which replaces Eq. (18.8). The quantity P, called percolation probability, is interpreted as the fraction of occupied sites belonging to the infinite cluster and Eq. (18.10) can be used to determine P once we know ρ_s, now interpreted as the density of finite clusters of size s.

Denoting by F the density of sites belonging to finite clusters, given by

$$F = \sum_{s=1}^{\infty} s\,\rho_s, \tag{18.11}$$

we see from (18.10) that

$$F + P = p. \tag{18.12}$$

Increasing the parameter p starting from zero, the model passes from a state in which there are only finite clusters ($P = 0$) to a state in which, in addition to finite clusters, there is an infinite cluster ($P \neq 0$). The presence of an infinite cluster characterizes the model as being in a percolating state, that is, in which occurs percolation. This transition is characterized as a phase transition in which P plays the role of order parameter of the percolating phase. When $p \leq p_c$, the order parameter P vanishes and $F = p$, that is, all occupied sites belong to finite clusters. When $p > p_c$, the order parameter becomes nonzero and only a fraction of occupied sites belong to finite clusters. In Table 18.1 we show the critical parameters p_c of site and bond percolation for several lattices obtained by various methods. In some cases the value of p_c is known exactly.

For an infinite system, the density of clusters f is given by

$$f = \sum_{s=1}^{\infty} \rho_s. \tag{18.13}$$

Table 18.1 Critical parameter for site and bond percolation for various lattices including the simple cubic (sc), body centered cubic (bcc) and face centered cubic (fcc). The errors in the numerical values with decimal point are smaller or equal to 10^{-n}, where n indicates the decimal position of the last digit (Source: 'percolation threshold', Wikipedia)

Lattice	d	z	Site	Bond
Honeycomb	2	3	0.69604	0.652703[a]
Square	2	4	0.592746	1/2
Triangular	2	6	1/2	0.347296[b]
Diamond	3	4	0.430	0.389
sc	3	6	0.311608	0.248812
bcc	3	8	0.2460	0.18029
fcc	3	12	0.19924	0.12016
Hypercubic	4	8	0.19689	0.16013
Hypercubic	5	10	0.14080	0.11817
Hypercubic	6	12	0.1090	0.09420

[a] $1 - 2\sin(\pi/18)$
[b] $2\sin(\pi/18)$

and the average sizes of the finite clusters S is given by

$$S = \sum_{s=1}^{\infty} s^2 \rho_s. \tag{18.14}$$

18.3 One-Dimensional Model

To determine the percolation properties of a one-dimensional chain, we start by calculation the quantity ρ_s from which other quantities can be determined. For a linear chain, the quantity ρ_s, which is the density of clusters of size s, is given by

$$\rho_s = p^s q^2. \tag{18.15}$$

This result is a consequence that a one-dimensional cluster of size s is defined by s consecutive occupied sites and two empty sites at the border. Replacing into (18.13) and performing the summation, we get

$$f = p(1 - p). \tag{18.16}$$

Replacing into (18.11) and performing the summation we get, for $p < 1$,

$$F = p, \tag{18.17}$$

which, replaced into Eq. (18.12), that defines the percolation probability P, reveals us that $P = 0$ for $p < 1$.

The result $P = 0$, for $p < 1$, show us that the one-dimensional model does not have a percolating phase, except in the trivial case in which all sites are occupied ($p = 1$). The absence of the infinite cluster for any $p \neq 1$ can be understood if we recall that if $p \neq 1$, then there is a nonzero probability ($q \neq 0$) that a site is empty. But being the lattice one-dimensional, the presence of empty sites is sufficient to destroy a possible infinite cluster.

Replacing (18.15) into (18.14) and performing the summation, we get

$$S = \frac{p(1+p)}{1-p},\tag{18.18}$$

whose behavior close to $p = 1$ is

$$S \sim (1-p)^{-1},\tag{18.19}$$

which means that the size of the clusters diverges when $p \to 1$.

18.4 Model in the Cayley Tree

A Cayley tree can be understood as a lattice of sites formed by concentric layers and constructed as follows. We start from a central site to which we connect z sites that constitute the first layer. Each site of the first layer is connect to $z - 1$ sites that make up the second layer. Each site of the second layer is connected to $z - 1$ sites that make up the third layer. And so on. Each site is connected to z neighbors, being connected to one site of the inner layer and $z - 1$ sites of the outer layer. A site is never connected to a site of the same layer.

On a Cayley tree, the average number of cluster of size s, per site, is

$$\rho_s = b_s p^s q^{(\sigma-1)s+2},\tag{18.20}$$

where $\sigma = z - 1$ is the ramification of the tree and z is called coordination number (number of neighbors). The constant b_s is a numerical coefficient which must be determined. The result (18.20) is a consequence of the fact that a cluster of size s is defined by s occupied sites and by a certain number of empty sites located at the border. In a Cayley tree this number depends only on s and is equal to $(z-2)s + 2$.

The replacement of (18.20) into (18.13), (18.11) and (18.14) gives, respectively,

$$f = q^2 \sum_{s=1}^{\infty} b_s p^s q^{(\sigma-1)s},\tag{18.21}$$

$$F = q^2 \sum_{s=1}^{\infty} s b_s \, p^s q^{(\sigma-1)s}, \tag{18.22}$$

$$S = q^2 \sum_{s=1}^{\infty} s^2 b_s \, p^s q^{(\sigma-1)s}. \tag{18.23}$$

The coefficients b_s must be such that $F = p$ for $p < p_c$ since in this case $P = 0$.
Next, we define the generating function $G(x)$ by

$$G(x) = \sum_{s=1}^{\infty} b_s x^s, \tag{18.24}$$

from which we get the following relations

$$x G'(x) = \sum_{s=0}^{\infty} s b_s x^s, \tag{18.25}$$

$$x G'(x) + x^2 G''(x) = \sum_{s=0}^{\infty} s^2 b_s x^s. \tag{18.26}$$

Comparing the expressions (18.21)–(18.23), we get the expressions

$$f = q^2 G(x), \tag{18.27}$$

$$F = q^2 x G'(x), \tag{18.28}$$

$$S = q^2 [x G'(x) + x^2 G''(x)], \tag{18.29}$$

where $x = p q^{\sigma-1}$.
Now, for $p < p_c$ we must have $F = p$ which, compared with (18.28), allows us
to write

$$G'(x) = q^{-(\sigma+1)}. \tag{18.30}$$

The right hand side of this equation must be interpreted as a function of the variable
x which is obtained by inverting the equation $x = (1-q)q^{\sigma-1} = q^{\sigma-1} - q^{\sigma}$.
Deriving (18.30) with respect to x, we get

$$G''(x) = -(\sigma+1)q^{-\sigma-2} \frac{dq}{dx} = -\frac{(\sigma+1)q^{-2\sigma}}{(\sigma-1) - \sigma q}. \tag{18.31}$$

Replacing the results (18.30) and (18.31) in (18.29), we get, after some algebraic manipulations and recalling that $q = 1 - p$,

$$S = \frac{p(1+p)}{1-\sigma p}, \tag{18.32}$$

valid for $p < p_c$. This expression diverges when $p = 1/\sigma$, marking the critical point. Therefore, the critical concentration of occupied sites for the percolation defined on a Cayley tree is

$$p_c = \frac{1}{\sigma} = \frac{1}{z-1}. \tag{18.33}$$

Hence,

$$S = \frac{p(1+p)}{\sigma(p_c - p)}. \tag{18.34}$$

and the behavior of S close to the critical point is given by

$$S \sim (p - p_c)^{-1}, \tag{18.35}$$

for $p < p_c$, that is, S diverges at the critical point.

To get f we start by integrating (18.30),

$$G(x) = \int q^{-(\sigma+1)} \frac{dx}{dq} dq = -\frac{1}{2}(\sigma - 1)q^{-2} + \sigma q^{-1} + C, \tag{18.36}$$

where C is a constant to be determined. Taking into account that $G = 0$ when $x = 0$, that is, when $q = 1$, then we conclude that $C = -(1 + \sigma)/2$. Hence,

$$G(x) = -\frac{1}{2}(\sigma - 1)q^{-2} + \sigma q^{-1} - \frac{1}{2}(1+\sigma), \tag{18.37}$$

which can be written in the form

$$G(x) = -\frac{1}{2}(\sigma - 1)\left(\frac{1}{q} - \frac{\sigma}{\sigma - 1}\right)^2 + \frac{1}{2}\frac{1}{\sigma - 1}. \tag{18.38}$$

From (18.27) and using this last result, we get

$$f = p - \frac{1+\sigma}{2}p^2. \tag{18.39}$$

valid for $p < p_c$.

Generating function   To obtain explicitly the generating function $G(x)$, we have to invert the expression $x = q^{\sigma-1} - q^{\sigma}$ and replace the result into (18.37). The inversion, however, cannot be made in an explicit way for any value of σ. Hence, we will obtain $G(x)$ for $\sigma = 2$, in which case the inverse function can be found explicitly. In this case, $x = q - q^2$, and the inversion gives two results

$$q = \frac{1}{2}\{1 \pm \sqrt{1 - 4x}\}. \tag{18.40}$$

Before doing the replacement of this result into (18.37), we must decide which of the two solutions we should use. Taking into account that the expression (18.37) is valid for $p < p_c$, that is, for $q > q_c$ and that in the present case $p_c = 1/2$ or $q_c = 1/2$, then we should use the result

$$q = \frac{1}{2}\{1 + \sqrt{1 - 4x}\}. \tag{18.41}$$

Replacing (18.41) into (18.37), we get

$$G(x) = \frac{1}{2} - \frac{1}{2}\left(\frac{1 - \sqrt{1 - 4x}}{2x} - 2\right)^2. \tag{18.42}$$

Similarly, we obtain $G'(x)$ substituting (18.41) in (18.30)

$$G'(x) = \left(\frac{1 - \sqrt{1 - 4x}}{2x}\right)^3, \tag{18.43}$$

and $G''(x)$ substituting (18.41) in (18.31)

$$G''(x) = \frac{3}{\sqrt{1 - 4x}}\left(\frac{1 - \sqrt{1 - 4x}}{2x}\right)^4. \tag{18.44}$$

Next, we will obtain $f(p)$, $F(p)$ and $S(p)$ for $p > p_c = 1/2$ since the results for $p < p_c$ have already been obtained in the previous section. Taking into account that $x = pq$, then $\sqrt{1 - 4x} = \sqrt{(1 - 2p)^2} = 2p - 1$, for $p > 1/2$, so that

$$G(x) = -\frac{3}{2} - \frac{1}{2p^2} + \frac{2}{p}, \tag{18.45}$$

$$G'(x) = \frac{1}{p^3}, \tag{18.46}$$

$$G''(x) = \frac{3}{(2p - 1)p^4}. \tag{18.47}$$

Hence, using (18.27)–(18.29), we get

$$f = q^2 \left(-\frac{3}{2} - \frac{1}{2p^2} + \frac{2}{p} \right), \tag{18.48}$$

$$F = \frac{q^3}{p^2}, \tag{18.49}$$

$$S = \frac{q^3}{p^2} \frac{2-p}{(2p-1)}, \tag{18.50}$$

which are valid for $p > p_c$.

The order parameter P is obtained from (18.12), $P = p - F$, that is,

$$P = \frac{p^3 - q^3}{p^2} = \frac{1}{p^2}(2p-1)(1-p+p^2), \tag{18.51}$$

for $p > p_c$. Close to the critical point, we get

$$P = 6(p - p_c), \tag{18.52}$$

$$S = \frac{3}{8} \frac{1}{|p - p_c|}, \tag{18.53}$$

valid for $p > p_c$.

Order parameter The percolation probability P, which plays the role of the order parameter of the percolating phase, can be obtained by a another reasoning. This is done by assuming that it is equal to the probability that the central site of the Cayley tree is connected, through active bonds, to a site of the outermost layer of the tree. An active bond is a connection between two nearest neighbor occupied sites. Denoting by Q_0 the probability that the central site is not connected to the outermost layer, then

$$Q_0 = q + pQ_1^z, \tag{18.54}$$

where Q_1 is the probability that a site of layer 1 is not connected to the outermost layer. Indeed, the central site is not connected to the outermost layer if it is either empty (probability q) or, in case it is occupied (with probability p), if all the z sites of layer 1 is not connected to the outermost layer (probability Q_1^z).

Next, we take a site of layer 1 and repeat the same reasoning above to get

$$Q_1 = q + pQ_2^\sigma, \tag{18.55}$$

where Q_2 is the probability that a site of layer 2 is not connected to the outermost layer. Notice that the exponent of Q_2 is σ. As the number of layer increases, Q_2 approaches Q_1 and these two quantities become identical in the limit of an infinite number of layers. Therefore, in this limit

$$Q_1 = q + pQ_1^\sigma. \tag{18.56}$$

In addition, in this limit $P = 1 - Q_0$. Therefore, from (18.56) we can obtain $Q_1(p)$ as a function of p and next determine $Q_0(p)$ using (18.54) and finally $P(p)$.

A solution of (18.56) is $Q_1 = 1$, which is independent of p. This trivial solution gives $Q_0 = 1$ so that $P = 0$. There is however another solution, that depends on p. Taking into account that $Q_0 = 1 - P$, we can use (18.54) to write Q_1 in terms of P and next replace in (18.56) to get

$$(1 - \frac{P}{p})^{1/z} = 1 - p + p(1 - \frac{P}{p})^{\sigma/z}. \tag{18.57}$$

This is an equation whose solution gives $P(p)$ as a function of p. To obtain the behavior of P for values of p close to the critical point, we expand both sides of this equation up to quadratic terms

$$P^2 - \frac{2zp}{q}(p - \frac{1}{\sigma})P = 0. \tag{18.58}$$

Therefore, the non-trivial solution is given by

$$P = \frac{2zp}{q}(p - \frac{1}{\sigma}), \tag{18.59}$$

which is valid only for $p > 1/\sigma$ because otherwise P will be negative. We see thus that P indeed vanishes continuously when $p = p_c = 1/\sigma$. Being this expression valid around the critical point, we may still write

$$P = 2\frac{\sigma + 1}{\sigma - 1}(p - p_c), \tag{18.60}$$

that is, $P \sim (p - p_c)$.

Density of clusters To determine the density of clusters ρ_s we should calculate the coefficient b_s, what can be done from the expansion of the generating function $G(x)$ in powers of x. This calculation can be done in an explicit way when $\sigma = 2$. For $p < p_c$, the expression for $G(x)$ is given by (18.42), which we write as

$$G(x) = \frac{1}{4x^2}\left[(1 - 4x)^{3/2} - (1 - 6x + 6x^2)\right]. \tag{18.61}$$

Using the binomial expansion, we get

$$G(x) = \sum_{s=1}^{\infty} \frac{3(2s)!}{(s+2)!s!} x^s, \tag{18.62}$$

from which we get

$$b_s = \frac{3(2s)!}{(s+2)!s!}, \tag{18.63}$$

valid for $s \geq 1$. Therefore, the density of clusters $\rho_s = b_s p^s q^{s+2}$ is given by

$$\rho_s = \frac{3(2s)!}{(s+2)!s!} p^s q^{s+2}, \tag{18.64}$$

valid for $s \geq 1$ and $p < p_c = 1/2$. For large s,

$$\rho_s = \frac{3}{\sqrt{\pi}} q^2 s^{-5/2} (4pq)^s, \tag{18.65}$$

valid for $p < 1/2$. Near the critical point

$$\rho_s = \frac{3}{4\sqrt{\pi}} s^{-5/2} e^{-cs}, \tag{18.66}$$

where

$$c = 4(p - p_c)^2 \tag{18.67}$$

For $p > p_c$, we can show that the above expressions for ρ_s are similar. It suffices to exchange p and q.

18.5 Dynamic Percolation

In this section we analyze a stochastic model whose stationary distribution is identified with percolation. More precisely, the stochastic process generates clusters that are identified with the percolation clusters. The dynamic percolation model is best formulated in the language of epidemiology and, in this sense, it can be understood as a model for the spread of an epidemic. It is defined as follows. Each site of a regular lattice, identified as an individual, can be found in one of three states: susceptible (S), infected (I) or exposed (E). Infected and exposed individuals remain forever in these states. A susceptible individual may become infected or exposed according to the rules introduced next.

At each time step a site is chosen at random. If it is in state I or in state E, the state of the site is not modified. If it is in state S, then (a) the site becomes I with probability pn_v/z, where n_v is the number of neighbors in state I and z is the number of neighbors or (b) the site becomes E with probability qn_v/z, where $q = 1 - p$. Another equivalent way of defining the stochastic process, and more interesting for our purposes, is as follows. At each time step, we choose a site at random, say site i. If it is in state S or E nothing happens. If it is in state I, then one neighbor, say j, is chosen at random. If j is in state I or E, nothing happens. If j is in state S, then j becomes I with probability p and becomes E with probability $q = 1 - p$.

Consider now a lattice full of sites in state S except for a single site which is in state I. Applying the rules above, we see that a cluster of sites in state I grows from the original site. The growth process stops when there is no site in state I that has a neighbor in state S. That is, if all neighbors of the sites in state I are found in state E.

To show that a cluster generated by the rules mentioned above is a possible cluster of site percolation, we consider another lattice, which we call replica, with the same structure and same number N of sites. Each site of the replica can be occupied with probability p or empty with probability $q = 1 - p$. A possible configuration of the replica is generated as follows. For each site i we generate a random number ξ_i identically distributed in the interval between 0 and 1. If $\xi_i \leq p$, the site i is occupied, otherwise it remains empty. In this way, several clusters are generated in the replica. A cluster is defined as a set of occupied sites such that each site has at least one site occupied.

Next we focus on a certain cluster of the replica and in a certain site of the cluster which we call site 0. Now we look at the original lattice and at the growth of a cluster from the site 0. Initially, all sites of the original lattice are in state S except site 0, which is in state I. We choose a site at random, say site i. If site i is in state I, we choose one of its neighbors, say site j. If j is in state S, then it becomes I if $\xi_j \leq p$ and E otherwise. According to these rules, all sites in state I form a cluster that contains the site 0. Moreover, the sites that are in state I correspond to occupied sites in the replica and those sites that are in state E correspond to empty sites in the replica. Since each site of the cluster was generated with probability p, then this cluster is a possible cluster of the site percolation model. Figure 18.2 shows a cluster generated by this dynamics in the square lattice at the critical point.

Critical exponents The critical behavior around the critical point for various quantities is characterized by critical exponents. Next we list some relevant quantities and the respective critical exponents. Order parameter, which is the fraction of sites belonging to the infinite cluster,

$$P \sim \varepsilon^\beta, \tag{18.68}$$

where $\varepsilon = |p - p_c|$ Variance of the order parameter

$$\chi \sim \varepsilon^{-\gamma}, \tag{18.69}$$

Spatial correlation length

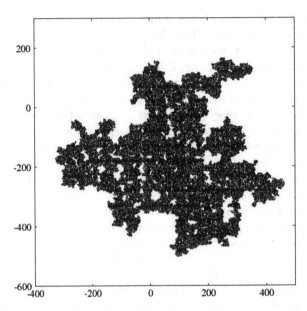

Fig. 18.2 Cluster with 82,062 sites, corresponding to site percolation on a square lattice, generated at the critical point according to the rules of dynamic percolation. The cluster has fractal structure of dimension $d_F = 91/48$

$$\xi \sim \varepsilon^{-\nu_\perp}. \tag{18.70}$$

Time correlation length

$$\xi_\| \sim \varepsilon^{-\nu_\|}. \tag{18.71}$$

Density of clusters

$$\rho_s \sim s^{-\tau} e^{-cs}, \qquad c \sim \varepsilon^{1/\sigma}. \tag{18.72}$$

Number of active sites

$$N_A \sim t^\theta, \tag{18.73}$$

at the critical point. Surviving probability

$$\mathcal{P} \sim t^{-\delta}. \tag{18.74}$$

Table 18.2 Critical exponents of the universality class of dynamic percolation, according to the compilation by Muñoz et al. (1999) and Henkel et al. (2008). The errors in the numerical value with decimal point are found in the last digit

d	β	γ	$\nu_\perp$	d_F	τ	$\nu_\parallel$	z	θ	δ
2	5/36	43/18	4/3	91/48	187/91	1.506	1.130	0.586	0.092
3	0.417	1.82	0.88	2.528	2.19	1.169	1.38	0.49	0.35
4	0.64		0.68		2.31		1.61	0.30	0.60
5	0.84		0.57		2.41		1.81	0.13	0.81
≥ 6	1	1	1/2	4	5/2	1	2	0	1

The fractal dimension d_F is defined by

$$N \sim L^{d_F}, \tag{18.75}$$

where N is the number of occupied sites inside a region of linear size L. The dynamic exponent z is defined by means of the following relation

$$\xi \sim t^{1/z}, \tag{18.76}$$

determined at the critical point.

In Table 18.2 we show the critical exponents obtained by various methods that include exact and numerical for the dynamic percolation universality class. When the dimension is larger than the upper critical dimension $d = 6$, the critical exponents are those we have obtained from the solution in the Cayley tree. In two dimensions the static critical exponents are known exactly.

References

F.C. Alcaraz, M. Droz, M. Henkel, V. Rittenberg, Reaction-diffusion processes, critical dynamics, and quantum chains. Ann. Phys. **230**, 250 (1994)

M.P. Allen, D.J. Tildesley, *Computer Simulation of Liquids* (Clarendon, Oxford, 1987)

G. Arfken, *Mathematical Methods for Physicists* (Academic, New York, 1970)

H.E. Avery, *Basic Reaction Kinetics and Mechanisms* (MacMillan, London, 1974)

N.T.J. Bailey, *The Mathematical Theory of Epidemics* (Hafner, New York, 1957)

G.A. Baker Jr., *Quantitative Theory of Critical Phenomena* (Academic, Boston, 1990)

M.S. Bartlett, *Stochastic Population Models in Ecology and Epidemiology* (Methuen, London, 1960)

C. Bennett, G. Grinstein, Role of irreversibility in stabilizing complex and nonergodic bahavior in locally interacting discrete systems. Phys. Rev. Lett. **55**, 657 (1985)

K. Binder (ed.), *Monte Carlo Method in Statistical Physics* (Springer, Berlin, 1979; 2nd edn., 1986)

K. Binder, D.H. Heermann, Monte Carlo simulation in statistical physics (Springer, Berlin, 1988; 2nd edn., 1992; 3rd edn., 1997; 4th edn., 2002; 5th edn., 2010)

J.J. Binney, N.J. Dowrick, A.J. Fisher, M.E.J. Newman, *The Theory of Critical Phenomena* (Clarendon, Oxford, 1992)

S.R. Broadbent, J.M. Hammersley, Percolation processes, I. Crystals and mazes. Proc. Camb. Phil. Soc. **53**, 629 (1957)

S. Chandrasekhar, Stochastic problems in physics and astronomy. Rev. Mod. Phys. **15**, 1–89 (1943)

W.T. Coffey, Yu. P. Kalmykov, J.T. Waldron, *The Langevin Equation* (World Scientific, Singapore, 2005)

L. Crochik, T. Tomé, Entropy production in the majority vote model. Phys. Rev. E **72**, 057103 (2005)

E.F. da Silva, M.J. de Oliveira, Critical discontinuous phase transition in the threshold contact process. J. Phys. A **44**, 135002 (2011)

S. Dattagupta, *Relaxation Phenomena in Condensed Matter Physics* (Academic, Orlando, 1987)

K.C. de Carvalho, T. Tomé, Self-organized patterns of coexistence out of a predator-prey cellular automaton. Int. J. Mod. Phys. C **17**, 1647 (2006)

S.R. de Groot, P. Mazur, *Non-equilibrium Thermodynamics* (North-Holland, Amsterdam, 1962)

J.R.G. de Mendona, M.J. de Oliveira, Stationary coverage of a stochastic adsorption-desorption process with diffusional relaxation. J. Stat. Phys. **92**, 651 (1998)

M.J. de Oliveira, Isotropic majority-vote model on a square lattice. J. Stat. Phys. **66**, 273 (1992)

M.J. de Oliveira, Numerical stochastic methods in statistical mechanics. Int. J. Mod. Phys. B **10**, 1313 (1996)

© Springer International Publishing Switzerland 2015

T. Tomé, M.J. de Oliveira, *Stochastic Dynamics and Irreversibility*, Graduate Texts in Physics, DOI 10.1007/978-3-319-11770-6

M.J. de Oliveira, Exact density profile of a stochastic reaction-diffusion process. Phys. Rev. E **60**, 2563 (1999)

M.J. de Oliveira, Diffusion-limited annihilation and the reunion of bounded walkers. Braz. J. Phys. **30**, 128 (2000)

M.J. de Oliveira, Linear Glauber model. Phys. Rev. E **67**, 066101 (2003)

M.J. de Oliveira, T. Tomé, Inhomogeneous random sequential adsorption on bipartite lattices. Phys. Rev. E **50**, 4523 (1994)

M.J. de Oliveira, T. Tomé, R. Dickman, Anisotropic random sequential adsorption of dimers on a square lattice. Phys. Rev. A **46**, 6294 (1992)

M.J. de Oliveira, J.F.F. Mendes, M.A. Santos, Non-equilibrium spin models with Ising universal behavior. J. Phys. A **26**, 2317 (1993)

D.R. de Souza, T. Tomé, Stochastic lattice gas model describing the dynamics of the SIRS epidemic process. Physica A **389**, 1142 (2010)

D.R. de Souza, T. Tomé, R.M. Ziff, A new scale-invariant ratio and finite-size scaling for the stochastic susceptible-infected-model. J. Stat. Mech. P02018 (2011)

R. Dickman, Nonequilibrium lattice models: series analysis of steady states. J. Stat. Phys. **55**, 997 (1989)

R. Dickman, T. Tomé, First-order phase transition in a one-dimensional nonequilibrium model. Phys. Rev. A **44**, 4833 (1991)

R. Dickman, R. Vidigal, Quasi-stationary distributions for stochastic processes with an absorbing state. J. Phys. A **35**, 1147 (2002)

R. Dickman, J.-S. Wang, I. Jensen, Random sequential adsorption: series and virial expansions. J. Chem. Phys. **94**, 8252 (1991)

E. Domany, W. Kinzel, Equivalence of cellular automata to Ising models and direct pecolation. Phys. Rev. Lett. **53**, 311 (1984)

P. Ehrenfest, T. Ehrenfest, Über zwei bekannte Einwände gegen das Boltzmannsche H-Theorem. Physikalische Zeitschrift **8**, 311–314 (1907)

A. Einstein, Über die von der molekularkinetischen Theorie der Wärme geforderte Bewegung von in ruhenden Flüssigkeiten suspendierten Teilchen. Annalen der Physik **17**, 549–560 (1905)

J.W. Evans, Exactly solvable irreversible processes on Bethe lattices. J. Math. Phys. **25**, 2527 (1984)

J.W. Evans, Random and cooperative sequential adsorption. Rev. Mod. Phys. **65**, 1281 (1993)

Y. Fan, J.K. Percus, Asymptotic coverage in random sequential adsorption on a lattice. Phys. Rev. A **44**, 5099 (1991)

B.U. Felderhofm, Spin relaxation of the Ising chain. Rep. Math. Phys. **1**, 215 (1970)

W. Feller, *An Introduction to Probability Theory and its Aplications*, 2vols. (Wiley, London, 1950)

P.J. Fernandez, *Introdução à Teoria das Probabilidades* (Livros Técnicos e Científicos, Rio de Janeiro, 1973)

M.E. Fisher, J.W. Essam, Some cluster size and percolation problems. J. Math. Phys. **2**, 609 (1961)

P.J. Flory, Intramolecular reaction between neighboring substituents of vinyl polymers. J. Am. Chem. Soc. **61**, 1518–1521 (1939)

A.D. Fokker, Die Mittlere Energie rotierender elektrischer Dipole im Strahlungsfeld. Annalen der Physik **43**, 810–820 (1914)

G. Frobenius, Über Matrizen aus nicht negativen Elementen. Sitzungsberichte der Königlich Preussischen Akademie der Wissenschaften 456–477 (1912)

F.R. Gantmacher, *Applications of the Theory of Matrices* (Interscience, New York, 1959)

C.W. Gardiner, *Handbook of Stochastic Methods for Physics, Chemistry and Natural Sciences* (Springer, Berlin, 1983; 4th edn., 2009)

R.J. Glauber, Time-dependent statistics of the Ising model. J. Math. Phys. **4**, 294 (1963)

B.V. Gnedenko, *The Theory of Probability* (Chelsea, New York, 1962)

P. Grassberger, On phase transitions in Schögl's second model. Z. Phys. B **47**, 365 (1982)

P. Grassberger, A. de la Torre, Reggeon field theory (Schögl's first model) on a lattice: Monte Carlo calculations of critical behavior. Ann. Phys. **122**, 373 (1979)

M.D. Grynberg, T.J. Newman, R.B. Stinchcombe, Exact solutions for stochastic adsorption-desorption models and catalytic surface processes. Phys. Rev. E **50**, 957 (1994)

G. Grinstein, C. Jayaprakash, Y. He, Statistical mechanics of probabilistic cellular automata. Phys. Rev. Lett. **55**, 2527 (1985)

J.M. Hammersley, Comparison of atom and bond percolation processes. J. Math. Phys. **2**, 728 (1961)

J.M. Hammersley, D.C. Handscomb, *Monte Carlo Methods* (Chapman and Hall, London, 1964)

H. Haken, *Synergetics, An Introduction* (Springer, Berlin, 1976)

T.E. Harris, Contact interactions on a lattice. Ann. Prob. **2**, 969 (1974)

D.W. Heermann, *Computer Simulation Methods in Theoretical Physics* (Springer, Berlin, 1986; 2nd edn., 1990)

M. Henkel, M. Pleimling, *Non-equilibrium Phase Transitions, Vol. 2: Ageing and Dynamical Scaling far from Equilibrium* (Springer, Dordrecht, 2010)

M. Henkel, H. Hinrichsen, S. Lubeck, *Non-equilibrium Phase Transitions, Vol. 1: Absorbing Phase Transitions* (Springer, Dordrecht, 2008)

T.L. Hill, *Free Energy Transduction and Biology* (Academic, New York, 1977)

T.L. Hill, *Free Energy Transduction and Biochemical Cycle Kinetics* (Springer, New York, 1989)

H. Hinrichsen, Non-equilibrium phenomena and phase transitions into absorbing states. Adv. Phys. **49**, 815 (2000)

B.R. James, *Probabilidade: um Curso em Nível Intermediário* (IMPA, Rio de Janeiro, 1981)

H.-K. Janssen, On the nonequilibrium phase transition in reaction-diffusion systems with an absorbing atationary state. Z. Phys. B **42**, 151 (1981)

J. Johnson, Thermal agitation of electricity in conductors. Phys. Rev. **32**, 97–109 (1928)

I. Jensen, R. Dickman, Time-dependent perturbation theory for non-equilibrium lattice models. J. Stat. Phys. **71**, 89 (1993)

M. Kac, Random walk and the theory of Brownian motion. Am. Math. Mont. **54**, 369–391 (1947)

M. Kac, *Statistical Independence in Probability, Analysis and Number Theory* (The Mathematical Association of America, Washington, DC, 1959)

S.A. Kauffman, Metabolic stability and epigenesis in randomly constructed genetic nets. J. Theor. Biol. **22**, 437 (1969)

J.G. Kemeny, J.L. Snell, *Finite Markov Chains* (Van Nostrand, Princeton, 1960)

A. Khintchine, Korrelationstheorie der stationären stochastischen Prozesse. Mathematische Annalen **109**, 604–615 (1934)

W. Kinzel, Phase transitions of cellular automata. Z. Phys. B **58**, 229 (1985)

A. Kolmogoroff, Über die analytischen Methoden in der Wahrscheinlichkeitsrechnung. Mathematische Annalen **104**, 415–458 (1931)

A. Kolmogoroff, Zur Theorie de Markoffschen Ketten. Mathematische Annalen **112**, 155–160 (1936)

K. Konno, *Phase Transitions of Interacting Particle Systems* (World Scientific, Singapore, 1994)

H.A. Kramers, Brownian motion in a field of force and the diffusion model of chemical reactions. Physica **7**, 284–304 (1940)

L.D. Landau, E.M. Lifshitz, *Statistical Physics* (Pergamon, Oxford, 1958)

P. Langevin, Sur la théorie du mouvement brownien. Comptes Rendus des Séances de l'Académie des Sciences **146**, 530–533 (1908)

D.S. Lemons, *An Introduction to Stochastic Processes in Physics* (Johns Hopkins University Press, Baltimore, 2002)

T.M. Liggett, *Interacting Particle Systems* (Springer, New York, 1985)

A.A. Lushinikov, Binary reaction $1 + 1 \rightarrow 0$ in one dimension. Sov. Phys. JETP **64**, 811 (1986)

A.A. Markov, Rasprostraneniye zakona bolshikh chisel na velichiny, zavisyashchiye drug ot druga. Izvestiya Fiziko-Matematicheskogo obschestva pri Kazanskom universitete **15**, 135–156 (1906)

A.A. Markov, Rasprostraneniye predelnykh teorem ischisleniya veroyatnostey na summu velichin, svyazannykh v tsep. Zapiski Imperatorskoy Akademii Nauk po Fiziko-matematicheskomu Otdeleniyu **22** (1908)

J. Marro, R. Dickman, *Nonequilibrium Phase Transitions* (Cambridge University Press, Cambridge, 1999)

D.A. McQuarrie, Stochastic approach to chemical kinetics. J. Appl. Prob. **4**, 413 (1967)

P. Meakin, J.L. Cardy, E. Loh Jr., D.J. Scalapino, Maximal coverage in random sequential absorption. J. Chem. Phys. **86**, 2380 (1987)

N. Metropolis, A.W. Rosenbluth, M.N. Rosenbluth, A.H. Teller, E. Teller, Equation of state calculations by fast computing machines. J. Chem. Phys. **21**, 1087 (1953)

E.W. Montroll, G.H. Weiss, Random walks on lattices, II. J. Math. Phys. **6**, 167 (1965)

O.G. Mouritsen, *Computer Studies of Phase Transitions and Critical Phenomena* (Springer, Berlin, 1984)

M.A. Muñoz, R. Dickman, A. Vespignani, S. Zaperi, Avalanche and spreading exponents in systems with absorbing state. Phys. Rev. E **59**, 6175 (1999)

G. Nicolis, I. Prigogine, *Self-Organization in Nonequilibrium Systems* (Wiley, New York, 1977)

R.M. Nisbet, W.C.S. Gurney, *Modelling Fluctuating Populations* (Wiley, New York, 1982)

R.S. Nord, J.W. Evans, Irreversible immobile random adsorption of dimers, trimers, . . . on a 2D lattices. J. Chem. Phys. **82**, 2795 (1985)

H. Nyquist, Thermal agitation of electric charge in conductors. Phys. Rev. **32**, 110–113 (1928)

L. Onsager, Reciprocal relations in irreversible processes, I. Phys. Rev. **37**, 405–426 (1931)

L.S. Ornstein, F. Zernike, Accidental deviations of density and opalescence at the critical point of a simple substance. Proc. R. Acad. Amst. **17**, 793–806 (1914)

W. Pauli, Über das H-Theorem vom Anwachsen der Entropie vom Standpunkt der neuen Quantenmechanik, in *Problem der modernen Physik, Arnold Sommmerfeld zum 60*, P. Debye (Hrsg.) (Hirzel, Leipzig, 1928), s. 30–45

A. Pelissetto, E. Vicari, Critical phenomena and renormalization-group theory. Phys. Rep. **368**, 549 (2002)

O. Perron, Zur Theorie der Matrices. Mathematische Annalen **64**, 248–263 (1907)

J. Perrin, L'agitation moléculaire et le mouvement brownien. Comptes Rendus des Séances de l'Académie des Sciences **146**, 967–970 (1908)

J. Perrin, *Les Atomes* (Alcan, Paris, 1913)

M. Planck, Über einen Satz der statistischen Dynamik und seine Erweiterung in der Quanten-theorie. Sitzungsberichte der Königlich Preussischen Akademie der Wissenschaften 324–341 (1917)

G. Pólya, Über eine Aufgabe der Wahrscheinlichkeitsrechnung betreffend die Irrfahrt im Strassen-netz. Mathematische Annalen **84**, 149–160 (1921)

L.E. Reichl, *A Modern Course in Statistical Mechanics* (University of Texas Press, Austin, 1980)

E. Renshaw, *Modelling Biological Populations in Space and Time* (Cambridge University Press, Cambridge, 1991)

S.O. Rice, Mathematical analysis of random noise. Bell Syst. Tech. J. **23**, 282–332 (1944); **24**, 46–156 (1945)

H. Risken, *The Fokker-Planck Equation, Methods of Solution and Applications* (Springer, Berlin, 1984; 2nd edn., 1989)

M.M.S. Sabag, M.J. de Oliveira, Conserved contact process in one to five dimensions. Phys. Rev. E **66**, 036115 (2002)

M. Sahimi, *Applications of Percolation Theory* (Taylor and Francis, London, 1994)

F. Schlögl, Chemical reaction models for non-equilibrium phase transitions. Z. Phys. **253**, 147 (1972)

J.E. Satulovsky, T. Tomé, Stochastic lattice gas model for a predator-prey system. Phys. Rev. E **49**, 5073 (1994)

J. Schnakenberg, Network theory of microscopic and macroscopic behavior of master equation systems. Rev. Mod. Phys. **48**, 571 (1976)

W. Schottky, Über spontane Stromschwankungen in verschiedenen Elektrizitätsleitern. Annalen der Physik **57**, 541–567 (1918)

G.M. Schütz, Reaction-diffusion processes of hard-core particles. J. Stat. Phys. **79**, 243 (1995)

V.K.S. Shante, S. Kirkpatrick, Introduction to percolation theory. Adv. Phys. **20**, 325 (1971)

M. Smoluchowski, Zur kinetischen Theorie de Brownschen Molekularbewegung und der Suspensionen. Annalen der Physik **21**, 756–780 (1906)

M. Smoluchowski, Über Brownsche Molekularbewegung unter Einwirkung äusserer Kräfte und deren Zusammenhang mit der verallgemeinerten Diffusionsgleichung. Annalen der Physik **48**, 1103–1112 (1915)

H.E. Stanley, *Introduction to Phase Transitions and Critical Phenomena* (Oxford University Press, New York, 1971)

D. Stauffer, Scaling theory of percolation clusters. Phys. Rep. **54**, 1 (1979)

D. Stauffer, *Introduction to Percolation Theory* (Taylor and Francis, London, 1985)

W. Sutherland, A dynamical theory for non-electrolytes and the molecular mass of albumin. Philos. Mag. **9**, 781–785 (1905)

C.J. Thompson, *Mathematical Statistical Mechanics* (Princeton University Press, Princeton, 1979)

T. Tomé, Spreading of damage in the Domany-Kinzel cellular automaton: a mean-field approach. Physica A **212**, 99 (1994)

T. Tomé, *Irreversibilidade: Modelos de Rede com Dinâmicas Estocásticas*, Tese de Livre-Docência, Instituto de Física, Universidade de São Paulo, 1996

T. Tomé, Entropy production in nonequilibrium systems described by a Fokker-Planck equation. Braz. J. Phys. **36**, 1285 (2006)

T. Tomé, M.J. de Oliveira, Self-organization in a kinetic Ising model. Phys. Rev. A **40**, 6643 (1989)

T. Tomé, M.J. de Oliveira, Stochastic mechanics of nonequilibrium systems. Braz. J. Phys. **27**, 525 (1997)

T. Tomé, M.J. de Oliveira, Short-time dynamics of critical nonequilibrium spin models. Phys. Rev. E **58**, 4242 (1998)

T. Tomé, M.J. de Oliveira, Nonequilibrium model for the contact process in an ensemble of constant particle number. Phys. Rev. Lett. **86**, 5643 (2001)

T. Tomé, M.J. de Oliveira, The role of noise in population dynamics. Phys. Rev. E **79**, 061128 (2009)

T. Tomé, M.J. de Oliveira, Entropy production in irreversible systems described by a Fokker-Planck equation. Phys. Rev. E **82**, 021120 (2010)

T. Tomé, M.J. de Oliveira, Susceptible-infected-recovered and susceptible-exposed-infected models. J. Phys. A **44**, 095005 (2011)

T. Tomé, M.J. de Oliveira, Entropy production in nonequilibrium systems at stationary states. Phys. Rev. Lett. **108**, 020601 (2012)

T. Tomé, R. Dickman, The Ziff-Gulari-Barshad model with CO desorption: an ising-like nonequilibrium critical point. Phys. Rev. E **47**, 948 (1993)

T. Tomé, J.R. Drugowich de Felício, Probabilistic cellular automaton describing a biological immune system. Phys. Rev. E **53**, 3976 (1996)

T. Tomé, M.J. de Oliveira, M.A. Santos, Non-equilibrium Ising model with competing Glauber dynamics. J. Phys. A **24**, 3677 (1991)

T. Tomé, A. Brunstein, M.J. de Oliveira, Symmetry and universality in nonequilibrium models. Physica A **283**, 107 (2000)

T. Tomé, A.L. Rodrigues, E. Arashiro, M.J. de Oliveira, The stochastic nature of predator-prey cycles. Comput. Phys. Commun. **180**, 536 (2009)

G.E. Ulenbeck, L.S. Ornstein, On the theory of the Brownian motion. Phys. Rev. **36**, 823–841 (1930)

N.G. van Kampen, A power series expansion of the master equation. Can. J. Phys. **39**, 551 (1961)

N.G. van Kampen, *Stochastic Processes in Physics and Chemistry* (North-Holland, Amsterdam, 1981; 2nd edn., 1992)

M.C. Wang, G.E. Uhlenbeck, On the theory of Brownian motion, II. Rev. Mod. Phys. **17**, 323–342 (1945)

N. Wax (ed.), *Selected Papers on Noise and Stochastic Processes* (Dover, New York, 1954)

B. Widom, Random sequential filling on intervals on a line. J. Chem. Phys. **58**, 4043 (1973)

B. Widom, Random sequential addition of hard spheres to a volume. J. Chem. Phys. **44**, 3888 (1966)

N. Wiener, The average of an analytic functional. PNAS **7**, 253–260 (1921)

W.W. Wood, J.D. Jacobson, Preliminary results from a recalculation of the Monte Carlo equation of state of hard spheres. J. Chem. Phys. **27**, 1207 (1957)

J.M. Yeomans, *Statistical Mechanics of Phase Transitions* (Clarendon, Oxford, 1992)

E.N. Yeremin, *The Foundations of Chemical Kinetics* (Mir, Moscow, 1979)

R.M. Ziff, E. Gulari, Y. Barshad, Kinetic phase transitions in an irreversible surface-reaction model. Phys. Rev. Lett. **56**, 2553 (1986)

Index

© Springer International Publishing Switzerland 2015
T. Tomé, M.J. de Oliveira, *Stochastic Dynamics
and Irreversibility*, Graduate Texts in Physics,
DOI 10.1007/978-3-319-11770-6

Printed in the United States
By Bookmasters